Phytopharmaceutical Technology

Phytopharmaceutical Technology

P. H. List and P. C. Schmidt

Institute for Pharmaceutical Technology
University of Marburg, Germany

Wiley India Pvt. Ltd.

Phytopharmaceutical Technology

Authorised reprint by Wiley India Pvt. Ltd., 4435-36/7, Ansari Road, Daryaganj, New Delhi-110002.

Reprint: 2010

Printed at: *Unique Color Carton, New Delhi*

ISBN : 978-81-265-2497-6

Contents

Chapter 5
Apparatus and Machinery

Foreword to Original German Edition

Phytopharmaceuticals are produced from fresh or dried plants, or parts of plants, by expression, distillation and other operations. Characteristically, the active constituents are obtained together with other materials present in the plant. The doctor and the patient expect the activity of the final product to be favourably influenced by the presence of these accompanying materials. To retain this quality in a phytopharmaceutical, it is not sufficient therefore to carry out a quantative extraction of what might generally be considered to be the active constituents. Rather, it must be ensured that the co-existing materials are also obtained in like kind and quantity from batch to batch. Since the active principles of many medicinal plants are still unknown, as is the nature of the co-existing material, a consistent quality of phytopharmaceutical can be attained only if all steps in the pre-processing of the plant material and in the manufacture of the final product are exactly described and strictly adhered to.

As in other technological areas, great changes have occurred over the last decade in the manufacture of phytopharmaceuticals. New processes have been developed or brought in from other areas. The attitudes of the legislators and also doctors and patients to these 'natural' medicines are now different from those of fifty years ago. As a result, the analytical methods used in quality assurance/control need to be refined. No monograph describing the preparation of these products is available. The authors have therefore tried to describe the whole area of phytopharmaceutics in such a way as to enable the reader to find manufacturing processes and quality control procedures suitable for his or her own special requirements. They have also endeavoured to provide unambiguous definitive statements covering standards and standardization procedures in order to help clarify the existing confusion of rulings. Generally, where products that are described in pharmacopoeal monographs are named in headings and tables, the old Latin descriptive terms are used (e.g. Extractum Belladonnae), but only English names are used in the text.

The authors should like to acknowledge the contributions made by the

following persons, companies, and institutes without which an appreciation
of the current state of technology and its scientific basis would not have been
possible:

Martin Bauer Co., Vestenbergsgreuth

Dr Ing K-H Brunner, Westfalia Separator
AG, Oelde

Doz Dr Bela Dános, Botanical Institute,
University of Budapest

Dr agr habil Chlodwig Franz, Technical
College, Müchen, Freising-Weihenstephan

O Greither and co-workers, Salushaus,
Bruckmühl

András Guylás, Szilasmenti MGTSZ
Kerepestarcsa

Prof Dr F W Hefendehl,
Bundesgesundheitsamt, Berlin

Dr W Herold and co-workers, Finzelbergs
Nachf, Andernach

Dr Gyözö Hortobágyi, Gedeon Richter,
Budapest

Dr C Lander, Bundesgesundheitsamt, Berlin

Dr H-G Menssen, A Nattermann & Co.,
Köln

Dr F Reiff and co-workers, E Merck,
Darmstadt

R Rinkel, L Heumann & Co., Nürnberg

Prof Dr H Schilcher, Institut für
Pharmakognosie und Phytochemie der FU
Berlin

W Spaich, Chemisch-Pharmazeutische
Fabrik Carl Müller und Staufen-Pharma
GmbH Göppingen

Prof Dr Dr h c E Stahl, Institut für
Pharmakognosie und Analytische
Phytochemie der Universität des
Saarlandes, Saarbrücken

Dr H Stumpf, Dr W Schwabe, Karlsruhe

Prof Dr Péter Tétényi, Director, Institute
Drug Plant Technology, Budapest

Dr Christa Ullrich, Weleda AG, Schwäbisch
Gmünd

We also thank the Deutschen Akademischen Austauschdienst (DAAD, the
German Academic Service for Information Exchange) for the support given
to one of us for a fact-finding journey to Hungary. Our thanks go to Frau J
Grossgart for her assistance in the writing and checking of our manuscript.
Not least, we thank the Publishers for their helpful collaboration.

January 1984   Paul Heinz List
                                                                 Peter C Schmidt

Notes on the English Edition
Phytopharmaceutical Technology is a translation of the German book *Technologie pflanzlicher
Arzneizubereitungen* with virtually no changes. The authors, Professors P H List and P C
Schmidt, presented a detailed description of relevant manufacturing processes, machinery,
control procedures and standards that is likely to remain a sound reference tool for many years
to come. A substantial demand was evident for this information to be available in English,
resulting in the present work. Special thanks are due to David Ellaby who translated the volume,
to Richard Schmidt who provided technical consultation, and to the original authors.

June 1989   Heyden & Son Limited

Trade Marks and Registered Trade Marks

Some product and corporation names mentioned in this book may be subject to restriction of use under trade name or trade mark enactments, which will vary from country to country. Readers should note, therefore, that free use of such names should not be inferred from their occurrence herein.

Pharmacopoeia abbreviations

2 AB-DDR	2nd Edition East German Pharmacopoeia
BP 80	British Pharmacopoeia 1980
DAB 6	West German Pharmacopoeia 6th Edition 1926
DAB 7	West German Pharmacopoeia 7th Edition 1968
DAB 8	West German Pharmacopoeia 8th Edition 1978
DAN IX	Danish Pharmacopoeia Edition IX, 1948
Helv. IV	Swiss Pharmacopoeia Edition IV, 1907
Helv. V	Swiss Pharmacopoeia Edition V, 1933
Helv. VI	Swiss Pharmacopoeia Edition VI, 1972
Helv. VI Suppl. 1	Swiss Pharmacopoeia Edition VI, 1973 Supplement
OAB 9	Austrian Pharmacopoeia 9th Edition, 1960
OAB 81	Austrian Pharmacopoeia 10th Edition, 1981
Ph. Eur.	European Pharmacopoeia Edition I, 1969
Ph. Eur. II	European Pharmacopoeia Edition II, 1971
Ph. Eur. III	European Pharmacopoeia Edition III, 1975
Ph. Franc. IXe Ed.	French Pharmacopoeia Edition IX, 1972–1974
USP XX	The United States Pharmacopoeia Twentieth Revision, 1980
NF XV	The National Formulary Fifteenth Edition, 1980

Chapter 1

Medicinal forms

1.1. Opening remarks

That branch of the science and practice of pharmacy concerned with the preparation of the dosage form of a drug is often termed pharmaceutical formulation technology or, more simply, pharmaceutics. The products of that technology are, of course, pharmaceuticals. Phytopharmaceuticals are those products derived from plant materials. A galenical in English usage is an obsolescent term used to describe a drug preparation of plant (or, to a lesser extent, animal) origin, but in certain European countries the term refers to the entire field of formulation technology [1.1].

Phytopharmaceuticals are produced from fresh or dried, or otherwise preserved, plants or parts of plants by expression, extraction, distillation, and other operations. Characteristically, the active constituents are obtained together with coexisting materials in the plant. Depending upon the methods used in the production of raw material and crude drug, and also upon subsequent stages in processing, the active constituents may either be obtained in their original natural form or they may become changed to a greater or lesser extent. Such changes may be physical, physico-chemical, or chemical in nature and in most cases may well be unintentional and therefore also uncontrolled, but may also be intentionally and purposefully induced. Pectic substances (Section 1.2.2.2.1) provide examples of these possibilities.

The manufacturing process will often include stages designed to remove so-called ballast material such as chlorophyll, tannins, resins, etc, and thus in general produce a selective enrichment of the active principles. Through the use of suitable procedures, such enrichment can be continued until finally a pure active constituent is isolated. With such a continuous transition from whole extract, through concentrated extract, to pure active constituent, it is difficult to delineate clearly the meaning of the term phytopharmaceutical. Certainly, pure natural products such as morphine, hyoscyamine, K-strophanthin and many others are not regarded as phytopharmaceuticals. Phytopharmaceuticals should always contain the active principles together with coexisting materials from the source plant, these additional materials having a greater or lesser beneficial influence upon the activity of the drug.

A phytopharmaceutical may frequently not represent the final dosage form administered to the patient. Dry extracts, for example, are further processed to produce powder mixtures, tablets, suppositories, and other dosage forms. They are considered to be intermediate or semi-finished products whose technical qualities are conducive to further processing. Also, one can list a number of phytopharmaceuticals that are used as pharmaceutical formulation aids, for example de-enzymized acacia gum and gum tragacanth.

The use of phytopharmaceuticals in medicine is termed phytotherapy [1.2]. Differences of opinion are often encountered concerning the use of this term and also the general value of this therapeutic approach, it often being repeated that the active principles of very many phytopharmaceuticals are still unknown. Moreover, in only a few cases have the beneficial effects of co-existing materials, which are themselves devoid of activity, been demonstrated experimentally, even though years of experience have shown their influence to be undeniable [1.3–1.7]. It is precisely this dearth of information that makes it so essential to prepare phytopharmaceuticals of the highest technical quality, and to standardize and stabilize these as far as possible.

1.2.   Medicinal forms from fresh plants

1.2.1.   FRUIT PULPS

Fruit pulps are solid or pasty viscid preparations which are to be taken internally. After removal of any hard components, the parts of the fruit to be used are crushed. Very dry fruits are made into a slurry with small quantities of water and and then pressed through a suitable sieve. The solids in the sieve are discarded and water is then evaporated until the mass reaches the desired consistency. Fruit pulps with certain additives are still used medicinally, though now only as purgatives, e.g. 'Neda-Früchtewürfel' ('Fruit Cubes') and 'Pastapalm' (see Rote Liste Nos 55062 and 55064, 1983).

Pharmacopoeias therefore tend now not to include pulps, although *DAB 6* does mention by way of example juniper berry juice, also known as cade juice, Succus Juniperi inspissatus. This product, however, should be regarded as a pulp.

Fruit pulps are important as intermediate products in the food industry, particularly in the soft drinks industry. In countries where they are grown various fruits (oranges, lemons, grapefruit, granadilla fruits and many different kinds of berries) are first coarsely crushed and the crude pulp obtained is frozen in relatively small casks (containers of 10–20 L capacity) and dispatched to the cold store. In the soft drinks factory or flavourings manufacturer the frozen crude material is ground on colloid mills (see Section 5.2.2) and then processed immediately either to a non-alcoholic fruit juice drink or, by spray-drying, to a further intermediate product (see Section 5.7.1).

1.2.2. JUICES. SUCCI

1.2.2.1. Terms and definitions

The term succi (Latin succus or sucus = juice) is used for a wide variety of phytopharmaceutical preparations manufactured or obtained by a diversity of methods. Hence juices expressed from fresh acidic fruits differ in nature and method of preparation from concentrated juices, Succi inspissati, such as are obtained by careful concentration of aqueous extracts, e.g. concentrated liquorice juice, Succus Liquiritiae. Other true plant juices such as aloes, opium or crude rubber are not classed as juices at all.

1.2.2.2. Press juices

Press juices or expressed juices are obtained by expression from fresh fruits or other juice-rich parts of plants, if necessary after addition of a little water and occasionally with stimulated fermentation.

When fresh parts of plants are pulverized and crushed, and especially when such pulps are left to stand, enzymatic processes (not possible in the intact plant where the enzyme and substrate are separated) take place, and original plant constituents are modified. Such alterations are desirable if they release originally bound substances (e.g. in woodruff) or if they impart to the product better properties for further processing (e.g. the breakdown of pectins). Alcoholic fermentation of sugar frequently occurs due to the ubiquitous presence of yeasts and this process is often accelerated by artificial addition of yeast. Certain enzymes may be added to assist the breakdown of pectins (see Section 1.2.2.2.1). Stabilization is necessary if uncontrollable enzymatic alterations of original plant constituents are to be prevented. This is achieved either by rapid drying of the fresh material (see 2.1) or by measures involving denaturation of the enzymes, e.g. by briefly heating the plants at over 70°C or by exposing them to the action of ethanol vapour (see Section 1.2.4).

1.2.2.2.1. Fruit juices

Fruit juices are generally obtained from fully ripe, usually acid-rich fruits such as raspberries, currants, cherries, etc. These fruits are all rich in pectins, making expression and filtration more difficult and causing the finished juices and syrups to gel (see Section 1.2.3). These pectins must therefore be destroyed. *DAB 6, Ph. Helv. V* and *USP XX–NF XV* describe one method and *Ph. Dan. 48* and *BP 80* describe a different one for decomposing pectins.

In the method described in *DAB 6*, for example, fresh raspberries are crushed, loosely covered and left to stand at room temperature with repeated stirring until 10 mL of a filtered sample of the juice can be mixed with 5 mL ethanol without becoming cloudy. The mass is then expressed and the juice left to stand in order to settle suspended matter, filtered and then processed to syrup.

When the pulp is left to stand, fermentation takes place, which, in the method described in *Ph. Helv. V*, is accelerated by prior addition of sugar and yeast.

Ph. Dan. 48 describes a method in which pectinase (3 g/kg fruit pulp) is ground with a small quantity of the fruit slurry and added to the crushed fresh, fully ripe washed fruits at 45°C. The mixture is left to stand at this temperature for 3 h with repeated stirring and, after a further addition of pectinase (3 g/kg), the pulp is left to stand for yet a further 3 h. This process is repeated until the mass can be easily expressed. Pectinase (3 g/kg expressed juice) is again added and the mixture left to stand for 3 h at 45°C. A sample of 10 mL of the cooled juice must remain clear when mixed with 5 mL alcohol.

Fruit juices are also included as monographs in the current Pharmacopoeias. *BP 80* describes the preparation of concentrated raspberry juice. After removal of the pectins by addition of pectinase the juice is filtered until clear and sucrose is added to bring its density to 1.050–1.060 g/mL, after which it is concentrated by evaporation to one-sixth of its original volume. Sodium metabisulphite, or some other suitable preservative, is then added.

Cherry juice, which is obtained from fresh cherries, the fruits of *Prunus cerasus* L., is registered in *USP XX–NF XV*. The washed fruits, from which the stalks but not the stones have been removed, are crushed without breaking the stones. Benzoic acid (0.1%) is dissolved in the fruit pulp, which is then left to stand at room temperature until a small filtered sample shows no cloudy turbidity after the addition of half its volume of alcohol, even after standing for 30 min. The juice is then expressed from the pulp and filtered. The difference in the method lies in the different way in which the pectins are destroyed.

Represented simply, a pectin molecule consists of ~ 100 D-galacturonic acid components linked to one another by α-$(1\rightarrow4)$-glycosidic bonds. One end of the molecule possesses reducing properties. Depending on the fruit from which the pectin is obtained, a varying percentage of the carboxyl groups will be esterified with methanol. The pectin is described as highly esterified when this proportion is more than 50% and as only partially esterified when the proportion is below this. The water solubility also decreases with diminishing degrees of esterification. Pectic acid or polygalacturonic acid is present when there is less than 10% esterification. Only its alkali metal salts are soluble in water.

Alcohol groups esterified with acetic acid are also important, especially in the pectins of beetroot and pears. The pectins in plant tissues are mainly present as protopectin in the middle lamellae, which cements the cells together. The polygalacturonic acid chains exist here as amorphous, water-insoluble substance. The individual moleules are inter-linked via divalent ions, predominantly Ca^{2+} and Mg^{2+}.

The spatial linking of the α-$(1\rightarrow4)$-glycosidically bound galacturonic acid units within the pectin molecule itself is axial–axial, the individual components in the chain being twisted towards each other. This aids water absorption, and hence swelling, and also facilitates the influx of water soluble substances. In the natural state the protoplast cell wall is surrounded by a film of water. This is of great technical importance. Enzymes can pass through the water layer and break down the middle lamella and the primary wall, i.e. macerate (in the biological sense) parenchymal tissue, and Ca^{2+} icns can enter to strengthen the tissue.

The greater the amount of cross-linking of the protopectin or the pectin in the middle lamella or in the primary cell wall, the stronger the plant tissue. Natural pectins are readily modified by technical processing.

Pectin esterases (PEs) are hydrolases that cleave the methyl esters of polygalacturonic acids, and are present in practically all higher plants. However, pectin-depolymerizing enzymes, pectin glycosidases (PGs), occur only in certain plants or parts of plants. The effects of PE and PG on products prepared from fresh plants or parts of plants (fruits) have been described in detail in the food technology literature [1.11].

One well known effect is the result of the uncontrolled action of PE in the manufacture of orange juices, where gels can form, especially in concentrates; in certain circumstances, this results in sedimentation of the suspended matter because of the close proximity of the particles to one another. If, however, orange-PE is left to act for a time on freshly expressed orange juice with simultaneous addition of small quantities of calcium salt (and very fine grinding of the fresh pulp), followed by complete heat-denaturation of all enzymes, it is possible to stabilize the homogeneous suspension of the solids, and the juice can remain stable for several years.

PE, or pectin de-esterified with PE, can also be used for clarifying lemon juices [1.12].

Enzyme preparations are used to increase the yield of fruit juices, especially of fruits rich in pectin such as blackcurrants, cherries and red grapes. These fruits contain both PE and PG. High yields of anthocyan in pigments, which in the natural state are bound to pectins, are particularly valued in red fruits.

Hydrolysis of the ester group by the PE specific to the particular plant occurs using the methods described in *DAB 6, Ph. Helv. V* or *USP XX–NF XV*, thereby producing free pectic acid. Pectinase (a commercial mixture of PE and PG), however, depolymerizes the polygalacturonic acid strands, as in the methods described in *Ph. Dan. 48* or *BP 80*. This is significant, as polygalacturonic acid, which remains after the first method, produces both poorly soluble and poorly filterable pectates upon addition of Ca^{2+} or Mg^{2+}.

PEs and PGs are mostly obtained from cultures of fungal moulds such as *Aspergillus niger, A. oryzae, A. ehrlichii, A. wentii, Penicillium chrysogenum, Botrytis cinerea*, etc. The quantity and composition of the pectolytic enzymes obtained from the individual mould species depends on the culture conditions, and especially on the type and concentration of the pectins present in the substrate. For example, there is a PG commercially available which, in addition to partially esterified pectin, preferentially cleaves insoluble proto-pectin and only slightly attacks highly esterified soluble pectin. This preferred hydrolysis of the protopectin cement in the middle lamellae results in rapid disintegration of most of the bindings which hold the cells of the plant parenchymal tissue together. This process can be assisted by a combination of PGs and suitable cellulases. Nectars, nectar concentrates and opaque juices of high quality which keep well can be prepared in this way from raw or blanched starting materials such as carrots, tomatoes, apples and other vegetables and fruits.

This type of plant preparation is important, particularly in food tech-nology and in reconstitutable products. At present the only enzyme prepara-tions used in the production of medicines from plants are pectin-cleaving enzymes.

Fruit juices are preserved either by steam-sterilization of the juices in the final sealed bottles or by sterile filtration and aseptic filling into presterilized bottles, where sorbic acid, benzoic acid or formic acid (or their salts) and other permitted preservatives are often added. Such additives become unnecessary when the fresh fruit juices are converted immediately into syrups, as the high sugar concentration produces a low a_w value, which prevents microbial growth (1.13].

1.2.2.3.　Juices not obtained by expression

These include juices, and process stages thereof, exuded from plants and parts of plants by 'bleeding' from wounds inflicted on plants naturally or artificially. They are generally collected in a partially dried state or as a liquid and then subjected to further processing. There are so many of these plant juices that only a few characteristic representatives are mentioned here.

1.2.2.3.1.　Aloes

The old name Succus Aloes inspissatus indicates that this preparation is regarded as a juice. Aloes can be found as a monograph in nearly all the current Pharmacopoeias.

There are many different varieties of aloes, most of which are named after their country of origin. They are obtained from various plants of the genus *Aloe*, which comprises about 250 species. Some Pharmacopoeias demand the use of Cape aloes, whereas others permit Cape, Curaçao, Sokotra and other *Aloe* species. Reference 1.14 gives information on the numerous species of aloes.

The recovery of aloes is described below using the example of Cape aloes. The shredded leaves are stacked in a loose heap up to ~ 1 m high, in a tub lined with animal skin and set in the ground, so that all their cut surfaces are turned towards the base of the tub. After several hours the juice which has flowed into the tub is collected in cannisters and, usually, concentrated at the place where it is obtained over open fires with stirring. The originally yellow/brown juice turns into a very dark brown viscous or thickly liquid mass. This is put into ordinary tin cans, where it solidifies upon cooling to a glassy mass which fractures like mussel shells.

The commercial varieties of aloes are basically distinguished by their external appearance:

1. *Aloe lucida*, which is glassy, translucent and glossy.
2. *Aloe hepatica*, which turns the colour of liver due to the deposition of crystalline aloin during the slow thickening and concentration, and which has a dull, matt surface.

These external characteristics give no information about the quality, which is determined solely by the aloin content.

1.2.2.3.2. Opium

Opium is the latex or milky juice exuded from, and drying out on, the capsule surface when the unripe seed capsules of *Papaver somniferum* L. are cut. One or more horizontal, oblique or vertical incisions are made in the lower half of the seed capsule. The incisions open the latex ducts but do not penetrate the inner wall of the seed capsule, so that the latex is not lost inwards. The incisions are made in the evening and the dried juice is scraped off before dawn to prevent the opium becoming too darkly coloured as a result of exposure to sunlight. The traditional techniques of cutting and collection differ from one producing country to another.

The crude opium obtained in this way is made into varying forms of specified weight, according to its country of origin. The importance of opium as a medicine is declining, as therapeutic interest has shifted to the pure alkaloids and their semisynthetic conversion products, the effects of which can be more easily controlled by the physician. The world demand for opium is nevertheless still very high. Morphine, the principal alkaloid of opium, is processed further to more potent or differently acting derivatives such as codeine, ethylmorphine and morphinan or apomorphine derivatives.

Opium is of technological interest in that its substantial lack of plant tissue parts (any tissue residues present originate from the poppy leaves used to wrap the opium loaves) makes special demands on the method of extraction. Opium can be considered as one example of a whole series of juice, resin or balsam drugs. Such drugs must be extracted either by maceration (see opium extract or opium tincture in *DAB 8*) or by maceration followed by percolation after the powdered drug plant has been mixed with an equal proportion of an inert material such as sea-sand [cf. Ref. 1.14].

1.2.2.3.3. Gum arabic. Gummi arabicum. Acaciae gummi *Ph. Eur. I.*

Gum arabic is mentioned here as a representative of a number of plant gums
which are used pharmaceutically. It is the air-hardened gum exuded naturally, or obtained by incision, from the trunk and branches of *Acacia senegal*
W. or other African species of *Acacia* (*Ph. Eur. I*).

In the main production areas the trees are artificially wounded to induce a
copious flow of gum. When the leaves begin to fall, strips of bark 3 cm wide
and at least 60 cm long are peeled off with an axe, care being taken not to
damage the cambium. Within 20–30 days lumps of gum are exuded on the
surface of the wound. These are collected, cleaned and graded by colour and
size at the markets or, if the trees are cultivated, on the producer's plantation.
The principal constituent of gum arabic is arabic acid, a branch-chained
polysaccharide consisting of D-galactose, L-arabinose, L-rhamnose, D-glucur-
onic acid and methylglucuronic acid. The arabic acid present in gum arabic
exists as a salt called arabin.

Gum arabic contains quantities of oxidases and peroxidases, making it
necessary for it to be deenzymated before it can be used pharmaceutically.
This is accomplished by sterilizing a freshly prepared solution and evaporating it to dryness under reduced pressure at a temperature not exceeding 60°C.
In other methods a solution of the starting material is filtered until clear and
then heated for 30 min while steam is passed through it. A single vigorous
boiling also suffices. The deenzymated product is also available commercially.

1.2.2.3.4. Myrrh. Myrrha

Myrrh is cited here as an example of a gum resin drug. It represents the air-
dried gum resin exuded naturally, or after injury, from the schizogenically
and subsequently lytically expanded secretory spaces in the bark of *Commiphora molmol* E. and other *Commiphora* species. The tincture is prepared, as
for opium, by the principle of maceration (*DAB 8*) or by percolation after
mixing with equal parts of sea-sand [1.14].

1.2.2.3.5. Peru balsam. Balsamum peruvianum *DAB 8*

Peru balsam is a secretion which is obtained from trees of *Myroxylon
balsamum* Harms. It does not occur naturally, but is formed only when the
trees are injured. An area of the outer layer of bark of ~ 15 × 25 cm situated
30 cm above the ground is removed from trees about 10 years old. The area is
then carbonized with a low-temperature flame for 4–5 min. The balsam
formed and exuded is absorbed by rags with which the wound has been
covered. This type of treatment can be repeated many times on a single tree
[1.14].

1.2.2.4. Artificial juices

This term covers 'juices' which do not occur as such in the plant and are not formed even after the plant has been wounded. Rather, they are obtained by solvent extraction (usually water) from plants or parts of plants, followed by concentration of the extracts. The most familiar example is liquorice, Succus liquiritiae.

1.2.2.4.1. Liquorice. Succus liquiritiae

The roots of freshly harvested liquorice plants, *Glycyrrhiza glabra* L., which are at least 4 years old are washed, cut up and ground in water to a fine, fibrous slurry. This is boiled for about 15 h, then strained and squeezed. The solids are then precipitated and the clear liquid is decanted. The liquid is concentrated and thickened in flat pans over small fires and the resultant viscid mass poured to form rod-shaped sticks. These are stamped with the manufacturer's mark, dried in air, polished by rubbing in liquorice solution and packed. Using modern production methods the extracts are now usually evaporated under reduced pressure.

1.2.3. SYRUPS. SIRUPI

Syrups are liquid preparations with a high sugar content which, when they contain juices or alcoholates, can be regarded as plant preparations. Being of an aromatic nature, they are mostly used as flavourings. Some dispensatories permit the addition of polyalcohols such as sorbitol and glycerol, starch syrup (glucose) or polyethyleneglycols to retard crystallization. A number of such syrups are mentioned in the current Pharmacopoeias. We mention here the following examples.

1.2.3.1. Blackcurrant syrup *BP 80*

Blackcurrant syrup can be prepared from the clarified juice of blackcurrants or from commercially available concentrated blackcurrant juice. The freshly prepared or commercial juice is adjusted to a density of 1.045 g/mL with water and 700 g of sucrose is dissolved in 560 mL of this adjusted juice. Benzoic acid, up to a maximum of 800 ppm (w/w), or sodium metabisulphite or some other suitable sulphite up to a maximum of 350 ppm (w/w) are permitted as preservatives. The addition of permitted food colourings is also allowed.

1.2.3.2. Raspberry syrup *BP 80*

Raspberry syrup can be prepared by diluting 1 volume of concentrated raspberry juice (see Section 1.2.2.2.1) with 11 volumes of sugar syrup (66.7% w/w). The addition of permitted food colourings is allowed.

1.2.3.3. Cherry syrup *USP XX–NF XV*

Cherry syrup can be prepared by dissolving 800 g of sucrose in 475 mL of cherry juice (see Section 1.2.2.1) by gentle heating in a water bath. The solution is then cooled and the froth which is formed removed. Ethanol (20 mL) and water are then added to bring the final volume to 1 L.

1.2.4. ALCOHOLATES. ALCOHOLATURAE

Alcoholates are prepared from fresh plants, or parts of plants, by maceration with ethanol. Alcoholates are generally recovered only from those plants in which the essential constituents would be completely or partially lost in a drying process.

The plants are ground up and macerated with cold ethanol (80–95%, depending on the dispensatory instructions followed). 'Stabilized alcoholates' are obtained by extraction from fresh plants, or parts of plants, by boiling under reflux. This denatures the plant enzymes which impair the preserving properties of the alcoholates.

Ph. Franc. IX describes Alcoolature de Citron (lemon) and Alcoolature d'Orange douce.

1.2.5. HOMOEOPATHIC PREPARATIONS FROM FRESH PLANTS

Most of the medicinal plant preparations used in homoeopathy are made from fresh plants, or parts of plants. Here the juice content and its composition are important as they govern the choice between various methods of preparing the 'mother tinctures'.

According to Hahnemann the fresh juice is taken as the unit of the strength of the medicine prepared from fresh plant material and hence mother tinctures are generally prepared by one of the methods from *HAB 1*:

Method 1,
  if the plants contain more than 70% expressible juice but neither ethereal oils, resins nor mucus.
Method 2a or 2b,
  if the plants contain less than 70% expressible juice, more than 60% moisture (loss upon drying) and no ethereal oils or resins.

Method 3a, 3b or 3c,
  if the plants contain ethereal oils and resins or less than 60% moisture (loss upon drying).

The individual methods can be found in the *Homoeopathic Pharmacopoeia* (*HAB 1 = Homöopathisches Arzneibuch*) and its supplements.

1.3. Dosage forms obtained from drugs

1.3.1. TERMS AND DEFINITIONS

In German pharmaceutical terminology 'Drogen', which for the sake of convenience is translated here as 'drugs', signifies dried whole plants or parts thereof, dried animals, parts of animals or secretions from animals. Only starting materials obtained from plants shall be discussed here.

The drugs are named after the part of the plant used. Dilg and Jüttner [1.1] produced a table of nomenclature which we have shortened here to plant drug names and reproduced as Table 1.1.

Table 1.1   Parts of drug plants

Plant part	Meaning	Example
Amylum	Starch	Amylum Tritici
Balsamum	Balsam	Balsamum Copaivae
Bulbus	Bulb	Bulbus Scillae
Cortex	Bark	Cortex Chinae
Flos	Flower	Flores Malvae
Folium	Leaf	Folia Uvae ursi
Folliculus	Pod (follicle)	Folliculi Sennae
Fructus	Fruit	Fructus Carvi
Gemma	Bud	Gemmae Populi
Glandula	Gland	Glandulae Lupuli
Gummi	Gum	Gummi arabicum (=Acacie Gummi)
Herba	Herb	Herba Thymi
Legumen	Capsule, pod	Legumina Phaseoli (=Cortex Phaseoli Fructus)
Lignum	Wood	Lignum Guajaci
Nux	Nut	Nux moschata
Pericarpium	Fruit skin, peel	Pericarpium Aurantii
Pix	Tar	Pix betulina
Pulpa	Fruit pulp	Pulpa Tamarindorum
Radix	Root	Radix Gentianae
Resina	Resin	Resina Jalapae
Rhizoma	Rootstock	Rhizoma Rhei
Semen	Seed	Semen Colchici
Siliqua	Pod, husk	Siliqua dulcis (=Fructus Ceratoniae)
Stigma	Leaf scar, stigma, spot	Stigmata Maidis
Stipes	Stem	Stipites Dulcamarae
Strobilus	Cone	Strobili Lupuli
Summitates [Pl.]	Twig	Summitates Sabinae
Tuber	Tuber	Tubera Salep
Turio	Shoot	Turiones Pini

In addition to the so-called whole drugs, which are coarse, uncut and unground pieces of starting materials [though finely particulate starting materials of a similar nature (e.g. starch) are also included], the degree of pulverization is also used for the naming of drugs, especially in the Pharmacopoeias.

Drugs in which the largest particles have a maximum diameter of 4 mm (coarsely shredded), 3 mm (moderately finely shredded) or 2 mm (finely shredded) are regarded as cut or shredded (concisus) drugs.

Drugs in which the largest particles have a maximum diameter of 0.75 mm (coarsely powdered), 0.30 mm (moderately finely powdered) or 0.15 mm (finely powdered) are regarded as powdered or pulverized drugs.

The granule or particle sizes for drugs given here do not differ significantly from those in the Pharamacopoeias; the various Pharmacopoeias likewise do not differ significantly from one another in the sizes they quote (see Section 4.1.1).

1.3.2. HERBAL TEAS. HERBAL REMEDIES

The simplest medicinal plant preparations are teas or infusions made from one or a mixture of drug plants. Some of the modern Pharmacopoeias contain precise instructions for their preparation. The following method of preparation is reproduced from *Helv. VI Suppl. 1*, 1973 as an example.

Herbal tea or infusion mixtures are mixtures of unground or suitably ground medicinal plants to which drug plant extracts, ethereal oils or medicinal substances can be added.

Unless otherwise prescribed in the monographs, the drug plants have to be suitably ground to the following particle sizes:

Leaves, flowers and herbs	500 μm
Leaves, flowers and herbs with leaf veins more than 300 μm thick	3150 μm
Fruits, seeds, woods, barks, roots and rhizomes	3150 μm

Fruits and seeds containing the ethereal oil in excretory cavities within the drug plant, e.g. Umbelliferae fruits and juniper berries, have to be crushed before use. All herbal remedies in individual-dose packs or in poultices must have a ground particle size of 1600 μm.

Any drug plant extracts or medicinal substances which are to be added must first be dissolved in water or ethanol and added by spraying or soaking either to the complete infusion mixture or to those drug plants in which the active substances will not be altered, or altered as little as possible, by the solvent. The solvent is then removed by evaporation at room temperature, or at most 40°C.

Infusion mixtures should be as homogeneous as possible. If they become slightly inhomogeneous they should be thoroughly mixed again in a suitable manner before being despatched.

Grinding to the required granule size is usually done by the drug importer or manufacturer on an industrial scale. The necessary mechanical processing is carried out in the pharmacy only when the herbal remedy has to be made up by the pharmacist or in the case of certain special drugs such as linseed and Umbelliferae fruits or other ethereal oil fruits as described above.

Laboratory toothed disc mills, also known as linseed crushers, which can be used to crush Umbelliferae fruits and juniper berries as well as linseed, are suitable for the small quantities involved here. The grinding should be done immediately before dispatch, in the case of ethereal oil drugs because of the loss of ethereal oil upon storage and in the case of linseed because of the tendency of the crushed product to become rancid.

Linseed is used as a mild laxative because of its swelling, mucus-forming constituents. The swelling, and hence the laxative effect, is only slight when linseeds are taken whole, but when crushed the cells are disintegrated, with the result that the mucus-forming constituents swell better and are assisted in their action by the exuded linseed oil. Crushing of the seeds also brings the cyanogenic glycosides contained in them into contact with the extracellular glycosidases, thus releasing hydrocyanic acid. Cyanide poisoning is unlikely with the usual dose of linseed, much less so as the hydrogen cyanide releasing enzyme linase is inactivated in the acidic medium of the stomach and any hydrocyanic acid which might possibly continue to be formed is detoxicated by the formation of its thiocyanate [1.15].

The drugs are cut or ground to the desired particle size on the industrial scale, although the transitions from one degree of 'fineness' to another are blurred (see Sections 4.1 and 5.2).

1.3.3. STANDARDIZED DRUG POWDERS. PULVERES NORMATI. PULVERES TITRATI

Accurate dosing of potent drugs requires the exact determination of their active substance content. Standardized powders are therefore used for dispensing and in pharmacies. *Ph. Eur.* and *DAB 8* contain a series of monographs which only briefly describe the preparation of drug powders, but which describe the value or content determination procedures in great detail. A minimum and a maximum value or content of active constituents is given in every case. *DAB 8* unfortunately gives different granule sizes of the powders those given in *Ph. Eur.*

DAB 8	300 µm	(for opium even 355 µm).
Ph. Eur.	180 µm	(also prescribed in the biological determination methods on cardioactive glycosides).

In all cases the active substance content of the powdered drug to be standarized is then determined for the given maximum granule or particle size. The Pharmacopoeia monographs stipulate that the determined value must lie either within or above the stated range of active substance content. In the latter case, where the powdered drug is too strong it must be adjusted to the required content by mixing with an inert material. The Pharmacopoeias recommend lactose or low-content drug powders as such diluents.

Table 1.2 lists the adjusted or standardized powders described as monographs in *Ph. Eur.* and *DAB 8*.

Table 1.2 Adjusted or standardized powders described in the pharmacopoeias

Standardized powder of	Pharmacopoeia	Active substance	Value determination[a]
Adonis	*DAB 8*	0.2% cymarine	Biological (value $W = 1.67–2.40$ mg/g drug)
Belladonna	*Ph. Eur. I*	0.28–0.32% 'hyoscyamine'	Chemical
Digitalis lanata	*DAB 8*	0.5% digoxin	Biological (value $W = 4.17–6.00$ mg/g drug)
Digitalis purpurea	*DAB 8*	1% digitoxin	Biological ($W = 7.50–13.33$ mg/g drug)
Hyoscyamus	*Ph. Eur. I*	0.05–0.07% 'hyoscyamine'	Chemical
Ipecacuanha	*Ph. Eur. I*	1.90–2.10% 'emetine'	Chemical
	DAB 8	0.2% convallatoxin	Biological ($W = 1.50–2.67$ mg/g drug)
Squill	*DAB 8*	0.2% proscillaridine	Biological ($W = 1.74–2.30$ mg/g drug)
Opium[b]	*DAB 8*	9.8–10.2% morphine	Chemical
D. stramonium	*Ph. Eur. I*	0.23–0.27% 'hyoscyamine'	Chemical

[a] The biological value determinations were carried out as directed in *DAB 8*, p. 29 'Determination of the active value of drugs with cardioactive glycosides'. Chemical methods are to be found in the monographs of the cited Pharmacopoeias. Emphasis of the active substance in the Table by '........' means that the total alkaloid content is calculated on the cited active substance.
[b] Standardized or adjusted opium is called 'Opium titratum' in *DAB 8*.

1.3.4. DRUG EXTRACTS

Drug extracts are preparations obtained by extracting drugs of a certain particle size with suitable extraction agents (menstrua). The extraction can be accomplished by the various methods described in detail under Section 4.2. The extract obtained after separation of the liquid from the drug residue is called a miscella. It may already represent the final liquid dosage form, e.g. as a so-called fluid extract, or be used as an intermediate product which is to be further processed as quickly as possible.

Five methods are given in *DAB 8* for the preparation of extracts. Three of them are used for the preparation of 'aqueous drug extracts' (see Section 1.3.4.1): the others, maceration and percolation, partly give ready-to-use medicinal preparations, e.g. fluid extracts and tinctures, and partly intermediate products, which are mostly further processed to dry extracts. Even the latter can be regarded as semi-finished products (see Section 1.1), as they are mostly used for the preparation of other dosage forms.

The Pharmacopoeias permit the use of other methods of preparation of 'extracts' if products with the same properties as those obtainable by the Pharmacopoeia methods can be obtained. This applies only to the extracts listed in the Pharmacopoeia.

1.3.4.1. Aqueous drug extracts

DAB 8 describes these types of medicinal preparations as dispensing medicines intended to be used as soon as possible. There are three distinct types: namely decoctions, infusions and macerates. The type of drug used makes it possible to make a choice between the various ways in which these different forms are prepared. Hence decoctions are generally prepared from root bark and wood, infusions from leaf, herb, and flower, and macerates are, for example, prepared from drug plants with a high mucus content.

Aqueous drug extracts, unless otherwise directed, are prepared according to *DAB 8* by the methods described in Sections 1.3.4.1.1–3 from 1 volume drug and 10 volumes water. Extracts of drugs obtainable only on prescription may be prepared only when the ratio of drug to water is stated in the prescription.

Unless otherwise directed in the drug monographs of the Pharmacopoeias, the drugs, depending on the type of plant parts are used for the extract in the following degrees of comminution:

Leaves, flowers and herbs	shredded (4000 µm)
Woods, barks and roots	shredded (2800 µm)
Fruits and seeds	(2000 µm)
Alkaloid-containing drugs	powdered (710 µm)

The apparatus used for the preparation of extracts must be made of inert materials. In addition to this official dosage of 'aqueous drug extracts', aqueous extracts have also been prepared on an industrial scale for a long time.

1.3.4.1.1.   Decoctions. Decocta *DAB 8*

DAB 8 permits decoctions to be prepared as follows:

> The drug, in the prescribed comminution, is put into water at a temperature above 90°C. The container is suspended in a water-bath and maintained at this temperature for 30 min, with repeated stirring. The mixture is then strained while still hot. If the prescribed weight of the decoction is not attained after gentle expression of the drug residue, the required quantity of boiling water is poured on to the residue which is again gently expressed. This second extract is added to the main decoction to give to the prescribed weight.

1.3.4.1.2.   Infusions. Infusa

Infusa are prepared as per *DAB 8* as follows:

> One part drug in the prescribed comminution is kneaded several times in a mortar with 3–5 parts water and left to stand for 15 min. The rest of the boiling water is then poured on to the mixture, which is suspended in a container in a water-bath and kept for 5 min, with

repeated stirring, at a temperature above 90°C. The mixture is covered and left to stand until cool. If the prescribed weight of infusion is not obtained after gentle expression of the drug residue, the required quantity of cold water is poured on to the residue which is gently squeezed out again. This second extract is added to the main infusion to make it up to the prescribed weight.

1.3.4.1.3.   Macerates. Macerata

Macerates are prepared as per *DAB 8* as follows:

The drug, comminuted as prescribed, is left to stand, with occasional stirring, for 30 min after the required quantity of water has been poured on to it at room temperature. The extract is then strained and made up to the prescribed weight with the rinsings.

1.3.4.1.4.   Aqueous drug extracts prepared on an industrial scale

Water, if necessary mixed with lower alcohols, plays an increasingly important role as menstruum for the technical preparation of drug extracts, especially where the end-product is to be a dried extract which is to be clearly soluble in water. Spray-drying (see Section 4.5.3.2) has made it possible to obtain powders which are easily soluble in water, and which since 1945 have become increasingly used in phytopharmaceutical preparations.

For technical reasons the types of preparation differ here from the directions in *DAB 8*, which are to be regarded as dispensing instructions. Batches of drug extract from 100 kg up to tons require extraction processes such as percolation or kinetic maceration (see Sections 4.2.2.2. and 4.2.3.2). The demand for ready-to-use medicines, which means further industrial processing, still remains.

Aqueous miscellae, especially when they are prepared at a low temperature in order to preserve the active substances, contain numerous microorganisms such as bacteria, fungal moulds and yeasts, which contaminate the plant starting material. The miscellae also act as an excellent nutrient medium for the replication of the microorganism. It must therefore be decided at this point whether measures are to be taken to reduce these microorganisms, and what these measures are to be. The following questions need to be posed in order to reach the correct decision:

How many organisms were there initially in the drug(s) used?
How many organisms are there in the miscella now?
Are the next processing stages suitable for reducing this organism count?
Is further processing possible immediately?
Are the constituents of the miscella resistant to heat?

The measures to be taken will be governed by the answers to these questions. Other negative properties of the miscella in addition to microbial contamination also have to be considered, like turbidity or opacity of the miscella, e.g. after standing for a short time, and the miscella may contain undesirable pigments (e.g. chlorophyll).

The necessary steps to be taken to purify the miscella are described in detail in Section 4.3.

1.3.4.2. Tinctures. Tincturae

According to *DAB 8*, tinctures are extracts from drug plants prepared with ethanol of varying concentration, ether, or mixtures of these, perhaps with certain additives, in such a way that 1 part of drug is extracted with more than 2 parts, but at most 10 parts, of extraction liquid. Solutions of dried extracts in ethanol of suitable concentration are also regarded as tinctures.

Tinctures having starting-drugs which are 'to be stored carefully' are prepared in the extraction ratio 1:10, and other tinctures are usually prepared in the ratio 1:5. The type and concentration of the extraction fluid used should be declared. All preparation processes should be carried out with apparatus made of inert materials resistant to the solvent and the constituents of the drug.

See Section 4.2 for the preparation methods 'Maceration' and 'Percolation' given in *DAB 8*; see Section 5.4 for the apparatus required for these.

Certain simple or composite solutions such as Tinctura Iodi and Nitromersol Tincture in *USP XX*, and others, were also formerly regarded as tinctures. This is still the case in *USP XX*. According to *DAB 8* the term tincture now applies only to medicinal plant preparations. *DAB 8* emphasizes only that tinctures should be 'protected from light' during storage. No shelf-life is given. Periodic examination for alterations in appearance, especially for turbidity or opacity and sedimentation is, however, recommended. Content determinations should be made repeatedly on potent tinctures. The tested tincture should be discarded if the determined active substance content has fallen below the minimum demanded in the Pharmacopoeia.

The statements made under 'Standardization and stability' (see Chapter 6) apply to the other tinctures.

1.3.4.3. Fluid extracts. Extracta fluida

Fluid extracts, like tinctures, are liquid preparations, the difference being that they are more concentrated. *DAB 8* recommend that at most 2 parts of fluid extract shall be made from 1 part of drug.

Other Pharmacopoeias, such as *Helv. VI* or *BP 80*, give no rigid limits for the ratio of drug to total extract, but permit the miscella obtained to be adjusted to a certain active substance content. Hence, for example, *Helv. VI* prescribes the adjustment of Extractum aurantii amari fluidum to a bitterness of 8–12 *Ph. Helv.* units, of Extractum capsici fluidum (which is however prepared with medicinal acetone) to a capsaicin content of 1.8–2.2%, and of Extractum chamomillae fluidum to a content of 0.20–0.30% ethereal oil.

BP 80, like *Helv. VI*, permits potent drugs to be adjusted to a defined active substance content, whereas the fluid extracts of other drugs are to be prepared in the ratio 1:1 to the drug used. Tinctures which, according to the cited Pharmacopoeias, may also be prepared from fluid extracts, generally have a tenth of the content of the fluid extracts. *ÖAB 81* also prescribes a similar procedure. The preparation is generally carried out by percolation or maceration.

An interesting example of the preparation of various types of extract from the same drug is found in *BP 80*, which describes the preparation of the forms:

Fluid extract (squill liquid extract), one part drug corresponding to one
   part fluid extract,
Tincture (Squill Tincture), ratio 1:10, and
Vinegar (Squill Vinegar), ratio 1:10,
from squill, Bulbus scillae maritimae (for Vinegars see Section 1.3.4.8).

1.3.4.4.   Thin extracts. Extracta tenua

Thin extracts, also known as Extracta tenua, are now regarded as obsolete. They are extracts which have been concentrated to a honey-like consistency. Certain solvent-free extracts have a natural honey-like consistency and have been classed as thin extracts; one example is Extractum Filicis maris of *ÖAB 9*.

1.3.4.5.   Viscous extracts. Thick extracts. Extracta spissa

Thick extracts are thickly liquid or viscous when warm, but are no longer fluid at room temperature. They are obtained from the miscellae by extensive but fairly careful concentration and are plastic masses containing varying quantities of residual moisture. They can be adjusted to a defined active substance content by the addition of calculated quantities of inert substances such as dextrin, lactose, etc. They have now been replaced almost completely by dried extracts because of their low stability and their susceptibility to microbial growth.

It should not be forgotten that 'dry' plastically workable extracts can be produced by concentration and drying of liquid extracts, especially when lipophilic extract substances get into the miscella via organic extraction agents, or when the eutectic point of the dry extract is made very low by a large number of components of a hydrophilic extract. Kneadable masses are then formed, especially when the extracts are dried by heating, and become friable, and hence grindable, only upon cooling.

The thick yeast extract made into a paste for the manufacture of pills and which is still official in *DAB 7* is mentioned as an example of a thick extract.

1.3.4.6. Dry extracts. Extracta sicca

Dry extracts are plant preparations obtained by the concentration and drying of liquid extracts under mild conditions. They should, if possible, be adjusted to a prescribed active substance content by the addition of inert adjuvant substances. There are two methods for doing this:

(1) The miscella is evaporated down to a thin extract under the careful conditions given in the particular Pharmacopoeia and this extract broth is then weighed. The dry residue and the active substance content of aliquot portions of it are determined. The thin extract is then adjusted to the required active substance content by the addition of the required quantity of dextrin or lactose. The required quantity of adjuvant substance is calculated by the formula

$$x = \frac{(100 - R)a}{b} - T$$

where a = weight of active substances (in g) in the extract broth, b = required content of active substances in dry extract (%), R = permitted residual moisture in dry extract (%), T = weight of the dry residue of the total extract broth (g), x = weight of adjuvant substance to be added (g).

After addition of the adjuvant substances the extract is dried to the desired residual moisture content.

(2) The miscella is evaporated to dryness; if only small quantities of it are available it is ground with the required quantity of adjuvant substance, which can also be calculated by the above equation. When large technical quantities are being processed, the dry extract, ground as fine as possible, must be mixed homogeneously with the calculated quantity of adjuvant substance in suitable mixers.

Dry extracts are usually very hygroscopic, and should therefore be ground and mixed under conditions which exclude moisture as much as possible. Intermediate and end-products must also be stored under dry conditions. Annealing or sealing of the products in suitable moisture-tight synthetic foils has proved a good method for this. If the foil manufacturers have not stated the permeability of their product by moisture, this can easily be determined by the method of List and Kassis [1.16]. This has the advantage that sufficiently moisture-tight, yet inexpensive foils, can be used.

1.3.4.7. Oily drug extracts. Medicinal oils. Olea medicata. Olea medicinalia

Medicinal oils are preparations containing medicinal substances dissolved or suspended in almond, peanut, olive, poppy seed, apricot kernel or peach kernel oil. Drying oils should not be used, although medium chain-length triglycerides (Miglyol 812 Neutral Oil) could be used to advantage. The medicinal substances can be in pure form or in the form of oily drug extracts. Only the latter are discussed here. Oily drug extracts are obtained by maceration or digestion of the ground drug with the oil.

This dosage form is now used relatively rarely. Because of their instability they should be prepared only in quantities intended for immediate consumption and stored in well-sealed glass containers in a cool, dark place.

Drugs considered for oily extracts are:

St John's wort
Henbane (*Hyoscyamus*)
Aconite (monkshood) root
Arnica blossom
Marigold
Camomile flowers
Rose petals
Tobacco leaves
Mullen flowers, etc.

1.3.4.8. Vinegars. Aceta

Vinegars are very old medicines which are now rarely used. They are obtained by extraction of drugs with dilute acetic acid. Such a preparation is found nowadays only in *BP 80*:

Squill Vinegar. Acetum Scillae

Squill, crushed 100 g

Dilute acetic acid 1000 mL

The drug is macerated for 7 days with acetic acid in a sealed vessel, with occasional stirring. The mixture is then strained, the residue squeezed out and the combined liquids boiled, left to stand for a further 7 days and filtered.

Density at 20°C is 1.025–1.045 g/mL.
Content of acetic acid: 5.4–6.3%.
Total extractive substances determined in 5 mL: at least 4.5% (w/v).

Chapter 2

Starting Material

The starting material for medicinal plant preparations consists of fresh plants or parts of plants and drugs. When the drug plant parts are preservable fruits, seeds, tubers or bulbs no clear distinction has to be made as to whether these are fresh plant parts or drugs. Generally speaking, they are drugs and are treated as such. In exceptional cases, for example for de-enzymation, they are treated as fresh plants. The processes such as are described under Section 2.1.1 commence as soon as such intact forms are cut, crushed or comminuted in any other way.

2.1. Recovery of drugs

2.1.1. BOTANICAL PRINCIPLES

The recovery of drugs, i.e. the processing of harvested or collected fresh plants or parts thereof to the dried whole drug, is the province of pharmacognosy (pharmaceutical phytology) and is described in numerous technical publications and monographs. Only certain fundamental facts of technical importance for further processing are mentioned here.

Hydrophilic active substances, such as alkaloids, glycosides, tannins, etc., are formed in the protoplasts of the living plant cell. From there they migrate through the inner plasma membrane, the tonoplasts, and into the vacuoles containing the cell fluid. Only *one* hydrophilic plant substance, cellulose, does not follow this route, but is secreted through the outer plasma membrane, the plasmalemma, to the exterior to form the basis of the cell membrane.

Lipophilic active substances, particularly ethereal oils, balsams and resins, are likewise formed in the plasma. In the simplest case they can be dispersed in the plasma, e.g. in rose-petals and in the flowers of the lime tree. In the oil cells of the Lauraceae, Zingiberaceae, etc. the plasma is gradually converted into ethereal oil, which then more or less completely fills the cell space. In some families (e.g. Rutaceae) the lysigenic excretion vacuoles are formed by the dissolution and fusion of a number of cells in a manner similar to that in which mucus cells occasionally expand into mucus spaces by dissolution of the cell walls. In other families, e.g. Umbelliferae, the ethereal oil is secreted from actively secreting adjacent cells into an intercellular space expanded by cell division, known as the schizogenic excretion vacuole. In the glandular hairs of the Labiatae and Compositae, this takes place in the extracellular space provided by the raised, swollen cuticular cavity. The hydrophilic active substances in fresh, living plants are therefore stored to a considerable extent in the aqueous cell fluid, whereas the lipophilic substances are rarely present in the hydrophilic plasma, but are usually found in separate cavities containing little water or almost none at all.

Immediately after harvesting, i.e. after separation of certain parts of the plant or after the whole plant, including roots, has been dug up, processes commence which mostly become externally apparent with the withering of the plants. Storage organs and preservable parts of plants such as tubers, seeds and certain dried fruits in principle undergo the same phenomena, although the modifying processes proceed very much more slowly here because of the low water content or because the structure of the skin of the fruit or the seed-case prevents evaporation. It is astonishing that still relatively little appears to be known about the processes which take place during withering of medical plants. A research team in the teaching department of Vegetable Husbandry of the Munich Technical College, which is also involved in the analysis of ethereal oils after harvesting, appears to be the only one concerned with the post-harvesting physiology of vegetables [1.8–1.10]. The consequences of rapid or gradual loss of water in various medicinal plants are of great importance for the processing of fresh harvested plants to drugs, i.e. for drying. Flück [2.1] logically describes the phenomenon as follows:

> The physical and physico-chemical state of the interior of the cell undergoes enormous change through loss of water. The gel state of the plasma is disturbed by dehydration, the active substances originally dissolved in the cell fluid partially precipitate and are attracted back to and into the structurally altered plasma. The enzymes predominantly localised in the plasma in the living plant come into contact with the active substances originally dissolved in the cell fluid as the drying-out process commences and, depending on the type of enzyme and substrate, lead to hydrolysis, oxidation (especially when air can also get in as the tissue dries out), polymerization and other changes. In some plants the active substances and the enzymes exist in different, spatially separated cells, with the result that these reactions take place only when tissue is destroyed during harvesting or during a further processing (e.g. in Cruciferae containing mustard oil, where the enzyme complex myrosin is present only in cells free of glycoside).

Most of the enzymes in the plant need adequate water to act. The water content in dry drugs is so low that decomposition reactions do not usually occur at all, or are retarded to such an extent that a certain degree of storage stability is guaranteed. It should not be forgotten that many enzymes survive the drying process and are reactivated when water/moisture is once again available.

The most important enzymes which break down substances for the production of drugs are:

Oxidases and peroxidases, which mainly oxidize phenols, unsaturated fatty acids, terpenes, etc.,

Hydrolases, which cleave esters and glycosides and break down polysaccharides,

Isomerases, which for example isomerize ergot alkaloids or other optically active substances.

There are two ways of protecting the drugs from the action of these enzymes:

(i) Drying as rapidly as possible,
(ii) Denaturing the enzymes

The latter is known as stabilization of drugs (see Section 1.2.4), and it is hardly ever practised on a large scale.

Enzymatic processes are not always undesirable. In many cases active and aromatic substances and other useful compounds are released only by such processes. The fermentation of tea to Thea nigra, the retting or steeping (fermentation) of fresh cocoa beans which produces the colour and aroma of the cocoa, and the release of coumarin from melilot (*Melilotus*) (or white clover, *Trifolium repens*) and woodruff (*Asperula odorata, Galium odoratum*) are mentioned as examples. The storage of alder buckthorn bark (Frangula cortex) for at least one year is recommended by the Pharmacopoeias for the enzymatic breakdown of constituents [2.2], and this can be considered as a process of this type.

There is therefore no generally applicable rule for the type of drying and storage processes to be carried out after harvesting.

2.1.2. Problems in the production of drugs of Pharmacopoeia quality

Schilcher [2.3] has listed and illustrated 11 of the most difficult problems in the production of drugs of Pharmacopoeia quality. Against the background of the very high consumption of drugs, this topic is of special importance. About 12 000 tons of drug plants were processed in West Germany in 1980 for herbal infusions alone. This does not include drugs processed to other dosage forms. Schilcher names the following possible problems:

(1) The drug is not as specified in the Pharmacopoeia because of deliberate or inadvertent substitution or adulteration.
(2) The plant parts do not correspond to those stated in the Pharmacopoeia.

(3)   The permitted level of 'foreign matter' laid down in *Ph. Eur. I* has been exceeded.
(4)   The maximum value of sulphate ash has not been adhered to.
(5)   The minimum content requirements for certain constituents either cannot be fulfilled for one year at all or cannot be fulfilled for quantities sufficient to meet demand.
(6)   The drug has been rendered unfit for use through infestation or microbial spoilage.
(7)   Microorganism contamination is too high.
(8)   Pesticide or preservative residues exceed permitted levels.
(9)   The quantities of heavy metal residues are in excess of those laid down for foodstuffs.
(10) Economic and industrial factors, for example standards, or sociological problems especially in the countries where the drugs are produced, make it difficult to meet the demands of the Pharmacopoeia, particularly with regard to providing a continuous supply.
(11) Legal regulations, e.g. the law on the import and export of wild animals and plants, cause almost insuperable difficulties when they are brought into force too quickly.

The author gives a number of striking examples of the cited problems. Shortly before this Schier had reported on drug adulterations in detail [2.4]. Drugs can be obtained from medicinal plants grown on farms or from collected wild medicinal plants. The latter have the advantage of being substantially free of pesticide residue, [2.5], though of course they are subject to the hidden risk of adulteration, confusion or mixing with other drugs, incorrect treatment and admixture of foreign material. A major obstacle to the collection of wild medicinal plants is the Washington Species Protection Agreement of 3 March 1973. The Agreement which has been in force in West Germany since 20 June 1976 aims at protecting certain endangered species of animals and plants by regulating international trade in them.

In its *Pharma Jahresbericht (Annual Drugs Report)* 1981/1982, pp. 49–50 [1.5] the West German Pharmaceutical Industries Association emphasizes that the manufacturers of herbal medicines were surprised by a draft of an order on the import and export of wild animals and plants submitted by the Federal Minister of Food, Agriculture and Forestry. The Association was asked for an opinion on it at a very late stage. As the draft order is still under discussion it is hoped that an exception clause can be made for medicinal plants.

Medicinal plants grown agriculturally are not subject to such restrictions. They also have a number of other advantages in that the risk of adulteration and confusion with other drug plants is almost totally eliminated. Breeding and perhaps even genetic techniques may make it possible to cultivate species which are more productive and resistant to disease, etc., and easier to process. Proper care of the ground in which they are grown can to a certain extent guarantee adequate harvest yields. One disadvantage is that the use of pesticides cannot be avoided. The type and quantity of pesticides used can of course be controlled with effort in order to keep the residue problem within tolerable limits.

In a press communication of 13 July 1982, A. Nattermann & Co., Cologne, informed the public that they were putting a phospholipid mixture on the market, obtained from plants, under the name 'Natipide', by means of which—probably through improvement of the wettability and spreadability of the formulations—up to 50% savings could be made on conventional plant pesticides [2.6].

A further potential source of plant starting-materials which has come under lively discussion in some quarters, namely tissue culture, would overcome all the previously mentioned disadvantages. However, this technique still requires a large amount of scientific research and will not be available to meet demands for some time yet.

2.1.3. PEST CONTROL

As stated under Section 2.1.2, there are various reasons for increasing the agricultural production of medicinal plants, and this involves the practice of all types of pest control.

The task of pest control is to ward off or destroy those plant or animal organisms which in any way, directly or indirectly, threaten human health or nutrition. It can be done with biological, physical or chemical agents. Detrimental effects on the biological equilibrium are to be avoided as far as possible by enforcing an integrated method of pest control in which biological pest control is coupled with physical and chemical methods.

Biological pest control uses the natural enemies of the pests to destroy them or promotes the colonization of the cultivation area by these enemies. Biological methods are also used for the protection of stored products. The simplest and certainly the oldest example is to keep cats in granaries and mills. 'Artificial' biologial methods are sometimes used; one example is the use of preparations of *Bacillus thuringiensis*, which selectively attack about 100 species of caterpillars but not useful insects.

The physical methods include mechanical weeding and the application of grease bands to fruit trees.

Biological and physical methods are, however, far exceeded by chemical pest control. Only with their aid have we succeeded in controlling pests so effectively that crop yields have increased considerably.

Chemical pesticides have of course been the cause of certain unpleasant side-effects, such as the occasional disturbance of the biological equilibrium (particularly after incorrect use of the chemical agent), the occurrence of resistance of pests to these pesticides in the case of pests which rapidly pass from one generation to the next, but especially the persistence of some of the pesticides or their metabolites in plants and other harvested material. This so-called residue problem is also a major factor in the drugs being considered here.

The areas of use of pest control are in the protection of growing plants, plant products in warehouses, materials, timber, wood, and also in general hygiene. The protection of plants, which is generally the most important in the cultivation of medicinal plants, includes all branches of agriculture such as the cultivation of arable crops, fruit, wine, medicinal plants, hops, vegetables and ornamental plants, as well as grassland, greenhouse crops and, not least, forests. Plant protection also includes the treatment of seeds and soil. The objective of warehouse protection is to deter and destroy pests such as insects, mites and rodents in harvested products. Pest control, or better still the keeping of pests away from drug warehouses, has an important part to play here.

2.1.3.1. Chemical pest control

Chemical pest control agents are used in the form of suitable preparations (formulations). Depending on the properties of the active substances and the products which they are to protect, they contain suitable carriers (e.g. kaolin, talc, silicic acid), adjuvants such as wetting agents and emulsifiers, organic solvents or propellant gases.

Various of active substance preparations exist, namely:

(a) Spray- or wettable powders: The active substance, together with the carrier material and wetting and dispersion agents, is finely ground to a particle-size of 1–20 μm and suspended in water prior to use.

(b) Powders: The active substance is finely ground to a particle-size of 10–50 μm in low concentration with the carrier material and any necessary addition of adhesives, and used without any further processing.

(c) Granulates: Here the active substance is adsorbed on to granulated carrier material (granule size 300 μm to several mm) and thus disseminated.

(d) Self-emulsifying concentrates: The active substance, to which emulsifiers have been added, is dissolved in as high a concentration as possible in an organic solvent, usually mineral oil. Immediately before use the concentrate is poured into the prescribed quantity of water, where it emulsifies spontaneously.

(e) Solutions in mineral oil fractions: These contain no emulsifiers and are applied to the plants to be protected either directly or as aerosols.

(f) Fumigants: The active substance is contained in a salt mixture acting as a 'trap' from which it vaporizes when heated. This type of preparation is used in closed rooms (store-rooms, greenhouses, etc.).

Depending on the pests to be controlled, the chemical agents can be divided into:

Fungicides
Herbicides (weed-killers)
Insecticides
Acaricides
Nematicides
Rodenticides
Bactericides
Other agents such as molluscicides, bird deterrents, rodent deterrents, etc.

The range of chemical pest control agents is not only very large but is also continually changing. Many of the substances listed here are no longer permitted in Germany or have been replaced by new compounds. DDT, for example, is no longer permitted in the western industrialized countries. However, it is still used in other countries, especially for the control of the malaria-transmitting *Anopheles* mosquito, and new DDT factories are still being built.

When testing drugs or drug plants for pesticide residues (see Section 2.2.1.10) it is more important to try to analyse for typical groups of pesticides rather than for individual compounds.

2.1.3.1.1. Fungicides

Fungicides are substances which kill fungal moulds and hence are used to prevent or control plant diseases caused by phytopathogenic fungi. They are used as seed or soil fungicidal disinfectants.

Seed disinfectants give prior protection to seeds against fungal disease pathogens which adhere to or are admixed with seed in the form of their dissemination organs such as spores, hyphae and scleratia. Disinfection of soil is much more difficult and expensive. There is also a group of fungicides called leaf fungicides, which are used for protecting leaves and fruits.

The substances principally used as fungicides are dithiocarbamates, organophosphorus compounds, pyridine derivatives, carboxin, thiabendazole and morpholine derivatives.

The following compounds have been and are still being used as leaf fungicides, some of them since the last century:

Bordeaux mixture (liquor), copper oxychloride,
colloidal sulphur, barium sulphide,
thiurams and dithiocarbamates,
N-(trihalogenomethylthio)-compounds,
nitro-compounds such as dichloronitroaniline,
dithia-compounds such as dithiadicyanoanthraquinone.

Some of these compounds possess both fungicidal and acaricidal activity.

2.1.3.1.2. Herbicides

Herbicides or weedkillers are chemical agents used for destroying plants or for preventing their growth. There are several types which differ in their mode of action, namely:

Total herbicides: which prevent or destroy every all plant growth.
Selective herbicides: which control the growth of undesirable plants (weeds) but do not affect cultivated plants.
Waterweed-killers: which act against aquatic plants in still and flowing water.
Harvesting aids: which, for example, cause the defoliation of cotton plants and hence make possible the mechanical harvesting of cotton.

Practically, only selective herbicides are used in the cultivation of medicinal plants. Table 2.1 shows the weed-killers which are now of economic importance (from Grimme [2.8]).

The compounds most commonly used in the cultivation of medicinal plants are carbamates, urea derivatives, growth promoters, di- and triazines and quaternary ammonium compounds.

2.1.3.1.3. Insect-controlling agents

This heading covers

Insecticides
Insect repellants
Insect attractants
Insect-sterilizing agents

Insecticides are by far the most important of this group. Attractants (e.g. sexual attractants) and insect-sterilizing agents have not yet been sufficiently tested and hence at present they can only be used experimentally.

Insecticides, unlike protective fungicides, are used only when the material to be protected is already infested with insects at various stages of development. The great variety of insecticides now available is partly due to the enormous number of insect species and partly to the occurrence of resistant insect strains. Efforts have also been made to create selective insecticides active only against harmful insects, but not against useful insects such as bees. A constant search is also being made for insecticides with the lowest possible toxicity to mammals.

The way in which insecticides are used is governed roughly by whether the target insects are biting insects such as beetles or caterpillars, or sucking insects such aphids and bugs. Insects can absorb the active substance by respiration, in their food or by contact. Insecticides are therefore described as respiratory toxins, food toxins and contact toxins. They can be classified by mode of action as

Table 2.1 Herbicides of current economic importance

Phenoxyacetic acid derivatives

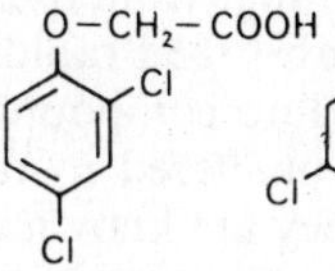

2,4-D 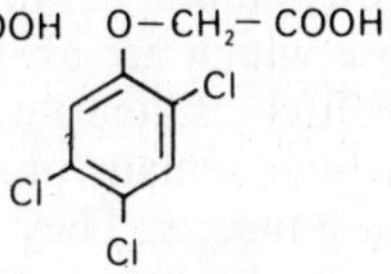   2,4,5-T 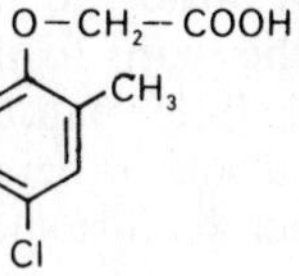   MCPA

Substituted phenols

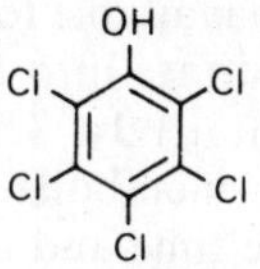

Pentachlorphenol
PCP

Substituted ureas

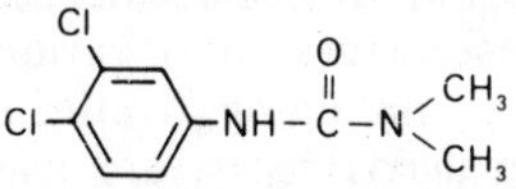

Diuron

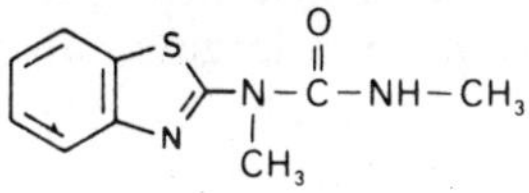

Methabenzthiazurone

s-Triazines

Simazine

Quats

Diquat

Paraquat

Carbamates and thiocarbamates

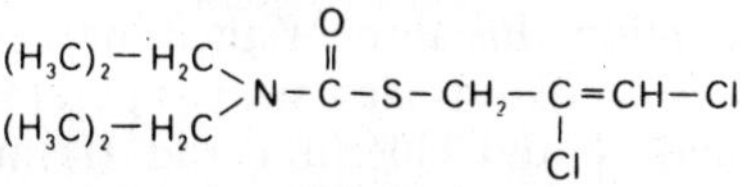

Diallate

Chlorinated aliphatic acids

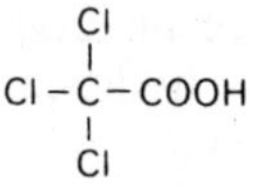

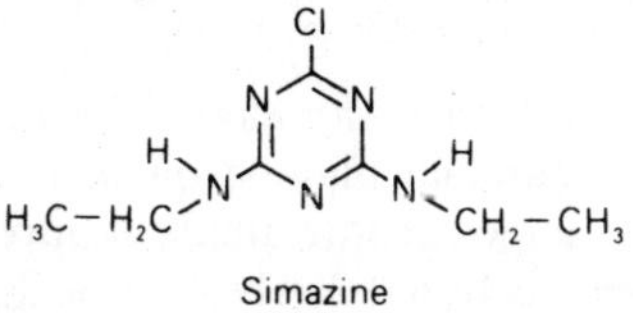

TCA 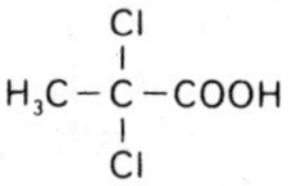   Dalapan

Toluidines

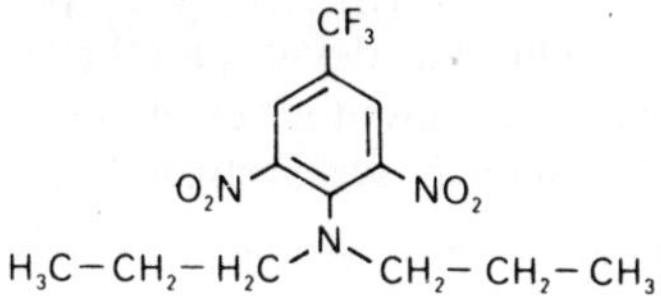

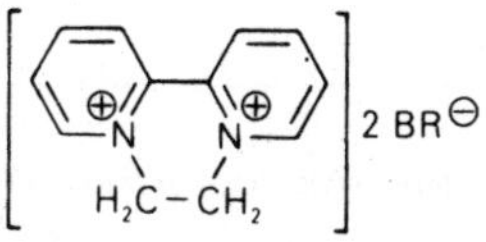

Trifuraline

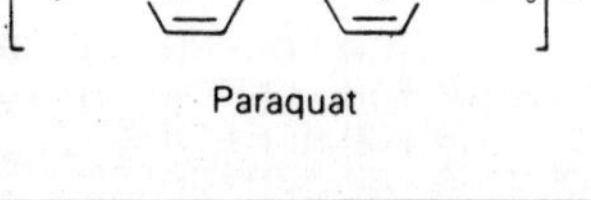

Individual herbicide

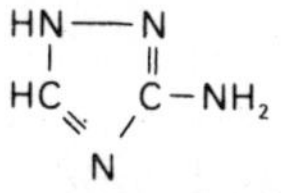

Amitrol

General cell toxins (protein-precipitants).
Enzyme toxins (phosphoric acid esters and N-alkylcarbamates).
Nerve toxins (e.g. chlorinated hydrocarbons).

The insecticides available for protecting medicinal plants range from those which do not penetrate into the plant to those which are absorbed rapidly and almost quantitatively. The latter can be differentiated into compounds which are rapidly metabolized and others which remain unaltered in the plant sap for some time and act against sucking insects. They are known as 'systemic insecticides'.

Inorganic insecticides have now largely been replaced by organic compounds because of their toxicity to mammals. They include arsenic compounds such as lead and calcium arsenate and Schweinfurt green, and fluorine compounds such as sodium fluoride, sodium fluorosilicate and cryolite. Because of their natural origin, plant insecticides from various *Tanacetum* species, such as pyrethrum (a crude extract containing pyrethrin I and II and cinerin I and II) and rotenone from derris roots, are used particularly in the hygiene sector. Their high price prevents their widespread use in the cultivation of medicinal plants. Although some insecticides have been withdrawn from the market, they are still available in large numbers and can be divided into four main groups:

(1)   Plant-derived insecticides.
(2)   Chlorinated hydrocarbons and related compounds, which also include DDT.
(3)   Organophosphoric acid esters and thiophosphoric acid esters.
(4)   Carbamidic acid esters (carbamates).

As there are so many synthetic insecticides, and in many cases it is not known which pesticide has been used by the medicinal plant grower, it is expedient when testing for pesticide residues to look for insecticide groups. Schilcher has devised methods for doing this (see Section 2.2.1.10). Table 2.2 shows representatives of the major groups of insecticides. For the chemistry of pesticides see Section 2.3.7.

Table 2.2   Principal groups of insecticides, with examples

Plant insecticides
Pyrethrum: Pyrethrum is a crude extract containing, principally, four esters, pyrethrin, I, pyrethrin II, cinerin I, and cinerin II)

Pyrethrin I: $R_1 = CH_3$; $R_2 = CH_2 - CH = CH - CH - CH_2$
Pyrethrin II: $R_1 = COOCH_3$; $R_2 = CH_2 - CH = CH - CH = CH_2$
Cinerin I: $R_1 = CH_3$ $R_2 = CH_2 - CH = CH - CH_3$
Cinerin II: $R_1 = COOCH_3$; $R_2 = CH_2 - CH = CH - CH_3$

Table 2.2 *Continued*

Nicotine: β-Pyridyl-α-methylpyrrolidine

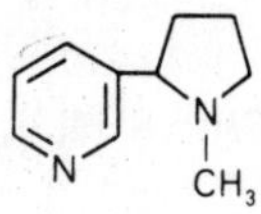

Rotenone: Derris

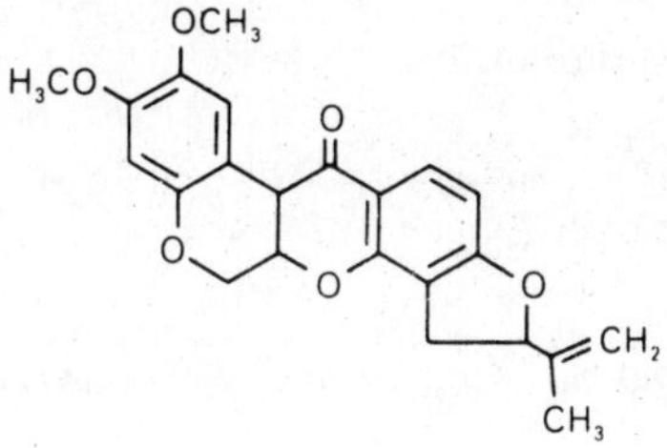

Chlorohydrocarbons and related compounds

Dichlorodiphenyltrichlorethane: DDT
Banned in many countries

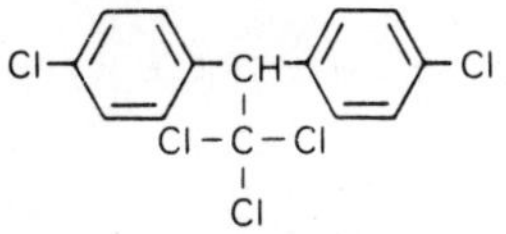

Hexachlorcyclohexane: Gammabenzeni Hexachlodium (WHO); HCH; Gammexane

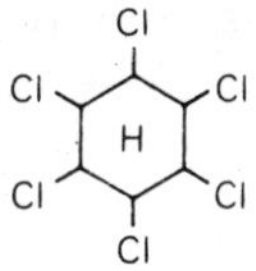

Organophosphorus compounds

O,O-Diethyl-O-(4-nitrophenyl)thionophosphate: Parathion; E-605

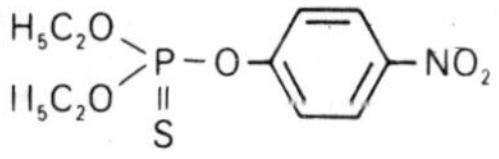

Methylparathion

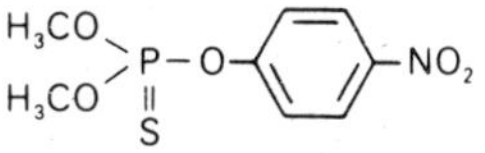

Dalf

Mp 35 36 °C
LD_{50}: 14 42 mg/kg in the rat

Fenitrophion

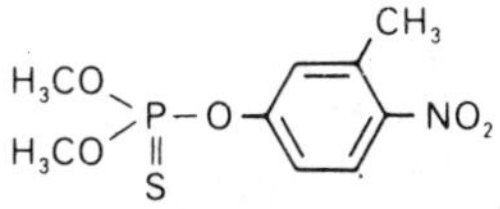

Folithoin; Sumithion

Bp at 0.1 mm Hg 109 °C
LD_{50}: 200 500 mg/kg in the rat

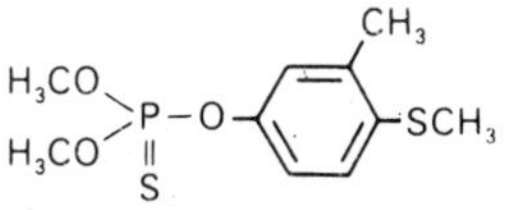

Baytex; Lebaycid

Table 2.2 *Continued*

O,O-Diethyl-O[2-(ethylthio)ethyl] phosphorothioate; Demeton; Systox

Mixture of 70%   30%

$$\begin{array}{l} H_5C_2O \\ \diagdown \\ P-O-CH_2-CH_2-S-C_2H_5 \\ \diagup \;\; \| \\ H_5C_2O \;\;\; S \end{array}$$

$$\begin{array}{l} H_5C_2O \\ \diagdown \\ P-S-CH_2-CH_2-S-C_2H_5 \\ \diagup \;\; \| \\ H_5C_2O \;\;\; O \end{array}$$

O,O-Diethyl-O[2-(ethylthio)ethyl] phosphorothioate; Methyldemeton; Metasystox

70%   30%

$$\begin{array}{l} H_3CO \\ \diagdown \\ P-O-CH_2-CH_2-S-C_2H_5 \\ \diagup \;\; \| \\ H_3CO \;\;\; S \end{array}$$

$$\begin{array}{l} H_3CO \\ \diagdown \\ P-S-CH_2-CH_2-S-C_2H_5 \\ \diagup \;\; \| \\ H_3CO \;\;\; O \end{array}$$

Metasystox (i)   Metasystox R

$$\begin{array}{l} H_3CO \\ \diagdown \\ P-S-CH_2-CH_2-S-C_2H_5 \\ \diagup \;\; \| \\ H_3CO \;\;\; O \end{array}$$

$$\begin{array}{l} H_3CO \\ \diagdown \\ P-S-CH_2-CH_2-S-C_2H_5 \\ \diagup \;\; \| \qquad\qquad\quad \| \\ H_3CO \;\;\; O \qquad\qquad\quad O \end{array}$$

Carbamates

3,5-Dimethyl-4-methylthiophenyl-N-methylcarbamate: Metmercapturon; Mesurol

2-Isopropoxyphenyl-N-methylcarbamate: Propoxur; Baygon; Blattanex; Unden

---

2.1.3.1.4. Acaricides

Acaricides are substances for controlling mites, particularly spider mites. These mites belong to the Arthropoda and multiply very rapidly (up to about 10 generations in one season of plant growth). They can cause considerable damage to the foliage of cultivated plants. Resistance to the usual chemicals soon appears as a result of their rapid multiplication and hence new acaricides must continually be developed.

Acaricidal substances occur among both fungicides and insecticides. Phosphoric and thiophosphoric acid esters, carbamidic acid esters and chlorinated aromatics with and without sulphur in the molecule exhibit acaricidal activity and are thus also found in the appropriate groups during testing for pesticide residues (see Section 2.2.1.10).

2.1.3.1.5. Nematicides

Nematicides are agents for the control of phytopathogenic threadworms (nematodes) living free in the soil or occurring in plants. Halogenated hydrocarbons, carbamidic and thiocarbamidic acid derivatives and thiophosphoric acid esters, the residues of which are also detected in the search for the corresponding groups, are also suitable nematicides (see Section 2.2.1.10).

2.1.3.2. Antibacterial and simultaneously antiverminous agents

2.1.3.2.1. Terms and definitions

Strictly speaking, these should come under the heading chemical pest control agents (Section 2.1.3.1). However, the measures described here are aimed at the treatment of drug plants after harvesting (rather than during growth) but before further processing. Long before chemical agents were used for pest control in the cultivation of medical plants and in the processing of harvested medicinal plants, food warehouses and transport depots and vehicles were being treated with chemical agents, usually pesticidal gases, for protection against or destruction of rodents and insects. The same applies to the destruction of parasites in dwellings and clothes. (Insecticides had already been in use for a long time for the destruction of head and body lice and for the control of itch mites). Hydrogen cyanide, methyl bromide, ethylene oxide and occasionally methyl formate are mainly used as toxic gases for pest control by fumigation, ethylene oxide being the most frequently used in the processing of harvested drug plants, although ethylene oxide is now prohibited as a pesticide and microbicide in quite a few countries.

Although ethylene oxide was formerly used principally for the destruction of vermin, the treatment of harvested drug plants with ethylene oxide is now directed primarily against microorganisms. With regard to the microbial purity of medicinal substances, which do not necessarily have to be sterile [2.9], harvested drug plants belong to category III, that is, they must be free of pathogenic bacteria and are permitted to contain at most 10^4 apathogenic bacteria/g and 100 fungal mould or yeast cells/g.

The limitation to 10^4 apathogenic bacteria represents a value which is still being debated, the demand is for a maximum of 10^3 bacteria/g. In view of the fact that bacterial count determinations obviously suffer a very large degree of error, a limitation of 10^3 bacteria/g would be meaningful for plant drugs taken in powder or tablet form of their natural state. Plant drugs which have been processed further should not contain more than 10^4 bacteria/g and here every effort should be made to ensure that the product finally handed over to the consumer contains not more than 10^3 apathogenic bacteria/g (mL) and no pathogenic bacteria (see also Section 2.2.1.9). Wallhäusser points out [2.10] that the demand for material 'free of pathogenic bacteria' far exceeds the capacity of the *in vitro* tests demanded in the Pharmacopoeia. The pathogenicity of bacteria cannot be definitely proved even by animal experiments. Moreover, no clear boundary can be drawn between pathogenicity and apathogenicity of microorganisms. There is therefore an increasing tendency to formulate from experience a catalogue of undesirable bacteria which includes certain Enterobacteriaceae such as *Salmonella* and *Shigella* species and, as a faecal indicator, *Escherichia coli*, as well as *Pseudomonas aeringosa* and *Staphylococcus aureus*.

As the fumigation conditions for reducing the bacterial count must be more rigorous than those for the destruction of vermin, the latter is only to be regarded as an attendant phenomenon of antimicrobial treatment.

2.1.3.2.2. Chemical and physical properties of ethylene oxide

Ethylene oxide may be obtained by the addition of hypochlorous acid to ethylene and the dissociation of hydrochloric acid from the resultant ethylene chlorohydrin with alkali at 100°C:

$$H_2C{=}CH_2 + HOCl \longrightarrow \underset{\underset{HO}{|}\quad\underset{Cl}{|}}{H_2C{-}CH_2}$$

$$HO{-}CH_2{-}CH_2Cl \xrightarrow[100°C]{Ca(OH)_2} H_2C\underset{\diagdown\diagup}{}CH_2 + CaCl_2 + H_2O.$$

or by direct oxidation of ethylene in the presence of silver catalysts:

$$2H_2C{=}CH_2 + O_2 \xrightarrow{Ag} 2H_2\overset{\diagup O \diagdown}{C{-}}CH_2$$

It is an easily liquefiable gas with a characteristic sweetish smell. It has the following physical characteristics:

Mol. wt.	44.05
Vapour density, relative to air	1.52
b.p.	10.5°C
m.p.	-112.5°C

Critical temperature   196.0°C
Critical pressure   73.3 bar
Viscosity at 0°C   0.32 mPa s

Density of liquid		Specific volume of liquid	
0°C	0.900 g/cm^3	0°C	1.111 cm^3/g
10°C	0.887 g/cm^3	10°C	1.127 cm^3/g
20°C	0.873 g/cm^3	20°C	1.145 cm^3/g
30°C	0.858 g/cm^3	30°C	1.165 cm^3/g
40°C	0.843 g/cm^3	40°C	1.186 cm^3/g
50°C	0.826 g/cm^3	50°C	1.210 cm^3/g

Liquid ethylene oxide expands enormously upon warming. The regulations on pressurized gases demand that 1 kg ethylene oxide must be allowed 1.3 L filling volume.

Heat of fusion (melting)	5.19 kJ/mol
Heat of vaporization	25.5 kJ/mol
Specific heat (liquid) at 0°C	85.9 kJ mol^{-1}°C^{-1}
at 25°C	87.7 kJ mol^{-1}°C^{-1}
Specific heat (gas) at 25°C	48.5 kJ mol^{-1}°C^{-1}
Heat of polymerization	92 kJ/mol
Heat of combustion	1308 kJ/mol
Flame point	−57°C
Decomposition temperature	571°C
Ignition temperature	ca. 430°C
Ignition group (VDE)	G 2
Explosion class (VDE)	2
Explosion limit in mixture with air	3–100 vol%
Max. workplace concentration	50 ppm
equivalent to	(cm^3 vapour/m^3 air) = 92 mg/m^3

2.1.3.2.3. Bactericidal action of ethylene oxide

Ethylene oxide is a very reactive compound because of the high strain in the oxiran ring. It reacts easily with nucleophilic compounds, by hydroxyethylation, according to the following scheme:

$$
R\!-\!\begin{cases} OH \\ COOH \\ NH_2 \\ NHR \\ CONH_2 \\ SH \end{cases} + \; CH_2\!\!-\!\!CH_2\; \longrightarrow \; R\!-\!\begin{cases} OCH_2CH_2OH \\ COOCH_2CH_2OH \\ NHCH_2CH_2OH \\ NRCH_2CH_2OH \\ CONHCH_2CH_2OH \\ SCH_2CH_2OH \end{cases}
$$

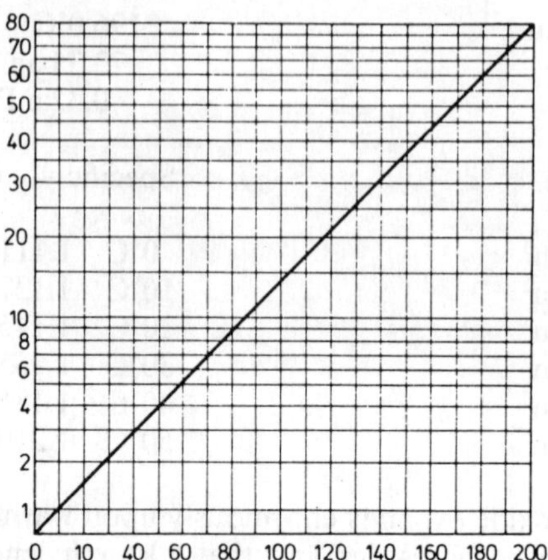

Figure 2.1 Vapour pressure of ethylene oxide.

The reaction is accelerated by acids and alkalis or by catalysts such as metal salts. In the presence of halogen ions it forms so-called hydrins (e.g. ethylene chlorhydrin) which in turn can also be regarded as alkylating agents. The toxicity of ethylene oxide makes it necesary for drug preparations gassed with this substance to be tested for residues of the gas (see Section 2.2.1.10). Proteins, being carriers of such functional groups, are the preferred reaction partners for ethylene oxide. This explains its action on microorganisms. In addition to bacteria, viruses, bacteriophages, yeasts and fungal moulds are also attacked ethylene oxide. Only spores possess partial resistance [2.11–16].

2.1.3.2.4.   Conditions for optimum ethylene oxide fumigation

The reactions of ethylene oxide with the constituents of drug plants, like its antimicrobial activity, also depend on a series of experimental conditions. The concentration of the gas in the fumigation area, the time for which it acts, the temperature and the prevailing atmospheric humidity or the water content of the material being treated are of great importance. The type of drug (leaf, herb or weed, root or bark drugs in intact, coarsely shredded or powdered form) has a definite though less significant influence.

Horn *et al.* [2.17] have worked out the following optimum conditions from experience of sterilization with ethylene oxide:

Gas concentration = 800–1200 mg/L.
Temperature = 50–60°C.
Relative humidity = 45–90%.
Action time = 2 h.

In practice the following three methods are mainly used:

(a) Sterilization at an excess pressure of 5–6 bar with 10–15% ethylene oxide, for about 1.5 h. This has the advantage that a non-combustible gas mixture of 10–15% ethox and 90–85% CO_2 can be used. However, the disadvantage is that pressure-resistant containers must be used.

(b) Sterilization at normal atmospheric pressure, 10–15% ethox, for 4–7 h. This method uses simple containers. However, it has the disadvantage that the material must be exposed to the gas for a long time due to slow diffusion.

(c) Sterilization at a pressure below atmospheric with 90–100% ethox. The low-pressure method is preferred, as higher and hence more effective gas concentrations can be used and thus the other parameters can be kept small. Either pure ethox or a mixture of 90% ethox and 10% CO_2 is used as sterilizing agent. This has the advantage that preevacuation of the chamber causes the inflowing gas to penetrate rapidly into all empty spaces within the material in the chamber; the absence of air inside the chamber eliminates the risk of explosion; as the pressure is still slightly below atmospheric pressure even after the gas has flowed into the chamber, no ethox can escape. A variant of this method is represented by the Degesch circulation method, which guarantees an even distribution of gas in the fumigation chamber.

Common to all these methods is the fact that after sterilization is complete any unused ethylene oxide is removed harmlessly from the chamber by evacuating the chamber, then flushing it out with air and evacuating it again. This procedure may have to be repeated to remove all of the ethylene oxide. There is no great risk of harmful quantities of ethylene oxide remaining in the fumigated drugs (i.e. no residue problem).

Most of the residues still present immediately after fumigation volatilize upon storage of the drugs before further processing or use. Moreover, any ethylene oxide which might still be present will be hydrolysed to hardly toxic ethylene glycol during extraction of the drugs with water or with organic liquids containing water. The same applies to any ethylene halohydrins (e.g. ethylene chlorohydrin) produced during the fumigation.

As the constituents of the fumigated materials, in this case drug plants, react with ethylene oxide to a varying extent according to their chemical nature and the chosen conditions, these very conditions must be carefully selected so that reactions with constituents, and particularly with active substances, do not take place to a detectable extent but so that an adequate microbicidal effect is attained.

It has been found that primary, secondary and even tertiary amino-groups react readily, which puts alkaloid drugs especially at risk. The most favourable fumigation conditions can be established by preliminary tests on drugs containing various active groups.

2.1.3.2.5.   Experimental protocol for determining the conditions for optimum ethylene oxide fumigation

2.1.3.2.5.1.   One-factor method

An investigation of all the factors involved in ethylene oxide fumigation would be very time-consuming and expensive if the classical 'one-factor method' were used. The most important rule of this method states that in each experiment only one parameter may be altered, as otherwise the measurement results cannot be assigned to the influence values. In the case of ethylene oxide fumigation this means that a total of five experiments (ten experiments if a repetition is to be made for each factor) are necessary for the four influence values (factors), i.e. ethylene oxide concentration, fumigation time, moisture content of the drug, and the fumigation temperature. Each factor may be varied here between only two levels.

If these factors are designated by the letters A, B, C and D, respectively, and the two levels by the indicies 1 and 2, the following equations are produced, e.g. for the alkaloid content (x) of a drug which is to be fumigated:

$$x(A) = x(A_2 - A_1) = x(A_2B_1C_1D_1 - A_1B_1C_1D_1)$$

$$x(B) = x(B_2 - B_1) = x(A_1B_2C_1D_1 - A_1B_1C_1D_1)$$

$$x(C) = x(C_2 - C_1) = x(A_1B_1C_2D_1 - A_1B_1C_1D_1)$$

$$x(D) = x(D_2 - D_1) = x(A_1B_1C_1D_2 - A_1B_1C_1D_1)$$

This method has the disadvantages that only some of the measurements are utilized for determining the effect of a variable, and that this 'planning of individual stages' only permits recognition of the effect of a single factor and does not enable interactions between the factors to be determined. Furthermore, all alterations of active substance content are always measured against the experimental combination $A_1B_1C_1D_1$. If this combination contains an error, this error will be transmitted to all further calculations.

2.1.3.2.5.2.   Factorial design

The disadvantages of the one-factor method can be avoided by using a statistical experiment planning method known as 'factorial design'. This was developed in England for use in agricultural fertilizer trials and has subsequently been adapted to the needs of the chemical industry.

The differences between conventional experiment planning and factorial design can be illustrated by the example of a simple reaction. The object is to investigate the effects of pressure and temperature on the yield of a reaction. The combinations shown in Table 2.3 are obtained with minimal experimen-

Table 2.3 Experimental parameters in the one-factor method		
Pressure	Temperature	
	T_1	T_2
P_1	(1)	(2)
P_2	(3)	

Table 2.4 Experimental parameters in factorial design		
Pressure	Temperature	
	T_1	T_2
P_1	(1)	(2)
P_2	(3)	(4)

tal material using the one-factor method, when T_1 and P_1 represent the lower and T_2 and P_2 the upper temperature and pressure levels.

The effect of altering the temperature is calculated by subtracting result (1) from result (2), and the effect of altering the pressure by subtracting result (1) from result (3). The accuracy of the result is increased by repeating each experiment, i.e. a total of six experiments must be carried out. However, if a factorial design is used for planning the experiments, a further experiment is necessary, i.e. T_2P_2 (4), as shown in Table 2.4.

Effect of temperature:
$\frac{1}{2}$ (result [(2) + (4)] − result [(1) + (3)])

Effect of pressure:
$\frac{1}{2}$ (result [(3) + (4)] − result [(1) + (2)])

Effect of reciprocal action (interaction):
$\frac{1}{2}$ (result [(4) + (1)] − result [(2) + (3)]).

The advantages of factorial design are obvious. Although four experiments have to be carried out here instead of three in the first experiment plan (or six with repetition), the effect of each parameter is obtained from two experiments (hence the division by 2) and information is also obtained on the interaction between the two factors which cannot be obtained with the one-factor method. The advantage of factorial design becomes even more obvious when additional factors are incorporated into the planning.

The theoretical principles of factorial design require that certain terms are to be defined and rules observed if the method is to be used correctly [2.18–20].

(i) Factor. By factor we mean any parameter which influences yield, quality or cost of a product and which is deliberately stipulated from experiment to experiment. A distinction is made between quantitative factors, which can be expressed by appropriate numerical values (e.g. pressure, temperature, concentration, etc.) and qualitative factors, for which no numerical values are possible (e.g. various materials, machines, etc.).

(ii) Level. The numerical adjustment of the quantitative factors or the arbitrary assignment of the qualitative factors to certain grades is defined as choice of level. Upper levels are indicated by a plus ($+$) sign or by the assignment of the lower case letters of the respective factors, lower levels by a minus ($-$) sign or the number (1), e.g. *a* means that factor *A* is at its upper level, all others are at the lower level, as in combination (1). In the choice of level it is essential that the difference between the upper and lower level should not be too large nor too small. In the first case there is the danger of the experiments being too close to the technically no longer feasible area. In the second case of a factor with a relatively small effect, there is the risk of the effect being no longer measurable, so that it could actually have been kept constant.

(iii) Effects. The effects describe the action of the influence of a factor on the investigation value. Here the experimental results must be in the form of numerical values. Scales have to be introduced when assessing the influences with subjective criteria, i.e. when it is not possible to give a numerical value to the results.

(iv) Interactions. Interactions produce superadditive results through the interaction of, and the reciprocal influences on, a number of factors (i.e. synergism of the effects).

Planning of experiments

A factorial experiment plan is designed on a very regular basis, starting from the experimental combination in which all factors are at the lower level [combination (1)]. The subsequent combinations are produced by multiplying these factors in alphabetical order with each immediately preceding combination. In this way it is possible to formulate large designs in the so-called standard form, which greatly facilitates calculation of the effects without the aid of computers (Table 2.5).

To avoid systematic errors the standard form in which the experiments are carried out must not be used as the sequence, as this must be random.

Deductions from the calculations of the effects and interactions

Definite statements on the influence of one or more factors on the result can be made when the numerical values of the effects and interactions are known. Calculation of these values is not complicated when the guidelines given below are followed.

(1) The effect of a factor (e.g. *A*) on yield is shown as the difference from the arithmetic means of the experiments carried out at the upper and lower level of this factor. In the 2^4 design

$$A = \tfrac{1}{8}[(a + ab + ac + abc + ad + abd + acd + abcd)$$
$$- ((1) + b + c + bc + d + bd + cd + bcd)].$$

Table 2.5 Example of Yates' method (2^3 design)

Standard form	Effects			
	A	B	C	D
(1)	−	−	−	−
a	+	−	−	−
b	−	+	−	−
ab	+	+	−	−
c	−	−	+	−
ac	+	−	+	−
bc	−	+	+	−
abc	+	+	+	−
d	−	−	−	+
ad	+	−	−	+
bd	−	+	−	+
abd	+	+	−	+
cd	−	−	+	+
acd	+	−	+	+
bcd	−	+	+	+
abcd	+	+	+	+

(2)   If the effect of a factor at the upper and lower levels of another factor varies in size it is assumed that the two factors are interacting. The interaction between factors A and B ($= AB$) is defined as half the difference between the effect of factor A at the upper level of B and the effect of A at the lower level of B. In the 2^4-design AB is calculated as follows:

a)   Effect of A at upper level of B:
$$A_2 = \tfrac{1}{4}(ab + abd + abc + abcd) - \tfrac{1}{4}(b + bc + bd + bcd)$$

b)   Effect of A at lower level of B:
$$A_1 = \tfrac{1}{4}(a + ac + ad + acd) - \tfrac{1}{4}((1) + c + d + cd)$$

c)   Difference $A_2 - A_1$
$$= \tfrac{1}{4}(ab + abc + abd + abcd - b - bc - bd - bcd)$$
$$- \tfrac{1}{4}(a + ac + ad + acd - (1) - c - d - cd)$$
$$= \tfrac{1}{4}(ab + abc + abd + abcd - b - bc - bd - bcd - a - ac - ad - acd + (1)$$
$$+ c + d + cd).$$

d)   The interaction amounts to half the difference:
$$AB = \tfrac{1}{8}[(1) + c + d + cd + ab + abc + abd + abcd)$$
$$- (b + bc + bd + bcd + a + ac + ad + acd)].$$

(3)   The threefold interaction (e.g. ABC) is defined as half the difference between the interaction AB with C at the upper and lower level, for example:

$$ABC = \tfrac{1}{8}[(a + b + c + abc + ad + bd + cd + abcd)$$
$$- ((1) = ab + ac + bc = d + abd + acd + bcd)]$$

Table 2.6   Factors influencing the effects of the gas

Standard form	A	B	C	D	AB	AC	AD	BC	BD	CD	ABC	ABD	ACD	BCD	ABCD
(1)	−	−	−	−	+	+	+	+	+	+	−	−	−	−	+
a	+	−	−	−	−	−	−	+	+	+	+	+	+	−	−
b	−	+	−	−	−	+	+	−	−	+	+	+	−	+	−
ab	+	+	−	−	+	−	−	−	−	+	−	−	+	+	+
c	−	−	+	−	+	−	+	−	+	−	+	−	+	+	−
ac	+	−	+	−	−	+	−	−	+	−	−	+	−	+	+
bc	−	+	+	−	−	−	+	+	−	−	−	+	+	−	+
abc	+	+	+	−	+	+	−	+	−	−	+	−	−	−	−
d	−	−	−	+	+	+	−	+	−	−	−	+	+	+	−
ad	+	−	−	+	−	−	+	+	−	−	+	−	−	+	+
bd	−	+	−	+	−	+	−	−	+	−	+	−	+	−	+
abd	+	+	−	+	+	−	+	−	+	−	−	+	−	−	−
cd	−	−	+	+	+	−	−	−	−	+	+	+	−	−	+
acd	+	−	+	+	−	+	+	−	−	+	−	−	+	−	−
bcd	−	+	+	+	−	−	−	+	+	+	−	−	−	+	−
abcd	+	+	+	+	+	+	+	+	+	+	+	+	+	+	+

Level of interaction columns span AB through ABCD; Effects columns are A, B, C, D.

(4)   In the fourfold interaction:

$$ABCD = \tfrac{1}{8}[((1) + ab + ac + bc + ad + bd + cd + abcd)$$
$$- (a + b + c + abc + d + abd + acd + bcd)].$$

Comparison of the standard form of design (Table 2.6) with the formulated equation systems clearly shows that the mathematical sign with which the individual experimental results are put into the calculation can also be deduced directly from the Table. The mathematical signs used for evaluating the interactions are obtained by multiplying the signs of the factors involved in the interaction, e.g. interaction $AB =$ sign of A ($-$) multiplied by sign of B ($-$) $= (+)$.

The rhythmic sequence of the change of sign reflects the symmetry of the experimental method. It is the prerequisite by which all effects and interactions can be determined independently of each other when all combinations are tested equally frequently.

The numerical values of the effects and interactions are calculated by adding or subtracting the experimental results according to their sign, which poses no great difficulty with programmable calculators or computers.

The results can, however, also be determined without the aid of computers using a method developed by Yates which uses the standard form of the experiment combinations. Two numbers, one of which is immediately below the other in the column, are added together and the one written below the other in the next column. The same pairs are then subtracted (the first from the second number) and written under the addition results. The process is repeated until there are as many columns as factors (hence four columns in the 2^4 design). The last column gives the effect (multiplied by 8 in the 2^4 design). The Yates method is demonstrated below by an example (2^3 design):

Fumigation conditions in relation to factorial design

Experiment combination	Measurements	Calculation process		
		(1)	(2)	(3)
(1)	9	13	23	51 Total
a	4	10	28	− 23 = 4 A
b	8	16	−11	− 7 = 4 B
ab	2	12	− 12	− 1 = 4 AB
c	11	− 5	− 3	5 = 4 C
ac	5	− 6	− 4	− 1 = 4 AC
bc	9	− 6	− 1	− 1 = 4 BC
abc	3	− 6	0	1 = 4 ABC
Sum	51			

The calculation process is checked by forming the sum of the measurements (= 51). This must agree with the number in the first line of column (3). As shown by Table 2.7, the value F is inversely proportional to the value of the random tests used for calculating the variances. When the variance is known, having been determined from a large number of experiments, a small squared sum of an effect is sufficient to verify significance. When the result is not significant, the absence of effect cannot therefore be established: only that the influence cannot be demonstrated in this experimental procedure.

The theoretical principles of factorial design illustrated here show that this is a method of planning experiments which makes it possible to determine the effects on, for example, yield, quality or costs of a product with a minimum of experiments and a relatively small probability of error. It should not be forgotten that the results obtained are (a) valid only within the chosen level limits and cannot therefore simply be generalized, and (b) subject to the chosen uncertainty/degree of unreliability or error.

Evaluation of the results

The effects and interactions determined in a factorial experiment plan must be checked for significance, as the variation of the experimental results is incorporated into the calculation. An effect must therefore differ from the experimental variation by a certain amount if it is to be regarded as significant. The variance of the measured value is used as a test criterion.

If this has not been calculated from earlier experiments, it is deduced from the results of the factorial experiment plan by determining the experimental variation using repetitions of the same experiment. The variance of the effects (formal variance) in a factorial experiment corresponds to the square of these effects. The formal variance is compared with the already known and/or calculated variance and tested for significance with the F-test. It is also possible to test for significance with the t-test.

If no information on variance is available it is usually assumed that the

Table 2.7 Fumigation conditions in relation to factorial design

f_2 \\ f_1	1	2	3	4	5	6	8	12	24	∞	f_2
1	161.45	199.50	215.72	224.57	230.17	233.97	238.89	243.91	249.04	254.32	1
2	18.512	18.999	19.163	19.248	19.298	19.329	19.371	19.414	19.453	19.496	2
3	10.129	9.552	9.276	9.118	9.014	8.941	8.844	8.744	8.638	8.527	3
4	7.710	6.945	6.591	6.388	6.257	6.164	6.041	5.912	5.774	5.628	4
5	6.607	5.786	5.410	5.192	5.050	4.950	4.818	4.678	4.327	4.365	5
6	5.987	5.143	4.756	4.534	4.388	4.284	4.147	4.000	3.841	3.669	6
7	5.591	4.737	4.347	4.121	3.972	3.866	3.725	3.574	3.410	3.230	7
8	5.317	4.459	4.067	3.838	3.688	3.580	3.438	3.284	3.116	2.928	8
9	5.117	4.256	3.863	3.633	3.482	3.374	3.230	3.073	2.900	2.707	9
10	4.965	4.103	3.708	3.478	3.326	3.217	3.072	2.913	2.737	2.538	10
11	4.844	3.982	3.587	3.357	3.204	3.094	2.948	2.788	2.609	2.405	11
12	4.747	3.885	3.490	3.259	3.106	2.999	2.848	2.686	2.505	2.296	12
13	4.667	3.805	3.410	3.179	3.025	2.915	2.767	2.604	2.420	2.207	13
14	4.600	3.739	3.344	3.112	2.958	2.848	2.699	2.534	2.349	2.131	14
15	4.543	3.683	3.287	3.056	2.901	2.790	2.641	2.475	2.288	2.066	15
16	4.494	3.634	3.239	3.007	2.853	2.741	2.591	2.424	2.235	2.010	16
17	4.451	3.592	3.197	2.965	2.810	2.699	2.548	2.381	2.190	1.961	17
18	4.414	3.555	3.160	2.928	2.773	2.661	2.510	2.342	2.150	1.917	18
19	4.381	3.522	3.127	2.895	2.740	2.629	2.477	2.308	2.114	1.878	19
20	4.351	3.493	3.098	2.866	2.711	2.599	2.447	2.278	2.083	1.843	20
21	4.325	3.467	3.072	2.840	2.685	2.573	2.421	2.250	2.054	1.812	21
22	4.301	3.443	3.049	2.817	2.661	2.549	2.397	2.226	2.028	1.783	22
23	4.279	3.422	3.028	2.795	2.640	2.528	2.375	2.203	2.005	1.757	23
24	4.260	3.403	3.009	2.777	2.621	2.508	2.355	2.183	1.984	1.733	24
25	4.242	3.385	2.991	2.759	2.603	2.490	2.337	2.165	1.965	1.711	25
26	4.225	3.369	2.975	2.743	2.587	2.474	2.321	2.148	1.947	1.691	26
27	4.210	3.354	2.961	2.728	2.572	2.459	2.305	2.132	1.930	1.672	27
28	4.196	3.340	2.947	2.714	2.558	2.445	2.292	2.118	1.915	1.654	28
29	4.183	3.328	2.934	2.702	2.545	2.432	2.278	2.104	1.901	1.638	29
30	4.171	3.316	2.922	2.690	2.534	2.421	2.266	2.092	1.887	1.622	30
40	4.085	3.232	2.839	2.606	2.449	2.336	2.180	2.004	1.793	1.509	40
60	4.001	3.151	2.758	2.525	2.368	2.254	2.097	2.918	1.700	1.389	60
120	3.920	3.072	2.680	2.447	2.290	2.175	2.016	2.834	1.608	1.254	120
∞	3.841	2.996	2.605	2.372	2.214	2.098	1.938	1.752	1.517	1.000	∞

smallest effect and/or a number of approximately equally small effects and interactions of the design are zero. The numerical values obtained are then attributable to experimental error. Significant interactions between several parameters are assumed to be very unlikely. In practice, the decision as to which effects and interactions are to be regarded as negligible or which can be disregarded must be made before the design is carried out if random selection of the smallest variances is to be guaranteed. This in turn is a condition for use of the F-table.

The formal variances of the effects and interactions are, as usual, calculated by squaring the numerical values and dividing by the number of experiments. The experimental variance is then estimated by multiplying the smallest value of the sum of the squares (SS) or the mean value of all small SS by the corresponding threshold value F from Table 2.7. This gives the significance value S. The condition for significance is that $SS > S$ (with 95% probability). A 5% probability of error is usual in quality assurance.

Formulation of an experiment plan for determining the optimum conditions for ethylene oxide fumigation

The following procedure is used in practice for formulating a factorial design:

(1)   Choice of factors.
(2)   Fixing of factor levels.
(3)   Establishment of the standard form of the experiment plan.

The choice of factors is of course crucial to the problem. The conditions under which a reaction of the constituents with ethylene oxide is no longer detectable have to be established. Here, the following factors must be considered:

Parameter	*Factor*
Gas concentration	*A*
Time of fumigation	*B*
Moisture content of drug	*C*
Temperature	*D*

The choice of the factor levels is also governed by the objective of the investigation. The levels in Table 2.8 have been fixed on the basis of the values of an antimicrobially active gas concentration known from experience and found in numerous publications.

Table 2.8   Fumigation conditions for a factorial design

Factors	Upper level ($+$)	Lower level ($-$)
A	1000 mg T-Gas/l	200 mg T-Gas/l
B	12 h	2 h
C	12.5%	9.95%
D	40°C	20°C

T-Gas is 90% ethylene oxide $+$ 10% CO_2.

If the standard form for a factorial design with four factors on two levels is chosen, then $2^4 = 16$ experiments must be carried out. The combinations are given in Table 2.6.

The experiment described here uses *Cinchona succiruba* costex pulvis as an example, since it is known that *Cinchona* bark loses a measurable quantity of alkaloid upon technical fumigation (although the reaction of cinchona alkaloids has still not been fully elucidated, the first stage of the modification may be a β-hydroxyethylation of the tertiary nitrogen in the quinuclidine ring). In accordance with general experimental protocol the actual moisture content of the drug under investigation was taken as the lower level, in this case 9.95%. The upper level of moisture content was chosen as 12.5%, and samples were adjusted to this value using a saturated solution of $SrCl_2.6H_2O$.

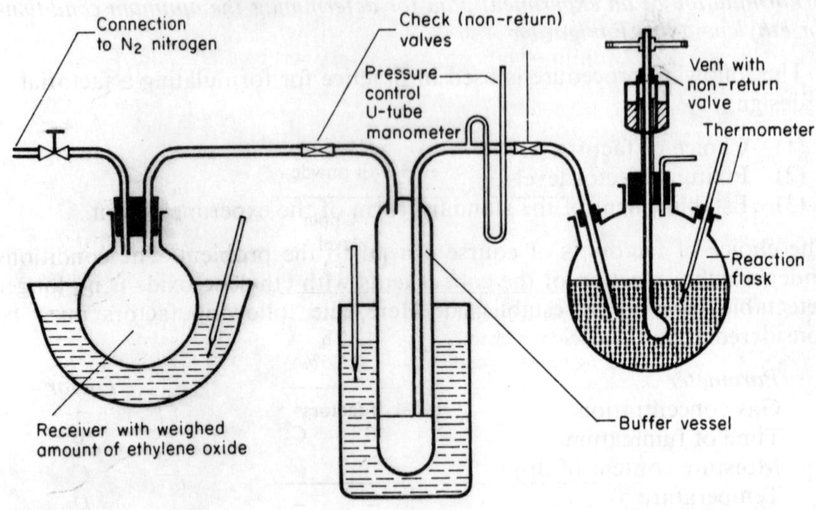

Figure 2.2 Apparatus for the fumigation of drugs with ethylene oxide.

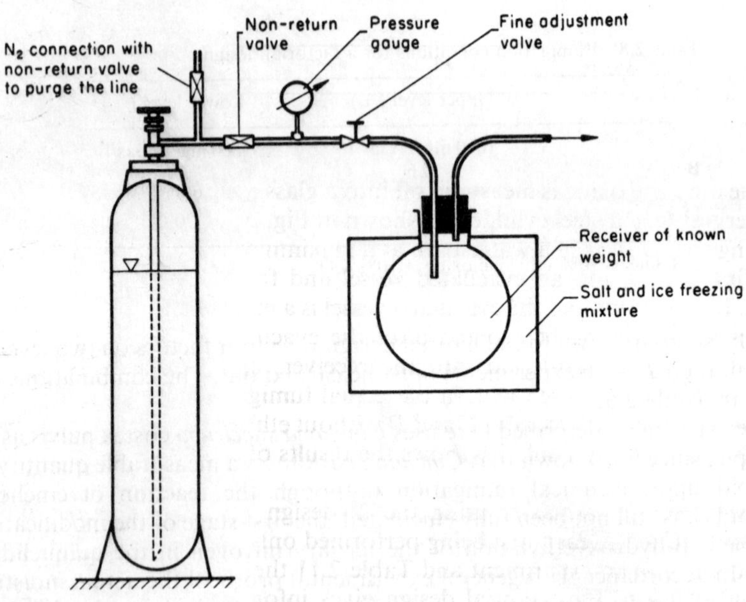

Figure 2.3 Drawing off a measured quantity of ethylene oxide.

The experimental fumigation of small quantities of drug is best carried out in an apparatus fulfilling laboratory safety conditions. An example is illustrated in Fig. 2.2.

Table 2.9 Blank experiments with powdered drug

Factors	Upper level (+)	Lower level (−)
D = Chamber temperature	40°C	20°C
B = Duration in chamber	12 h	2 h
C = Drug moisture content	12.5%	9.95%

Experiment combination	Level of factors			Alkaloid content (mg/g drug)
	D	B	C	
(1)	−	−	−	77.33
d	+	−	−	77.10
b	−	+	−	77.23
db	+	+	−	77.01
c	−	−	+	77.29
dc	+	−	+	76.91
bc	−	+	+	77.16
dbc	+	+	+	77.21
Mean:				77.15

The ethylene oxide is measured off into a glass round-bottomed flask from a reservoir (e.g. a steel cylinder) as shown in Fig. 2.3. The gas is vaporized by placing the receiver in a water-bath at a maximum temperature of 40°C. The gas first passes into an evacuated vessel and from there into the reaction flask. Between this and the evacuated vessel is a manometer. There are check-valves in the tubing before and after the evacuated vessel to prevent any backflow of reaction-liquid into the receiver, although check-valves and stirrers can be dispensed with in the actual fumigation of the drug.

The influence of factors B, C and D without ethylene oxide is best tested by blank experiments. Table 2.9 shows the results of such a test with *Cinchona* bark.

The actual experiments using the 2^4 design are then carried out in a random sequence, each one being performed only once. Table 2.10 gives the results of such an experiment and Table 2.11 their evaluation.

Evaluation of the factorial design gives information on those variables responsible for reducing alkaloid content. The effects and interactions are then calculated from combinations of the experimental results. Table 2.6 shows the sequence and signs. These values must be tested for significance.

Table 2.10 Results of factorial design (powdered drug)

Factors	Upper level $(+)$	Lower level $(-)$
A = Ethylene oxide concn	1000 mg/l T-Gas	200 mg/l T-Gas
B = Fumigation time	12 h	2 h
C = Drug moisture content	12.5%	9.95%
D = Fumigation temperature	40°C	20°C

Experiment No.	Experiment combination	A	B	C	D	Alkaloid content (mg/g drug)	% Variation[a]
8	(1)	−	−	−	−	76.91	− 0.31
4	a	+	−	−	−	75.38	− 2.29
12	b	−	+	−	−	75.40	− 2.27
16	ab	+	+	−	−	73.14	− 5.19
6	c	−	−	+	−	77.36[b]	− 0.27
2	ac	+	−	+	−	75.70[b]	− 1.87
9	bc	−	+	+	−	76.21[b]	− 1.22
13	abc	+	+	+	−	71.26[b]	− 7.62
11	d	−	−	−	+	76.86	− 0.37
3	ad	+	−	−	+	73.84	− 4.75
10	bd	−	+	−	+	68.23	− 11.56
15	abd	+	+	−	+	64.42	− 16.50
5	cd	−	−	+	+	77.16[b]	− 0.02
1	acd	+	−	+	+	71.15[b]	− 7.77
7	bcd	−	+	+	+	73.42[b]	− 4.86
14	abcd	+	+	+	+	59.37[b]	− 23.04

[a] % Variation from mean value X = 77.15 mg alkaloid per g drug calculated as quinine and cinchonin.

[b] Corrected values.

If it is assumed that the interactions between three or four factors were improbable and attributable to experimental error, and the mean value of these squared sums is multiplied by F (Table 2.7), corresponding to the number of these squared sums ($n = 5$), the significance S is obtained:

$$
\begin{array}{ll}
ABC & 6.014 \\
ABD & 1.446 \\
ACD & 6.77 \\
BCD & 1.19 \\
ABCD & 1.374 \\
\text{Sum total} & 16.794 \quad \tfrac{1}{5} = 3.358
\end{array}
$$

$F_{95\%} f_1 - 1, f_2 = 5, F = 6.61$ (Table 2.7)
$S = 3.358 \times 6.61 = 22.2.$

If 95% probability is to be taken as significant the square sums must be greater than $S = 22.2$.

Significance is indicated in column IV of Table 2.11 by a plus sign.

Table 2.11 Evaluation of results of factorial design (powdered drug)

I Factor	II (Effect 8)	III $\dfrac{(\text{Effect } 8)^2}{16}$	IV Significance
A	−37.29	86.9	+
B	−42.91	115.079	+
C	− 2.55	0.406	+
	−36.91	85.146	±
AB	−12.85	10.32	−
AC	−16.05	16.100	−
AD	−16.49	16.995	−
BC	+ 0.689	0.0297	−
BD	−24.23	36.693	+
CD	− 1.949	0.237	−
ABC	− 9.81	6.014	−
ABD	− 4.81	1.446	−
ACD	−10.41	6.77	−
BCD	+ 4.37	1.19	−
ABCD	− 4.69	1.374	−

The bacterial count determinations on samples treated by factorial design are illustrated in Table 2.12. These show that with this method it is possible to recognize the factors responsible for reduction of the alkaloid content and to determine the fumigation conditions which guarantee an adequate anti-

Table 2.12 Typical bacterial counts after treatments by factorial design methods

Expt. combi- nation	Level of factors A	B	C	D	Weight of T-Gas (mg)	Alkaloid content (mg/g drug)	% Deviation[a]	Bacterial count/g after 48 h at 35°C
(1)	−	−	−	−	−	77.10	−0.06	>1100
c	−	−	+	−	−	76.86[b]	−0.37	>1100
d	−	−	−	+	−	77.23	+0.1	>1100
cd	−	−	+	+	−	77.38[b]	+0.3	>1100
(1)	−	−	−	−	135.5	76.89	−0.41	>1100
(1)	−	−	−	−	136.0	77.04	−0.14	>1100
c	−	−	+	−	132.0	76.43[b]	−0.93	>1100
c	−	−	+	−	134.0	77.0 [b]	−0.2	>1100
d	−	−	−	+	134.5	76.47	−0.88	>1100
d	−	−	−	+	133.0	76.78	−0.47	>1100
cd	−	−	+	+	131.5	76.74[b]	−0.52	75
cd	−	−	+	+	132.0	77.52[b]	+0.48	150
cd	−	−	+	+	135.0	76.69[b]	−0.59	43

[a] % variation from mean value X = 77.15 mg alkaloid per g drug calculated as quinine and cinchonin.

Corrected values.

microbial action and yet at the same time achieve the greatest possible conservation of the active substances.

The following optimum fumigation conditions are given for *Cinchona* bark:

Ethylene oxide concentration	200 mg/L
Fumigation time	2 h
Drug moisture content	12.5%
Reaction temperature	40°C

To reduce the number of preliminary experiments it is recommended (and also correct) that the active constituents of drugs are classified as sensitive or less sensitive to technical fumigation where, as has been shown, the alkaloid drugs are the most sensitive.

(An example of the formulation of a factorial experimental plan for the extraction of drugs is in Section 4.2.3.2.3).

2.1.3.2.6. Evaluation of ethylene oxide fumigation

Ethylene oxide is particularly suitable for the necessary antibacterial and disinfestation treatment of harvested drug plants, which have very high natural bacterial counts and are often susceptible to a high degree of infestation by insects (at all stages of development) and other animals.

However, it is also a fact that ethylene oxide not only acts on living organisms but also reacts with constituents of the treated drug plants, thereby β-hydroxyethylating functional groups bearing active hydrogen atoms. Primary, secondary and tertiary amines are particularly susceptible to attack. Products of altered solubility are formed, i.e. the reaction products are more hydrophilic than the original substances and therefore have different pharmacological properties.

Many attempts have been made to replace the fumigation of drugs with alkylating agents (i.e. ethylene oxide) by other methods. Treatment with pressurized steam or with ionizing radiation fails when large volumes of drugs are being processed. Truebenbach has shown [2.21] that optimal conditions which give a satisfactory reduction of the bacterial count together with the greatest possible preservation of the constituent substances can be calculated for the performance of the process with the aid of factorial experiment planning. In this case disinfestation is essentially a by-product.

No residue problems arise when the process is carried out correctly. Schilcher has shown that in large-scale investigations no measurable quantities of ethylene oxide or ethylene chlorohydrin remain, even in oily drugs such as linseed [2.22]. This is very important, as the only further mechanical processing to which linseed is subjected before it is administered is simple crushing. Drug plants and drugs which are processed to different medicinal preparations after fumigation raise no problems.

2.2. Analysis of drugs

2.2.1. QUALITY ASSURANCE OF IMPORTED PRODUCTS

Most imported drugs should be put in a quarantine store, where they should remain until the quality assurance laboratory releases them for further processing. The store, like any other store, should be constructed so that no detrimental alterations can occur to the produce during storage. Adequate ventilation and suitable storage conditions must be ensured. Bundles of drug plants which have become damp must be quickly dried out to prevent growth of moulds. Unchecked transfer from the quarantine store to the factory store must be made impossible.

The 'goods inward' check extends to a series of pharmacognostic, physical, chemical and possibly biological and microbiological investigations, i.e.:

(1)   Macro- and microscopic examination for identity.
(2)   Possible thin-layer chromatographic examination for identity.
(3)   Examination for foreign inorganic and organic impurities.
(4)   Determination of drying loss and water content.
(5)   Determination of ash.
(6)   Determination of crude fibre.
(7)   Determination of extractable components.
(8)   Determination of the active substance(s) (insofar as they are known).
(9)   Determination of microbial contamination and of the absence of pathogenic bacteria.
(10)   Examination for pesticide residues.

2.2.1.1. Sampling

All these assays require statistically trustworthy sampling. *USP XX* gives the following directions for this.

(a)   It is recommended that samples of plant drug materials in powdered or crushed form (or plant drug materials which in their natural state already exist as small components, e.g. Umbelliferae fruits) which are available in large quantities, and the components of which have a diameter of not more than 1 cm, should be taken with a sampler which is pushed down from the top to the bottom of the container. If the total weight of the amount of drug material under investigation is less than 100 kg, at least two samples must be taken at different points. At least 250 g must be taken in this way (if the total quantity of drug material under examination comes from numerous containers, the number of containers to be tested is shown in Table 2.13).

If the total weight is more than 100 kg, more samples should be taken by the method described above, mixed and quartered. Remove two diagonally opposed quarters. They will not be used for the analysis. The two remaining quarters are again mixed and quartered repeatedly until two of these quarters

Table 2.13 Selection of test containers from bulk supply

Number of containers supplied	Number of containers to be tested
1–10	1–3
11–25	3–4
26–50	4–6
51–75	6–8
76–100	8–10
> 100	min. 10

together weigh about, but not less than, 125 g. They represent the sample to be investigated.

(b)   In the case of drug plant materials with components larger than 1 cm it is recommended that the samples be taken by hand. If the total weight of drug material to be tested is less than 100 kg, at least 500 g shall be the sample quantity to be investigated; this quantity is to be taken from different parts of the container. The procedure described in (a) should be adopted if the dry material comes from numerous containers. If the total weight is more than 100 kg, more samples are to be taken in the same way, mixed, quartered as above and divided in such a way (see (a)) that two of these quarters together weight about, but not less than, 250 g.

The examinations described in Sections 2.2.–10 are carried out on the weight-reduced test samples. It is to be decided in each individual case whether a normal selection of test methods suffices, or whether the list of tests should be extended. We give the directions in *USP XX* for some of these

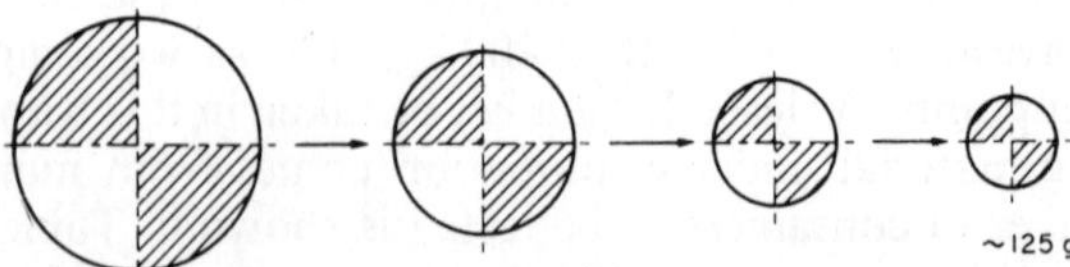

Figure 2.4   Repeated quartering of a sample until a weight below 125 g is attained.

tests. If the drugs are official drugs made up in the pharmacy the directions of the respective pharmacopoeias apply. Non-official drugs should be processed according to the state of scientific knowledge as given by supplementary collections of directions (DAC, Hager and others) or publications.

The general directions of *USP XX* are given in the following:

Analysis sample (USP XX)
An adequate representative quantity is taken from the sample obtained as described above. If the drug material has not been ground or powdered, the quantity taken is finely ground until it can pass through a sieve with a mesh-width of 0.84 mm. If the sample cannot be ground, its granule-size should be made as small as possible. The ground material is mixed by rolling on paper or on a test cloth, and the required analysis sample taken from it.

2.2.1.2. Foreign organic impurities (*USP XX*)

Between 25–500 g of the sample to be investigated is weighed and spread out in a thin layer. Foreign organic materials should be picked out by hand as far as possible. The quantity of such material removed is weighed and expressed as a percentage of the investigated quantity of the sample.

In processing factories, this type of examination is extended to the entire quantity of material being processed, if it is to be processed to infusion preparations, i.e. it is not limited just to the amount of sample taken. In this case the hand-picking process is usually carried out after shredding of the drug plants (see Section 5.2.1). Before shredding, dust and other lighter contaminants (e.g. feathers, blown seeds) and specifically heavier contaminants such as sand, stones, metal fragments, etc., are removed by blast-sifting. The material is then shredded and put in a thin layer on a conveyor belt which passes checkers who pick out foreign impurities and bad drug plant parts by hand.

2.2.1.3. Total ash (*USP XX*)

Between 2–4 g of the air-dried analysis sample is weighed in a porcelain crucible and incinerated by gentle heating. The mass should not be heated above slight red-heat and should not be charred. The ash is cooled and weighed. If carbon-free ash cannot be obtained, the cooled contents of the crucible are extracted with hot water, and any undissolved material is collected on an ashless filter and again incinerated, together with the residue left after evaporation of the aqueous extract, in the same crucible. The washings are then added, the water cautiously evaporated off and the residue heated to dull red heat, cooled, weighed and the ash content calculated as a percentage of the original weight of the material.

2.2.1.4. Acid-insoluble ash (*USP XX*)

The total ash obtained is boiled for 5 min with 25 mL 3N hydrochloric acid. The insoluble residue is collected in a filter crucible or on an ashless filter paper, washed with hot water, dried and incinerated. The proportion of acid-insoluble ash is determined as a percentage of the original weight of the material.

Other Pharmacopoeias permit the determination of the 'sulphate ash' remaining after incineration of the residue left after addition and evaporation of sulphuric acid.

Determination of ash content cannot always be used for detecting soil contamination. The ash contents of phytopharmaceuticals can, for example, also indicate whether the leaves or roots of medicinal plants were used as starting-materials, though even here it cannot be assumed that roots would produce the higher ash values. For example, the leaves of Belladonna drug plants give ash contents of 11–17% (usually 12–13%), whereas the roots yield only 4–7% ash [2.23].

2.2.1.5. Crude fibre (*USP XX*)

About 2 g of the drug sample, accurately weighed, is extracted with ether. Then 200 mL of 1.25% sulphuric acid is added to the extracted drug and the whole mixture boiled for 30 min under reflux in a 500 mL flask. The mixture is then filtered through a hardened filter and the residue washed with boiling water until free of acid. All of the residue is rinsed back into the flask with 200 mL of boiling 1.25% sodium hydroxide solution and again boiled under reflux for 30 min. The liquid is then quickly filtered through a tared filter and the residue on the filter is washed with boiling water until neutral, dried at 110°C to constant weight, then incinerated, likewise to constant weight. The difference between the weight of the dried residue and that of the incidental residue represents the weight of the crude fibre. It is expressed as percentage of the original weight of the material.

The determination of crude fibre represents a criterion much used in food and agricultural chemistry for the proportion of food or fodder which the human or animal body cannot digest and utilize. It must, however, be remembered that it is not an absolute value and that the crude fibre, in the ideal case identical with cellulose, therefore varies a great deal both qualitatively and quantitatively according to the method of determination used. The quality determined also depends on the degree to which the material has been ground up.

The crude fibre content is, for example, an indicator of the proportion of husk in cacao products (see textbooks on foodstuffs chemistry).

2.2.1.6. Determination of extractable fraction (*USP XX*)

This determination is vitally important for the further processing of plant drugs. It permits the selection of a suitable solvent (see Chapter 3). The extracts obtained with the various solvents can be conveniently used, for example, for thin-layer chromatographic investigations of constituents.

2.2.1.6.1. Alcohol-soluble extractive substances

A 2 g sample prepared as above is weighed in a tared, dried paper extraction thimble using a weighing glass with a ground glass cover. Sodium hydroxide (200 mg) is put in the distillation flask of a Soxhlet apparatus together with the required quantity of alcohol and the sample is extracted for 5 h. The paper thimble with the insoluble residue is then dried for 30 min at 105°C and weighed. The value obtained in the determination of the water content by azeotropic distillation (see Section 2.2.1.7) is reduced in proportion to the weight of the analysis sample and the result subtracted from the weight of the sample. The difference between this result and the weight of the drug residue gives the quantity of alcohol-soluble extractive substance.

2.2.1.6.2. Dilute alcohol-soluble extractive substances

A 2 g sample is macerated in an Erlenmeyer flask with 70 mL dilute alcohol. The mixture is shaken every 30 min during the first 8 h, then left to stand for a further 16 h without shaking, and then filtered. The flask and residue are washed out with small portions of dilute alcohol until the total liquid amounts to 100.0 mL, and 50.0 mL of the mixed filtrate is evaporated in a tared evaporating dish on a water-bath. The residue is dried to constant weight at 105°C and weighed. The percentage of substances extractable with dilute alcohol is calculated from the quantity obtained.

2.2.1.6.3. Hexane-soluble extractive substances

A 2 g sample, accurately weighed, is extracted with n-hexane for 20 h in a Soxhlet apparatus. The hexane extract is left to evaporate of its own accord in a tared evaporating dish. The residue is then dried for 18 h over phosphorus pentoxide and weighed. The hexane-soluble proportion is calculated as a percentage of the weighed sample.

2.2.1.6.4. Non-volatile ether-soluble extractive substances

The procedure is as described in Section 2.2.1.6.5., 'Volatile ether-soluble extractive substances'. The weight of the extract after drying in a desiccator and then at 105°C to constant weight represents the proportion of non-volatile ether-soluble extractive substance.

2.2.1.6.5.   Volatile ether-soluble extractive substances

A 2 g sample dried for at least 12 h over phosphorus pentoxide is weighed accurately and extracted in a Soxhlet apparatus with dry ether for 20 h. The ether extract is left to evaporate of its own accord in a tared evaporating dish. The residue is then dried for 18 h over phosphorus pentoxide and the total content of ether-soluble extractive substances is determined. The residue is then gradually warmed and finally dried to constant weight at 105°C. The weight loss represents the proportion of volatile ether-soluble extractive substance.

2.2.1.6.6.   Water-soluble extractive substances

The procedure is as described in Section 2.2.1.6.2., 'Dilute alcohol-soluble extractive substances', using water instead of dilute alcohol.

2.2.1.7.   Determination of drying loss and water content (*Ph. Eur. I*)

Drying loss is the weight loss in percentage (w/w) produced, according to the monographs, by drying the sample in a desiccator at atmospheric pressure and room temperature over phosphorus pentoxide, or at room temperature and reduced pressure (below 20 Torr or 26.6 mbar) or at a higher temperature and reduced pressure (below 20 Torr or 26.6 mbar) after a given time, or in a drying oven at 100–105°C (or as specified).

The water content is determined as described in *Ph. Eur. I* and other Pharmacopoeias, e.g. *USP XX*, by azeotropic distillation in accurately described apparatus.

2.2.1.8.   Determination of the active substance(s)

When a drug to be examined is described as a so-called officinal drug in an official Pharmacopoeia or listed as a non-officinal drug in a supplementary publication such as the Deutscher Arzneimittel Codex (*DAC*), the British Pharmaceutical Codex (*BPC*) or the National Formulary (*NF*), these also always describe value determination methods and tests for identification and purity. If such information is not given, other directions in the scientific literature (e.g. Hager) can be followed.

Value determinations are difficult when the active substance(s) of a phytopharmaceutical is/are not known. Certain frequently occurring principal substances characteristic of the phytopharmaceutical should, if possible, be determined in such cases as quantitatively as possible as described in Section 6.2 (Standardization of Phytopharmaceuticals) so that a constant batch-to-batch quality can be guaranteed. Utilization of the principal substances for quality determination beyond this appears problematical.

Careful execution of the checks described under Sections 2.2.1.1–7, 2.2.1.9 and 2.2.1.10, however, offers the guarantee of always having drugs of identical quality even when one or more of the active substances is not known.

2.2.1.9. Determination of microbial infestation and demonstration of the absence of pathogenic organisms

Since it became known in the 1960s that patients could contract serious infections from oral and ophthalmic drugs contaminated with *Salmonella* and pseudomonads, great efforts have been made both by scientists and by the authorities to put limitations on the degree of microbial contamination, even in medicinal preparations which do not obligatorily need to be sterile. The principal demand is for the demonstrated absence of pathogenic bacteria and so-called indicator-bacteria, e.g. *Escherichia coli*.

The permitted number of apathogenic bacteria in oral medicaments (10^3 bacteria/g) is very much lower than in foodstuffs and should, in our opinion, perhaps relate to the finished product. The limit in natural starting-materials should be 10^4 bacteria/g and 100 mould- or yeast-cells/g, especially when a reduction of the number of organisms can be expected from further processing.

2.2.1.9.1. Microbial status of non-sterile pharmaceutical products (*USP XX*)

Very few raw materials for the manufacture of pharmaceutical products are sterile when supplied by the factory. They require special treatment if they are to be brought to a microbial state acceptable for consumption. Here the adherence to strict hygienic precautions in the production of medicinal preparations is required for the effective reduction of the number of possible bacterial species and of individual organisms ... it may be necessary and expedient to carry out additional in-process microbiological checks at various processing stages, the 'goods inwards' check being particularly important here. Raw materials of animal, vegetable and even of mineral origin are frequently carriers of numerous, possibly pathogenic, microorganisms which can be passed on to subsequent processing stages. All starting materials and their subsidiary products must be stored in such a way that multiplication of any bacteria originally present is impossible.

The type and frequency of the tests depends on the product. Some Pharmacopoeia monographs demand the absence of one or more indicator microorganisms such as *E. coli*, *Salmonella* species, *Staphylococcus aureus* and *Pseudomonas aeruginosa*. The monographs prescribe a definite limitation of the number of aerobic bacteria in other products. The particular significance of microorganisms in non-sterile pharmaceutical products must be judged according to the use and nature of the product and the possible risks to the consumer.

We suggest that certain pharmaceuticals, such as drugs of vegetable and animal origin, and certain minerals should be regularly checked for certain microbial contaminants such as *Salmonella* species.

2.2.1.9.2. Bacterial count determination (Microbial Limit Tests *USP XX*)

This section contains tests for the determination of the number of replicable (viable) microorganisms and for the demonstration of the absence of certain species of microbes in all types of pharmaceutical products and materials, from crude starting-materials to finished products ... Strict adherence to aseptic working conditions must be ensured in the preparation and execution of the tests. Unless stated otherwise, 'incubation' means the storage of the material in an incubator at 30–35°C for 24–48 h. The word 'growth' is used in the sense of expressing the presence and expected multiplication of live microorganisms.

Further information in *USP XX* on Microbial Limit Tests concerns the composition of a phosphate buffer solution and of media containing 18 nutrients. It also includes directions for the determination of the total number of aerobic microorganisms. Only the serial dilution test, but not the plate-pouring method, is used in the investigation of drug plant materials because of their insolubility in the nutrient media. The analysis sample (10 g) described under Section 2.2.1.1 is to be used for each test.

Tests for the absence of *S. aureus* and *P. aeruginosa* and also of *Salmonella* species and *E. coli* are described in further sections.

We do not have the space here to reproduce the individual directions and we therefore refer the reader to the *USP XX* itself.

2.2.1.9.3. Microbiological examination [2.23]

(1) Preparation of sample
     The sample (10 g) (see Section 2.2.1.1) is extracted under aseptic conditions by vigorous shaking for ∼ 30 min in 90 mL sterile physiological sodium chloride (saline) solution. The microbial contamination of 1 mL of this stock solution is roughly determined with the Biotest Hygiene Check System and then the suitable dilution immediately prepared without the usual serial dilutions.

(2) Total bacterial count determination and bacterial differentiation

(a) Total bacterial count:
Plate-Count Agar Merck No. 5463 and Foodstuffs Bacterial Count Agar Merck No. 10231.
Incubation: 48 h at 30–32°C.

(b) Coliform bacteria:
Brila-broth (= Brilliant Green-bile-lactose broth) Merck No. 5454.
Incubation: 48 h at 37°C.
T 7-Agar Merck No. 5471 with bromothymol blue and DEV ENDO-Agar Merck No. 10684.
Incubation of both tests: 24 h at 37°C.

(c) *E. coli* in 0.1–1.0 g:
Brila broth Merck No. 5454 and tryptone-water Merck No. 10859 with Kovacs indole reagent.
Incubation of both tests: 48 h at 44°C.

(d) Enterobacteriaceae:
Testsystems 'Enterotube' [Roche] and API Laboratory System API 20 E.

(e) *Pseudomonas* species:
Enrichment with selenite-cystine enrichment broth Merck No. 7709.
Isolation with GSP agar Merck No. 10230, Leifson agar Merck No. 2896 and XLD agar Merck No. 5287.
Differentiation on GSP agar.

(f) Pathogenic staphylococci:
Enrichment in staphylococcal enrichment broth Merck No. 7892.
Isolation on Vogel–Johnson staphylococcal selective agar Merck No. 5405.
Incubation: Each 36 h at 37°C and identification by plasma coagulase test, DNase test and phosphatase test.

(g) Salmonellae:
Pre-enrichment with tetrathionate broth Merck No. 5285, selenite broth Merck No. 7717 and selenite-cystine solution Merck No. 7709, isolation on *Salmonella–Shigella* agar Merck No. 7667 and BPLS agar Merck No. 7232. Selective detection on iron-triple sugar agar Merck No. 3915, Kligler agar Merck No. 3913 and Costin lysin decarboxylase sulphohydrase test nutrient Merck No. 5266, additional identification by 'Enterotube' and API 20 E.

(h) Yeasts and fungal moulds:
On malt extract agar Merck No. 5398 and wort agar Merck No. 5448 by 2–5 days incubation at room temperature, differentiation on Czapek-Dox agar Merck No. 5460.

(i) Clostridia:
Enrichment on Brewer anaerobe agar Merck No. 5452, identification on SPS agar Merck No. 10235 and thioglycolate nutrient medium Merck No. 8191, incubation in an anaerobe pot.

(k) *Bacillus mesentericus*:
The cell coating from PC agar Merck No. 5463 is heat-treated at 60°C for 10 min and then inoculated on to PC and malt extract agar Merck No. 5398.
Incubation: 48 h at 37°C on PC agar and for 72 h at 32°C on malt extract agar.

(l) Heat-resistant bacteria:
The sample is treated in 100 mL casein–soy peptone broth Merck No. 5459 for 15 min with steam at ~ 100°C and then the aerobic and anaerobic spore-formers are detected at 100°C.

2.2.1.10. Examination for pesticide residues and heavy metals

The vast number of pesticides, and the fact that it is usually not known whether, which and in what quantity pesticides have been used by the drug plant grower, make it expedient to test for typical representatives of the major groups of pesticides. Schilcher has published the following directions, including tests for heavy metals and fumigation residues [Section 2.3.22]. In Table 2.14 he summarizes the pesticides which may be found in drug plant material and their preparations.

2.2.1.10.1. Examination for chlorinated hydrocarbons and organophosphoric acid esters

Chemicals
Acetonitrile, cyclohexane, dichloromethane, petrol (40–60°C), mostly for residue analysis—analytical grade sodium sulphate heated to red heat at 600°C for 4 h—Florisil (60–100 mesh ASTM) dried for 8 h at 250°C, cooled over silica gel in the desiccator and deactivated with 5% water [5 mL water (double-distilled) per 100 g Florisil].

Preparation of sample
50 g, or in the case of plants containing fat or oil 100 g, powdered representative average samples of the plant drug material (Section 2.2.1.1) are extracted three times in succession with acetonitrile quantis satis, each time for 15 min in an ultrasonic bath. The extraction vessel must be carefully sealed. The individual extracts are filtered through a fluted filter paper over calcined sodium sulphate, and the flask and filter are rinsed out with 50 mL acetonitrile. The combined extraction solutions are evaporated to dryness in a rotary

Table 2.14 Possible pesticide residues and impurities in drug plant materials

Pesticides (biocides)	Agents for protecting stored materials against vermin and undesirable microorganisms (bacteria, moulds, viruses)	Environmental pollution	Microorganisms
(A) Agents for control of animal pests: *Insecticides*, acaricides, molluscicides, rodenticides (B) Agents for control of plant diseases: *Fungicides*, nematicides (C) Agents for control of weeds and parasitic higher plants: *herbicides* (*weedkillers*)	(A) Insecticides (B) Disinfestation—disinfectant agents and their reaction products and metabolites: ethylene oxide, methylene bromide, phosphine, etc.	(A) Natural and synthetic radionuclides: potassium-40, carbon-14, tritium, radium-226, radon-222, lead-210, polonium-210, strontium-90 (B) Toxic trace elements: lead, cadmium, mercury, thallium, etc.	(A) Ubiquitous bacterial contamination (B) Undesirable decay-inducing microorganisms (especially moulds) (C) Prohibited disease pathogens

evaporator at $\sim 30°$C. The acetonitrile-free residue is dissolved in a mixture of 10 vols dichloromethane and 90 vols petrol and put on a prepared Florisil column (diameter ~ 2.5 cm, length 30–35 cm). The main fraction of chlorinated hydrocarbons is eluted with ~ 100 mL of a mixture of 15 vols dichloromethane and 85 vols petrol. The receiving vessel (a 250 mL round-bottomed flask) is then changed and the second elution with ~ 100 mL of a mixture of 25 vols dichloromethane and 75 vols petrol is collected. The rest of the chlorinated hydrocarbons are found in the second eluate. If the material is also being tested for phosphoric acid esters a third elution is carried out with ~ 80 mL of a mixture of 40 vols dichloromethane and 60 vols petrol, the column being finally rinsed through with 50 mL pure dichloromethane.

Residues of carotenoids, chlorophyll and ethereal oils, which adversely affect the analyses, must be removed from various plant drug materials by activated charcoal. A second purification on Florisil is often necesary for material containing fats and oils. The removal of interfering substances by means of concentrated sulphuric acid usually produces colourless eluates, though at the same time also causing some loss of pesticide. The eluates are concentrated to dryness separately in a rotary evaporator at a maximum temperature of 40°C. The bulb of the evaporator is attached to a 5 mL measuring flask, and concentrated eluate is then made up to this mark with cyclohexane at 20°C. A 1–5 mL portion of this solution is injected into the gas chromatograph.

Gas chromatography
HP 5750 G gas chromatograph with ECD-^{63}Ni and FID.

Temperatures: Injector block 240°C, detector 290°C, furnace running isothermally 190°C and in programme 150–200°C (2°C/min).

Carrier gas: Purified helium, 30 mL/min.

Flushing gas: Argon/methane 90:10, 100–130 mL/min.

Advance speed of paper: 6.35 mm/min.

Columns:
(1)   3.8% SE 30 on Chromosorb W AW-DMCS (80–100 mesh), 1.8 m glass column
(2)   5% or 15% QF 1 on Gaschrom Q (80–100 mesh), 1.8 m glass column.
(3)   6% OV 17 + 3% Dexil on Chromosorb W AW DMCS (80–100 mesh), 3.6 m glass column.

Columns 1 and 2 are used often, whereas column 3 is used only in doubtful cases. The quantitative evaluation is made either on the area, or the height of the peak, in comparison with comparative substances and internal or external standards.

2.2.1.10.2.   Examination for carbamate pesticides in the ppm-range with HPLC

Preparation of sample
Before HPLC determination the drug is extracted with dichloromethane, the solvent completely removed in a rotary evaporator and the residue dissolved in the mobile phase.

HPLC parameters
Separating-columns:

(1)   LiChrosorb RP-8 (10 µm), mobile phase = methanol/water (9:1).
(2)   LiChrosorb Si100 (10 µ,), mobile phase = 2,2,4-trimethylpentane/dioxan (8:2).

Satisfactory separation of carbamate pesticides in drug plant materials containing fats or oils and/or a large amount of chlorophyll and carotenoids is possible only when columns 1 and 2 are placed in series (two-column operation). Column 2 proved very useful in the usual routine.

Flow-rate: 1.25 mL/min and 30–40 bar
Detector: UV detector at 245 nm and 275 nm.
Paper advance speed: 600 mm/h
Evaluation: Quantitative evaluation on peak-height.

2.2.1.10.3. Determination of lead, cadium and mercury by atomic absorption spectrophotometry (AAS)

Preparation of sample

Large quantities (10–50 g) of the powdered drug material are solubilized by the Sperling method with a mixture of perchloric and nitric acids, smaller quantities (up to 300 mg) are solubilized by the Tölg method and determined by AAS. The few analyses carried out with a low-temperature incinerator (Plasma Processor 200-E, Technics) gave very good reproducible values, particularly with material which is difficult to incinerate to ash, e.g. Semen lini (linseed), Semen cucurbitae. The conditions are given in Table 2.15.

Table 2.15 AAS conditions (apparatus: Zeiss FMD 3/PMQ 3/MB 3 and Pye Unicam SP1900)

	Lead	Cadmium	Mercury
Standard solutions	0.05–2 µg/mL	0.005–0.5 µg/mL	5×10^{-5}–5×10^{-3} mg
Wavelength	283.3 nm	228.8 nm	253.7 nm
Light source	Hollow cathode lamp	Hollow cathode lamp	Hollow cathode lamp
Lamp current	15 mA	7 mA	15–20 mA
Slit	0.3 mm	0.4 mm	0.5 mm
Combustion gas/oxidizer	Acetylene/ compressed air	Acetylene/ compressed air	Flameless

2.2.1.10.4. Examination for fumigation residues

The drug material is shaken with 10–50 mL analytical acetone for 24 h at room temperature. The acetone extract is then filtered through a fluted filter paper and made up to 10 mL. 3–8 µL of the clear filtrate is injected directly into the gas chromatograph.

Column:	15% polypropyleneglycol on Chromosorb W-AW, 60–80 mesh, length 4 m, diameter $\frac{1}{8}$ in., glass.
Temperatures:	Column 60°C isothermic, detector FID 150°C, injector 120°C.
Carrier gas:	15 mL N_2/min. Combustion gases: air 150 mL/min, H_2 15 mL/min.
Paper advance speed:	100 cm/h.

Overall retention times; ethylene oxide 3.18 min, methyl formate 3.54 min.

Determination of ethylene chlorohydrin:

A 1–10 g sample of plant material or drug preparation is extracted as described above and 3–8 µL of the extract is injected directly into the gas chromatograph.

Column: 5% Carbowax 20 M on Chromosorb T, 40–60 mesh, length 4 m, diameter $\frac{1}{4}$ in., glass.

Temperatures: Column programmed 70–100°C (heated at 5°C/min), detector FID 200°C, injector 160°C.

Carrier gas: 30 mL N_2/min.

Combustion gases: Air 300 mL/min, H_2 30 mL/min.

Paper advance speed: 100 cm/h.

Evaluation and calculation for 5.1 and 5.2:

The peak-area is calculated by the formula: height × width at half of maximum intensity. The residue quantities R in ppm are as shown in Table 2.16.

Table 2.16   Residues (ppm)

Sample (g)	Ethylene oxide	Ethylene chlorohydrin	Methyl formate
2	1	6	2
5	0.4	2.4	0.0
10	0.2	1.2	0.45

$$R = \frac{F_A K 1000}{F_B E}$$

F_A = Peak-area of investigation solution (mm^2)

F_B = Peak-area of standard calibration solution (mm^2)

K = concentration of standard calibration solution (mg/10 mL acetone)

E = weight of sample under investigation (g/10 mL acetone).

2.2.1.11. Legal situation regarding pesticide residues and heavy metals

As no *ruling on maximum quantity* for pesticides, or *approximate values* for toxic heavy metals, or data on maximum values for bacterial counts is given in the Pharmacopoeias, the law on these is generally extremely vague. The Maximum Quantities Order (Germany) of 13 June 1978 (Bundesegesetzblatt I, p. 718) exists for foodstuffs, the so-called ZEBS approximate values for lead, cadmium and mercury are published by the Zentrale Erfassungs- und Bewertungsstelle für Umweltchemikalien (Central Environmental Chemicals Detectection Laboratory) and finally there are also microbiological regulations for certain foodstuffs in the LMBG (Lebensmittelbundes-Gesetzblatt = West German Federal Food Regulations Gazette). Ennet therefore recently expressed the view that it would certainly be in the interests of all concerned to establish a legal ruling similar to that for foodstuffs. If the wording of the general directions of *Ph. Eur. III*, p. 19, were strictly adhered to, legal regulations additional to the existing provisions would of course then not be absolutely necesary, in fact they would be superfluous. *Ph. Eur. III* states: 'The legal provisions do not cover all possible contaminants. Hence for example it is not assumed that an unusual contaminant which is not detected by the given examination methods is permitted when reason and good pharmacuetical practice demand its absence.'

Chapter 3

Extraction agents

3.1. Preliminary remarks, terms and definitions†

Extraction agents used in the preparation of phytopharmaceuticals must be suitable for dissolving the important therapeutic drug constituents and thus for separating them from the substances containing the drugs which are to be extracted. In pharmaceutical technology the extraction agent or solvent is known as a menstruum and the extract solution separated from the residual insoluble drug plant material is called a miscella. Menstrua are therefore solvents which readily dissolve, and must often also have a certain selectivity for, the extracted substances; hence, for example, bitter principles, mucins, pigments, resins, etc., should if at all possible not be dissolved in the extraction of vegetable oils. According to the customary definition in this technology solvents are, under normal conditions, volatile, usually organic liquids capable of dissolving other gaseous, liquid or solid substances without either themselves or the dissolved substance being chemically altered [3.1]. In pharmaceutical technology this is not always the case, as solvents such as oils which are not volatile under normal conditions are also used.

The only inorganic solvents used in pharmacy are water and CO_2 (see Section 4.2.4), although liquids such as ammonia, sulphur dioxide and hydrocyanic acid solutions are used for special purposes in pharmaceutical processing.

Water, pure organic liquids and mixtures of organic liquids with water or with other organic liquids are used as extraction solvents. These organic liquids are nearly always hydrocarbons and their derivatives such as halogenated hydrocarbons, alcohols, esters, ketones, ethers, oils, etc.

Selectivity, ease of handling, economy, protection of the environment and safety are major factors to be considered in the choice of a suitable solvent or of a mixture of several solvents. The last two factors in particular are becoming increasingly important due to the legislators' sharpened attention to them. This has gone so far as some manufacturers making serious

---

†Most of the numerical values mentioned in Chapter 3 have been taken from Houben-Weyl "Methoden der organischen Chemie" Volume 1/2 "Allgemeine Laboratoriumspraxis" (General laboratory practice), Part 2, fourth Revised Edition, Thieme Verlag, Stuttgart 1959.

attempts to design extraction processes so that the smallest possible number of 'safe' liquids, e.g. water and lower alcohols and, in special areas, super-critical gases (see Section 4.2.4), are used.

It must not be forgotten here that traditional plant extracts must satisfy the requirements of the Pharmacopoeias. This is usually possible only when the menstrua specified are used. A degree of choice is offered by using cheaper methanol instead of expensive ethanol (on which duties have to be paid) if its polarity is equal to that of the required ethanol–water mixture. As a rule of thumb, a 10% higher methanol concentration roughly corresponds to the required ethanol concentration.

One essential consideration in the choice of suitable solvents is the question of whether the menstruum remains wholly or partly in the end-product. If this is the case, physiologically harmless solvents must be used. If the end-product is free of solvent, as is the case with dried extracts, or if it still contains only harmless components from the originally used solvent mixture after further processing of the miscella, then for reasons of selectivity and economy the choice can be made without considering the physiological properties of the mixture used.

3.2.　Pharmaceutical solvents

The range of solvents available for the extraction of processed medicinal plant material and fresh medicinal plants is not large. Mixtures of organic solvents such as ether and ethanol, or mixtures of organic solvents such as alcohols with water, are used to produce certain effects. Azeotropic mixtures represent a special type of mixture. These are mixtures of two or more liquids which boil at a temperature characteristic of the mixture, i.e. the components are present in the vapour in the same relative concentration as they are in the liquid (they therefore cannot be separated by distillation or rectification). In drug extraction by the Soxhlet process, where the solvent is repeatedly recycled on to the extraction material by constant distillation, solvent mixtures can only be used if they are azeoptropic. Here it must be ensured that the composition of an azeotropically-boiling mixture is pressure-dependent. As drug extractions are frequently carried out at sub-atmospheric pressures, the liquid mixtures used must be suitable for this. Furthermore, the composition of a mixture regarded as azeotropic can be altered, for example, by water being retained during soaking and subsequent swelling of the drug material. The mixture remaining in circulation no longer boils azeotropically. Some binary and ternary azeotropic mixtures which can be used pharmaceutically are listed in Section 3.3.

3.2.1. ALIPHATIC HYDROCARBONS

Certain petroleum fractions are used for the extraction of lipophilic substances such as fats and waxes, often only for pre-extraction so that larger yields of extract can then be obtained with polar solvents. These fractions differ in their boiling ranges as follows:

Petroleum ether	b.p. 30–50°C
Light petrol	b.p. 60–95°C
Ligroin	b.p. 80–110°C
Cleaning petrol	b.p. 100–140°C
White spirit	b.p. 160–196°C

Pure hexane or heptane is used only for analytical purposes, e.g. spectroscopy.

n-Hexane

Mol. wt 86.17; $b.p._{760}$ 68.74°C; f.p. −95.34°C; D_4^{20} 0.65937; n_D^{20} 1.37486. Solubility in water 0.014 wt % (15.5°C), 0.0111 wt % (20°C). Azeotrope with water 61.55°C.

The MWC† value of the petrol fractions is 500 ppm. Because of the high flammability of petrols and the explosiveness of their vapour–air mixtures one must work with them using suitable protection against explosion.

n-Heptane

Mol. wt 100.198; $b.p._{760}$ 98.427°C; f.p. −90.601°C. D_4^{15} 0.68798; n_D^{20} 1.38765. Solubility in water 0.005 wt % (15.5°C). Solubility of water in n-heptane 0.0151 wt % (25°C).

n-Octane

Mol. wt 114.224; b.p. 125.665°C; f.p. −56.798°C; D_4^{20} 0.70252; n_D^{20} 1.39743. Solubility in water 0.0142 wt % (20.0°C).

Cyclohexane

Mol. wt 84.156; b.p. 80.738°C; f.p. 6.554°C; D_4^{20} 0.77855; n_D^{20} 1.42623. Azeotrope with water, composition 91% cyclohexane, $b.p._{760}$ 68.95°C. Solubility in water 0.010 wt. % (20°C).

---

†The German Commission for the Testing of Health-damaging Industrial Substances defines the MWC value (maximum workplace concentration) of a gaseous, vaporized or powdered industrial substance as that concentration in the air of the workplace, measured as respiratory level, which after careful testing is not generally expected to damage the health of the persons employed in the workplace even when they are exposed to it for 8 h per day. The MWC of gases and vapours is given in parts per million (ppm) (cm^3 gas/m^3 air) for a temperature of 20°C and a pressure of 1013 mbar. For pure gases the MWC can, and for dusts and sprays must, be given in mg/m^3.)

3.2.2. AROMATIC HYDROCARBONS

Because of their high flammability, the explosiveness of their vapour–air mixtures, their toxicity and especially their easy absorption through the skin, aromatic hydrocarbons are now used only in special cases for the manufacture of phytopharmaceuticals. In the MWC tables substances absorbed by the skin are marked with a letter 'H'.

Benzene

Mol. wt 78.108; b.p.$_{760}$ 80.10°C; f.p. 5.53°C; D_4^{10} 0.88947; n_D^{20} 1.50110. Azeotrope with water, composition 91.17% benzene, b.p. 69.25°C. MWC 25 ppm$\simeq$80 mg/m^3.

Toluene

Mol. wt 92.134; b.p.$_{760}$ 110.623°C; f.p. -94.991°C; D_4^{10} 0.87615; n_D^{20} 1.49693. MWC 200 ppm $\simeq$750 mg/m^3.

3.2.3. CHLORINATED HYDROCARBONS

Chlorinated hydrocarbons were for a long time the preferred extraction solvents for lipophilic substances. Those listed below are non-flammable and do not form explosive mixtures with air, though because of their toxicity and the fact that they are not permitted in effluent (being detrimental to biological effluent purification) they are now used only with great hesitation. Of the three compounds, i.e. dichloromethane (methylene chloride), trichloromethane (chloroform) and tetrachloromethane (carbon tetrachloride), methylene chloride is the one still most frequently used.

Dichloromethane, methylene chloride

Mol. wt 84.940; b.p.$_{760}$ 39.95°C; f.p. -96.7°C; D_4^{15} 1.33479; n_D^{20} 1.42456. Azeotrope with water, composition 98.5% methylene chloride, b.p. 38.1°C. MWC 200 ppm $\simeq$1720 mg/m^3.

Trichloromethane, chloroform

Mol. wt 119.389; b.p.$_{760}$ 61.15°C; f.p. -63.55°C. D_4^{20} 1.4892; n_D^{15} 1.44858. Azeotrope with water, composition 97.8% chloroform, b.p.$_{760}$ 52.12°C. MWC 50 p.p.m $\simeq$240 mg/m^3. 0.5–1% ethanol is added to commercial chloroform because it tends to decompose to phosgene in the presence of oxygen and light.

Carbon tetrachloride

Mol. wt 153.838, b.p.$_{760}$ 76.75°C; f.p. -22.99°C; D_4^{15} 1.60370; n_D^{15} 1.46305 Azeotrope with water, composition 95.0% carbon tetrachloride, b.p.$_{760}$ 66°C. MWC 10 ppm $\simeq$65 mg/m^3

3.2.4. ALCOHOLS

Only the lower alcohols, i.e. methanol, ethanol, propanol-1, propanol-2, and occasionally the butyl alcohols are used for drug extraction.

Methanol, methyl alcohol

Mol. wt 32.042; b.p.$_{760}$ 64.51°C; f.p. -97.49°C; D_4^{15} 0.79609; n_D^{15} 1.33057. No azeotrope with water. MWC 200 ppm $\simeq$ 260 mg/m^3.

Ethanol, ethyl alcohol

Mol. wt 46.086; b.p.$_{760}$ 78.325°C; f.p. -114.5°C; D_4^{15} 0.79360; D_4^{20} 0.78934; n_D^{20} 1.36139. Azeotrope with water, composition 95.6 wt % ethanol, b.p.$_{760}$ 78.2°C. MWC 1000 ppm $\simeq$ 1900 mg/m^3.

n-Propanol, propanol-1, n-propyl alcohol

Mol. wt 60.094. b.p.$_{760}$ 97.15°C; f.p. -126.2°C; D_4^{15} 0.80749; n_D^{20} 1.38556. Azeotrope with water, composition 70.9 wt % n-propanol, b.p.$_{760}$ 87.76°C.

Isopropanol is technically more important and less toxic.

Mol. wt 60.094; b.p.$_{760}$ 82.40°C; f.p. -89.5°C; D_4^{15} 0.78916; n_D^{25} 1.3747. Azeotrope with water, composition 87.4 wt % isopropanol, b.p. 80.3°C. MWC 400 ppm $\simeq$ 988 mg/m^3. Technically more important and less toxic than n-propanol.

Butanols, butyl alcohols

Table 3.1 Solvent related parameters of butyl alcohols

Property	n-Butanol Butanol-1 Butyl alcohol $CH_3-(CH_2)_2-CH_2OH$	i-Butanol Isobutyl alcohol 2-Methylpropanol-1 $(CH_3)_2CH-CH_2OH$	s-Butanol s-Butyl alcohol Butanol-2 $CH_3-CH_2-CH(OH)-CH_3$
Mol. wt	74.120	74.120	74.120
b. pizo	117.726°C	107.89°C	99.529°C
f.p.	-89.53°C	-108°C	-114.7°C
D4/5	0.81337	0.80576	0.81092
n_D^{15}	1.40118	1.39768	1.39946
Azeotrope	b.p.$_{760}$ 92.7°C	b.p.$_{760}$ 89.8°C	b.p.$_{760}$ 87.5°C
With water	57.7%	67%	72.7%
	n-butanol	i-butanol	s-butanol

3.2.5. KETONES

Of the numerous ketones used as solvents in chemical technology, only acetone and occasionally methyl ethyl ketone are used for drug extraction.

Acetone

Mol. wt 58.087; b.p. 56.24°C; f.p. -95.35°C; D_4^{20} 0.79079 (99.70%); n_D^{20} 1.35880 (99.70%). No azeotrope with water. MWC 1000 ppm $\simeq$ 2400 mg/m^3.

Butanone, Methyl ethyl ketone

Mol. wt 72.104; b.p.$_{760}$ 79.50°C; f.p. −87.30°C; D_4^{20} 0.80473; n_D^{20} 1.37850. Azeotrope with water, composition 88.73% butanone, b.p.$_{760}$ 73.41°C. MWC 200 ppm ≃590 mg/m³.

3.2.6. CARBOXYLIC ACIDS

Acetic acid is used in small quantities in the preparation of 'vinegar' dosage forms. Although this type of medicinal presentation is regarded as obsolete, the physical constants of pure acetic acid are given here for reference.

Acetic acid

Mol. wt 60.052; b.p.$_{760}$ 117.72°C (99.78%); f.p. 16.63°C; D_4^{20} 1.04923; n_D^{20} 1.27160. MWC 10 ppm ≃25 mg/m³.

3.2.7. ESTERS

Ethyl acetate is used as an extraction solvent for phytopharmaceuticals and especially for obtaining active substance concentrates or pure natural substances.

Ethyl acetate, acetic acid ethyl ester, acetic ester

Mol. wt 88.104; b.p.$_{760}$ 77.114°C; f.p. −83.79°C; D_4^{20} 0.90063; n_D^{20} 1.37239. Ethyl acetate and water are miscible 8.00 g ethyl acetate/100 g water and 3.30 g water/100 g ethyl acetate, at 25°C. MWC 400 ppm ≃1400 mg/m³.

3.2.8. ETHERS

All ethers used as solvents in the laboratory and in industry tend to form peroxides to a greater or lesser extent, which, when present in a certain concentration, can decompose explosively. Before they are used, ethers must therefore be tested for peroxides which, if present, must be removed. Only when ethers have been proved to contain no peroxides may they be distilled to dryness. The use of ethers as solvents for easily oxidizable substances also requires the prior removal of peroxides. Practically the only ether used as a solvent in galenics is diethyl ether.

Diethyl ether, ether

Mol. wt 74.120; b.p.$_{760}$ 34.481°C; D_4^{15} 0.71925; n_D^{20} 1.35272. Commercial diethyl ether may contain impurities like ethanol, acetaldehyde and water, and peroxides. MWC 400 ppm ≃1200 mg/m³.

3.2.9. WATER

As emphasized in Section 3.1, water is the most important of all extraction
solvents. It is used either alone or mixed with organic solvents, principally
lower alcohols. The question of the purity of water for the preparation of
plant extracts and what type of purification should be used has often been
discussed. In view of the fact that drug plants contain many mineral
substances of varying solubility and in greatly varying quantities, it would be
irrational to demand the use of distilled or demineralized water. It has been
shown that drinking water which meets all the hygienic requirements
represents water of the quality required. Only the requirements for drinking
water are therefore given (see also the appropriate literature, Reference 3.3).

Drinking water is water suitable for human consumption when it fulfils at
least the following requirements:

(1) Drinking water and water used in the foodstuffs industry must always
    be free of disease pathogens and substances harmful to health.
(2) Water intended for human consumption must be naturally as free of
    bacteria as possible and of an acceptable taste.
(3) Drinking water must be colourless or at any rate not definitely coloured,
    clear, cool and free of any foreign smell or taste.
(4) Water for domestic and industrial use should not contain too many
    salts, particularly hardeners, iron, manganese and organic substances
    (peat or humus).
(5) Drinking water should if possible not cause corrosion.

The following is also stated in the World Health Organization Standards
concerning the radiological consistency: 'It is important to keep the radioactivity in drinking water as low as possible. The following limit values are
acceptable for a drinking water consumed by man throughout his lifetime: α-
radiations 1 pC/L, β-radiations 10 pC/L. Water with a higher content of total
radioactivity can however still be drunk if it contains less dangerous
isotopes.'

Drinking water is generally the basis for the other qualities of water given
in the Pharmacopoeias.

3.2.10. OILS

Oily drug extracts are now rarely made in the pharmacy. Older Pharmacopoeias prescribe the use of oils which are suitable for consumption, have only
a slight colour, smell and taste, are not rancid, and do not become turbid
within 24 h at 10°C. Almond, olive, sesame, peanut, cottonseed and maize
oils can be used. They must be of Pharmacopoeia quality. Drying-oils are not
used. However, the semisynthetic Mygliol 812 Neutral Oil, Triglycerida
medicatenalia *DAB 8*, is available for this purpose. It fulfils the above-
mentioned requirements and, furthermore, is miscible with ethanol, is of low
viscosity and dissolves numerous natural substances.

3.3. Solvent mixtures

3.3.1. WATER-SOLUBILITY OF SOLVENTS

Solvent–water mixtures depend on the degree of miscibility of the two components. There exist fairly large mixture gaps between the solubility of water in solvents and that of solvents in water. Miscibility also varies with temperature. The values given in Table 3.2 relate to a temperature of 20°C.

Table 3.2 Miscibilities in water of various solvents at 20°C (wt %)

Solvent	Solvent miscibility with water	Water miscibility with solvent
n-Hexane	0.01	—
Cyclohexane	<0.01	0.01
Benzene	0.07	0.06
Toluene	0.06 (30°C)	0.05
Methylene chloride	1.32 (25°C)	0.20 (25°C)
Chloroform	0.8	0.2
Carbon tetrachloride	0.03	0.03
Methanol	∞	∞
Ethanol	∞	∞
n-Propanol	∞	∞
Isopropanol	∞	∞
n-Butanol	7.7	20.1
s-Butanol	20.1	30.3
Isobutanol	10	15
t-Butanol	∞	∞
Acetone	∞	∞
Methyl ethyl ketone	26.8	11.8
Acetic acid	∞	∞
Ethyl acetate	8.7	3.3
Diethyl ether	6.9	1.3

3.3.2. BINARY AND TERNARY AZEOTROPIC MIXTURES

The dissolution properties of an extraction agent can often be substantially improved by a combination of solvents of varying polarity. To utilize this phenomenon fully it is recommended that the composition of the menstruum be chosen so that a binary or ternary azeotropic mixture is produced. This has the advantage that upon concentration of the extracts the solvent boils constantly and the condensate, perhaps after a small correction by replacement of components preferentially retained in the drug residue, can be reused. Some such pharmaceutically utilizable binary and ternary mixtures are listed in Table 3.3.

Table 3.3 (Azeotropic) mixtures of constant boiling point

(a) Binary mixtures

Component 1	wt %	Component 2	wt %	b.p.$_{760}$ (°C)
n-Hexane	81	Benzene	19	68.9
n-Hexane	28	Chloroform	72	60.0
n-Hexane	79	Ethanol	21	58.7
n-Hexane	96	n-Propanol	4	65.7
n-Hexane	78	Isopropanol	22	61.0
Chloroform	72	n-Hexane	28	60
Chloroform	88	Methanol	12	53.5
Chloroform	93	Ethanol	7	59.4
Carbon tetrachloride	79	Methanol	21	55.7
Methanol	27	n-Hexane	73	50.0
Methanol	37	Cyclohexane	63	54.2
Methanol	39	Benzene	61	48.3
Methanol	21	Carbon tetrachloride	79	55.7
Methanol	12	Chloroform	88	53.5
Methanol	86	Acetone	14	55.7
Methanol	19	Ethyl acetate	81	54
Methanol	92[1]	Ethyl acetate	8[1]	62.3
Ethanol	95.57	Water	4.43	78.15
Ethanol	21	n-Hexane	79	58.6
Ethanol	30.6	Ethyl acetate	69.4	71.8
Ethanol	40	Methyl ethyl ketone	60	74.8
Isopropanol	23	Ethyl acetate	77	74.8
Methyl ethyl ketone	86	Methanol	14	55.9
Methyl ethyl ketone	89	Water	11	73.6
Methyl ethyl ketone	37	Benzene	63	78.3
Methyl ethyl ketone	29	Carbon tetrachloride	71	73.8
Diethyl ether	40	Ethanol	60	74.8
Ethyl acetate	95[1]	Water	5[1]	34.2
Ethyl acetate	81	Methanol	19	54
Ethyl acetate	69	Ethanol	31	71.8

(b) Ternary mixtures[2]

Component 1	mol %	Component 2	mol %	C't 3	mol %	b.p.$_{760}$ (°C)
Ethanol	23	Carbon				
Ethanol	12.4 (9)	tetrachloride	57.6	Water	19.4	61.8
Ethanol	22.8	Ethyl acetate	60.1 (81)	Water	27.5 (8)	70.3
Ethanol	22.2	Benzene	53.9	Water	23.3	64.9
Isopropanol	(18.7)	Cyclohexane	54.3	Water	23.5	62.1
Isopropanol	19.2	Benzene	(73.8)	Water	(7.5)	66.5
n-Propanol	18	Cyclohexane	54.8	Water	26.0	64.2
n-Propanol	8.9	Carbon	54.8	Water	26.9	65.4
n-Propanol	10.3	tetrachloride	62.8	Water	29.3	68.5
		Cyclohexane	60.3	Water	29.4	66.5

[1] These values are mol %.
[2] Values in parentheses are wt %.

Chapter 4

Processing

4.1 Comminution, grading and classification of drug plant materials

4.1.1. DEFINITIONS

4.1.1.1. Comminution or pulverization

This means the fragmentation of a substance into small particles by mechanical forces.

Several hundred thousand million kWh, in all about 6% of the world's energy production, is spent annually worldwide on comminution processes. As only about 1% of the energy consumed actually achieves the comminution whilst the rest goes to waste, the comminution process can be seen as an energetically inefficient operation. Summarized descriptions of comminution are given in the literature, e.g. by Ullmann [4.1] and in related works on process technology, e.g. by Adolphi [4.2].

Various types of comminution can be distinguished on the basis of the physical properties (i.e. 'hardness' and 'toughness') of the materials being processed (Table 4.1).

Table 4.1 Types of comminution

Type of comminution	Type of force	Examples of material
Hard comminution	Force between two surfaces by pressure, friction, hammering	Quartz, cement, glass
Medium hard comminution	Force between two surfaces by pressure, friction and impact	Gypsum, coal, salts
Soft comminution	Force on a surface by cutting, pressure and impact	Talc, grain

Another method of characterizing the different types of comminution uses the maximum granule diameter, d_{max} (Table 4.2). The ratio of any granulation parameter before and after comminution is defined as the degree of comminution Z, where the parameters are used so that $Z \geqq 1$, e.g.

$$Z_{d'} = \frac{d_0'}{d_1'}$$

Table 4.2 Comminution types by granule diameter

Type of comminution	d_{max}
Coarse fracturing	> 50 mm
Fine fracturing	5–50 mm
Rough grinding	0.5–5 mm
Fine grinding	50–500 μm
Very fine grinding	1–50 μm

where d' = granulation parameter of the RRSB distribution. (The RRSB distribution is an empirically determined function for graphical representation of granule size distributions. A distribution following this function produces a straight line in the RRSB granulation graph. The granule-size of the total residue at 36.8% is designated the granulation parameter d'.)

$$Z_{d_{80}} = \frac{d_{80_0}}{d_{80_1}}$$

Where d_{80} = granule-size of the throughput 80% of the distribution

$$Z_O = \frac{A_1}{A_0}$$

where A = (specific) surface area, d = particle diameter, index 0 = before grinding, and index 1 = after grinding.

The Pharmacopoeias make statements about the degree of comminution of drugs which are used in addition to the technical definitions. A fundamental distinction is to be made here between the Pharmacopoeias which fix the degree of comminution solely by stating the sieve mesh width and those which make additional descriptive statements.

The first group includes *Ph. Eur. I* and *Ph. Helv. VI* and the second group the *2.AB-DDR*, *USP XX* and *BP 80*. *Ph. Eur. I* uses the sieve sizes recommended by the International Standards Organization (ISO) (Table 4.3).

Table 4.3 Sieve sizes after *Ph. Eur. I*

Sieve number (mesh in μm)	Standard deviation of mesh (%)	Greatest mesh variation (%)
8000	±2	+ 3
5600	±3	+ 5
4000	±3	+ 7.5
2800	±3	+10
2000	±3	+10
1400	±3	+10
1000	±5	+15
710	±5	+17.5
500	±5	+20
420	±5	+25
355	±5	+25
300	±5	+25
250	±5	+25
180	±6	+35
125	±6	+40
90	±6	+40

Ph. Helv. VI follows a similar scheme (Table 4.4), but uses other sieve mesh widths, although the sieves of *Ph. Eur. I* are permitted here. The use of circular hole sieves is allowed, though the diameter of the holes with the same sieve number must be 1.25 times the mesh-width of a sieve with a square mesh.

Table 4.4 Sieve meshes after *Ph. Helv. VI*

Sieve number	Permitted standard deviation of mesh (%)	Permitted variation in greatest mesh (%)
9000	±2	+ 3
5000	±3	+ 5
3150	±3	+10
1600	±3	+10
800	±5	+15
500	±5	+20
315	±5	+25
250	±5	+25
200	±6	+35
160	±6	+40
100	±6	+40

The degree of comminution must, according to *Ph. Helv. VI*, fulfil certain requirements (Table 4.5).

Table 4.5   Degree of comminution after *Ph. Helv. VI*

9000:	All material must pass through a 9000 sieve
5000:	All material must pass through a 5000 sieve
3150:	All material must pass through a 3150 sieve
1600:	All material must pass through a 1600 sieve
800/250:	All material must pass through a 800 sieve and not more than 20% through a 250 sieve
500-250:	All material must pass through a 500 sieve and not more than 40% through a 250 sieve
315/200:	All material must pass through a 315 sieve and not more than 40% through a 200 sieve
250/160:	All material must pass through a 250 sieve and not more than 40% through a 160 sieve
200/100:	All material must pass through a 200 sieve and not more than 40% through a 100 sieve
160/100:	All material must pass through a 160 sieve and not more than 60% through a 100 sieve
100:	All material must pass through a 100 sieve

The production of a certain proportion of fine material is unavoidable in the shredding of drug plant material. *Ph. Helv. VI* recommends that this material should be removed using a sieve with the next smaller mesh size. This has to be done after the raw drug plant material has been transported or when it is to be used for the preparation of infusion mixtures.

Pharmacopoeias which give descriptive statements, in addition to mesh widths, include *2.AB-DDR*, *USP XX* and *BP 80*.

2.AB-DDR (Table 4.6) states the degree of comminution as the mesh width of the sieve through which all the comminuted substance will still just pass.

Table 4.6   Degree of comminution after *2 AB-DDR*

Sieve no.	Mesh	Degree of comminution
I	6.3 mm	Coarsely ground
II	3.15 mm	Moderately coarsely ground
III	2 mm	Moderately finely ground
IV	1.6 mm	Finely ground
V	1 mm	Very finely ground
VI	0.8 mm	Coarsely powdered
VII	0.5 mm	Moderately coarsely powdered
VIII	0.315 mm	Moderately finely powdered
IX	0.16 mm	Finely powdered
X	0.05 mm	Very finely powdered

2.AB-DDR demands that 100% of the particles pass through the sieve. *USP XX* and *BP 80* also give additional lower limits (Table 4.7).

Table 4.7 Degree of comminution after *USP XX* and *BP 1980*

Designation	Particle size (μm) *USP XX*[a]	Particle size (μm) *BP 80*
Very coarse	100% through 2380 μm sieve ≤20% through 250 μm sieve	
Coarse	100% through 840 μm sieve ≤40% through 250 μm sieve	100% through 1700 μm sieve ≤40% through 355 μm sieve
Moderately coarse	100% through 420 μm sieve ≤40% through 177 μm sieve	100% through 710 μm sieve ≤40% through 250 μm sieve
Moderately fine		100% through 355 μm sieve ≤40% through 180 μm sieve
Fine	100% through 250 μm sieve ≤40% through 149 μm sieve	100% through 180 μm sieve
Very fine	100% through 177 μm sieve	100% through 125 μm sieve

Some of the differences between these two Pharmacopoeias are considerable, i.e. a difference by a factor of 3 in the number of particles per gram is given for a spherical particle of density 1 for the designation 'very fine' based on the limiting granule sizes of 177 μm and 125 μm permitted by the *USP XX* and the *BP 80* respectively. The relevant surface areas differ about 6-fold. The designations of *DAB 7* and *DAB 8* (Tables 4.8 and 4.9) are still often used in the drug trade.

Table 4.8 Degree of comminution after *DAB 7*

Sieve no.	Mesh per DIN 4188, page 1 (February 1957)	Degree of comminution
0	10.00 mm	Very coarsely shredded
1	4.00 mm	Coarsely shredded
2	3.15 mm	Moderately finely shredded
3	2.00 mm	Finely shredded
4	0.80 mm	Coarsely powdered
5	0.315 mm	Moderately finely powdered
6	0.169 mm	Finely powdered
7	0.100 mm	Very finely powdered

Table 4.9 Degree of comminution after *DAB 6*

Sieve no	Mesh (mm)	Designation	
		English	Latin
1	4	Coarsely shredded	Concisus grossus
2	3	Moderately finely shredded	Concisus
3	2	Finely shredded	Minutim concisus
4	0.75	Coarsely powdered	Pulvis grossus
5	0.30	Moderately finely powdered	Pulvis
6	0.15	Finely powdered	Pulvis subtilis

4.1.1.2. Sorting or grading

Strictly speaking, this is a classification process. However, whereas in actual classification comminuted materials are separated into different classes, in grading whole pieces are graded by various characteristics, e.g. eggs are graded by size, the ripeness of tomatoes by colour and apples by size and appearance (rust spots).

As far as phytopharmaceuticals are concerned grading is important, particularly for those infusion preparations in which quite large pieces of plant parts are used. This is the case with seeds, berries and fruits, e.g. hazel nuts, almonds and pistachio nuts in the foodstuffs industry and karaya gum, gum arabic, fennel fruits and juniper berries in the pharmaceutical industry. The separation of ergot from cereal grain after harvesting is also carried out on sorting machines.

4.1.1.3 Classification

This means the separation of the particles of a substance into various 'classes' according to a physical characteristic. The size, weight, volume, shape, specific gravity and the sedimentation rate of the particles can be considered as such characteristics. Classification by particle size for example takes place during sieving, which is important for drugs, and also during blast sifting.

4.1.2. PRINCIPLES OF COMMINUTION

The first attempts at a mathematical description of the process of comminution were made as long ago as the last century. Various equations are given for the expended grinding energy:

$$W = C \left(\frac{1}{d_1} - \frac{1}{d_0} \right) \qquad \text{(Rittinger, 1867)}$$

$$W = C(\log d_0 - \log d_1) \quad \text{(Kick, 1885)}$$

where W = grinding-energy expended, C = constant, d_0 = initial granule size (m), and d_1 = granule-size of resultant fine material (m).

According to Rittinger, the resultant surface area is proportional to the grinding energy, whereas according to Kick it is the alteration of the volume which is proportional.

These equations apply strictly to the comminution of individual granules. Their application to aggregates and to the conditions of comminution by machine appears dubious. In 1952 Bond [4.3] produced a useful equation for the grinding of mineral substances in ball mills:

$$\frac{W}{m} = \alpha_i \left(\frac{1}{\sqrt{d_{80_1}}} - \frac{1}{\sqrt{d_{80_0}}} \right)$$

where $\frac{W}{m}$ = size-related energy expended (J kg^{-1}), α_i = constant [modified 'work index' (m$^{\frac{1}{2}}$ kg^{-1}) as per Bond], d_{80_0} = size (m) of particles 80% of which pass through sieve before grinding, and d_{80_1} = size (m) of particles, 80% of which pass through the sieve after grinding.

Table 4.10 shows the α_i values of certain materials determined in a laboratory ball mill.

Table 4.10 Work indexes for various materials

Substance	α_i (10^{-3} J m$^{\frac{1}{2}}$ kg^{-1})
Basalt	0.22
Cement clinker	0.15
Coal	0.12
Coke	0.23
Glass	0.03
Iron ore	0.17
Chalk	0.18
Quartz	0.15

Unfortunately, the given laws of comminution and the examples of drug substances can only be used to a very limited extent. Hard seeds, such as the cola nut or ginger, behave in a similar manner to mineral substances. However, the more fibrous the drug material, the less applicable will be the given relationships. Lauer (4.5) shows this by the different energy requirements for the comminution of sugar, bark and gelatins, the gelatins being particularly difficult to comminute (Fig. 4.1).

Sugar represents a brittle or friable substance, bark a fibrous substance and gelatin a viscid, ductile, difficultly comminutable substance. Three important conclusions can be drawn from Fig. 4.1:

(1) The energy requirement increases as the substance is more finely ground, being particularly high for very fine grinding.
(2) The energy requirement for brittle or friable substances, i.e. which can be ground easily, is much lower than for tough, viscid, ductile ones. The differences can be as much as 50-fold even when the granule size of the various substances is the same!
(3) There is a lower limit to which both friable and tough substances can be ground, i.e. they cannot be 'finely' ground even when enormous amounts of mechanical energy are supplied.

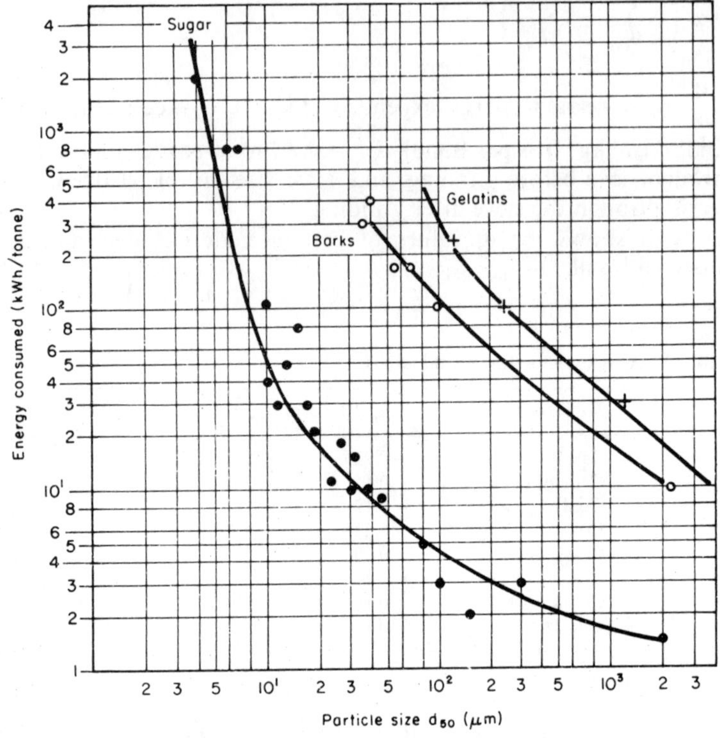

Figure 4.1 Specific energy requirements for the comminution of pharmaceutical material as a function of particle size.

In the case of a friable substance this limit is ~ 3–4 μm. For tough substances it is substantially higher, Fig. 4.1 shows it to be ~ 40–50 μm for bark and ~ 80–100 μm for gelatins.

Beushausen [4.4] gives general directions for the comminution of various substances by impact stamping, pressure, grinding friction, percussion and shredding (Fig. 4.2).

The more important types of force used in comminution are shown in Fig. 4.2. According to this, percussion and shredding should be the most suitable, which correlates with the actual types of machine used for the comminution of drug material.

4.1.3. PRINCIPLES OF CLASSIFICATION

Sieving and blast sifting are used as classification processes for the preparation of drugs. Only the principles of these two processes will be discussed

Property (structure)	Hammer action	Pressure	Friction	Impact	Cutting
Hard (abrasive)					
Medium hard					
Soft					
Brittle					
Resilient					
Tough					
Fibrous					
Poor thermal stability					

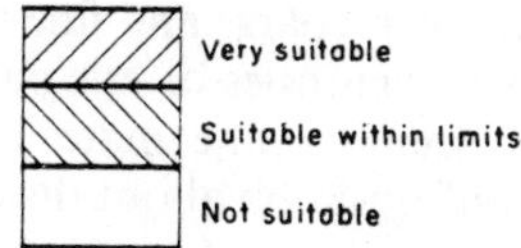

Figure 4.2 Methods for fine grinding and their relation to the characteristics of the material.

here. Detailed descriptions are to be found in the literature [4.6 and 4.2]. Of the sieving processes, only dry sieving is used for drugs; wet processes are not used. *Sieving* is predominantly a process for the separation (classification) of granular aggregates.

A distinction is made between *analytical sieving* and *preparative sieving*. The former is used to determine the granular composition of an aggregate and is done in batches. Continuous sieving is used almost exclusively in production.

During sieving, an aggregate is separated by its particle size into at least two fractions, i.e. coarse and fine material. When a 1 mm sieve is used all particles < 1 mm should theoretically appear in the fine material and all particles > 1 mm in the coarse material. However, in practice fine material particles get into the coarse material and conversely particles of coarse material appear in the fine material, resulting in the 'overlapping' shown in Fig. 4.4. These particles are jointly known as 'bad grain'.

As an overlap area is unavoidable, the way in which the desired separation limit can be defined is of decisive importance in the classification process.

The preparative separation limit, d_p $(= d_{50})$, is the granule size at which the frequency curves of coarse and fine material intersect.

The analytical separation limit, d_a, is the granule size at which the areas under the fine and coarse material curves are equal.

The overlap granule size d_o is the granule size at which equal quantities of 'bad grain' are found in coarse and fine material.

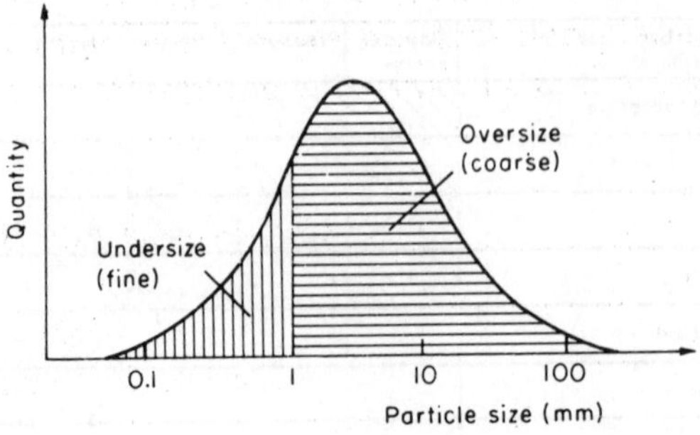

Figure 4.3   Size separation of an aggregate on a sieve.

In technical classification this overlap granule size is frequently defined as the separation limit using distribution curves of coarse and fine material and determining the granule size at which equal quantities of fine particles occur in coarse material and coarse particles occur in fine material. However, determining the separation limit gives no information about the sharpness of the separation (Fig. 4.5).

For definition and determination of the various degrees of sharpness of separation we refer to the specialist literature [4.7]. The characterization and definition of the sharpness of separation are to be regulated by a DIN Standard [4.8]. The sharpness of separation of a sieving process is influenced by the type of movement and the construction of the sieve used, and by the properties and the quantity of the material put through the sieve per time unit.

The sieving process can be divided into the following phases:

Charging and shaking of the sieve	(a)
Selection	(b)
Ejection of coarse material	(c)
Delivery of fine material	(d)

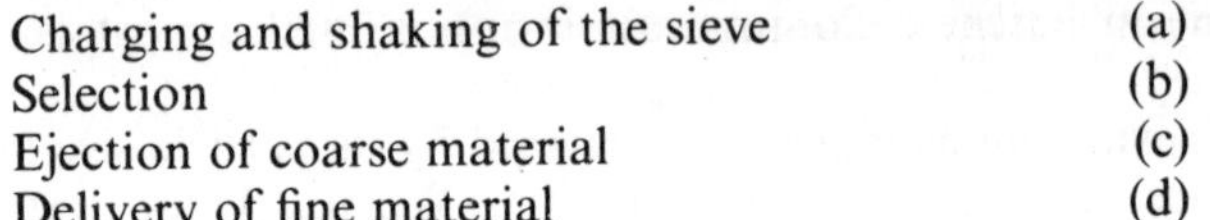

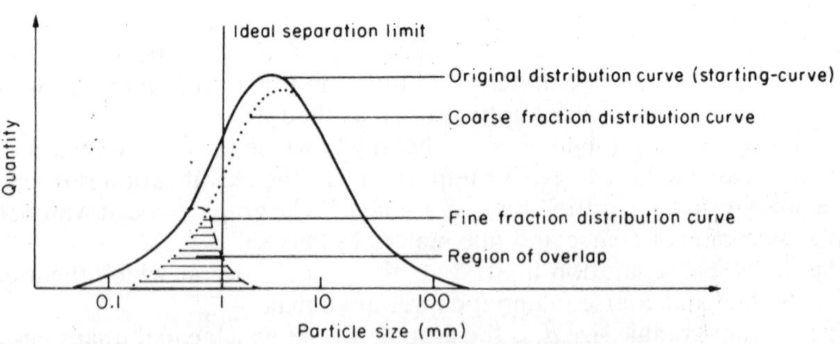

Figure 4.4   Overlap of coarse and fine materials.

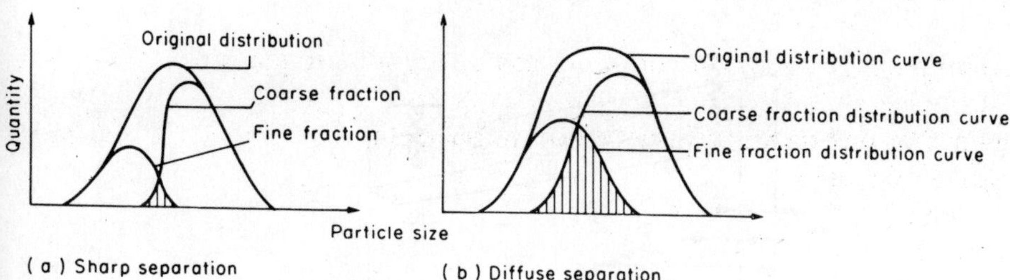

Figure 4.5   Variation in separation efficiency.

These conditions are illustrated in Fig. 4.6 [taken from Ref. 4.6]. In the first stage the fine granules pass through the material to the bottom of the sieve. This depends on the composition of the material and is facilitated by 'loosening' the material.

The second stage is the passage of the fine material through the surface of the sieve, where for this process there is a probability of only < 1, which is dependent on the type and quantity of the flow of material of granule diameter d and on the sieve mesh width l.

The probability of passage is governed by the ratio d/l. There are in theory three ranges:

(a)   Fine material ($d/l < 0.8$) which has a high probability of passing through the sieve without risk of clogging.
(b)   Borderline material ($0.8 \leqslant d/l \leqslant 1.5$) which, in the region $0.8 \leqslant d/l \leqslant 1.0$, has a slight probability of passage but a high risk of clogging.
(c)   The probability of passage is nil for material with $d/l > 5$.

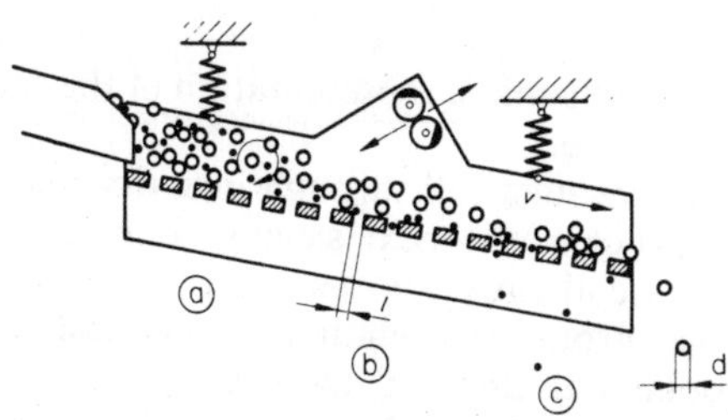

Figure 4.6   The sieve process: (a) charging and shaking; (b) selection; (c) ejection of coarse material and discharge of fine material. d is the particle diameter and l is the sieve mesh dimension.

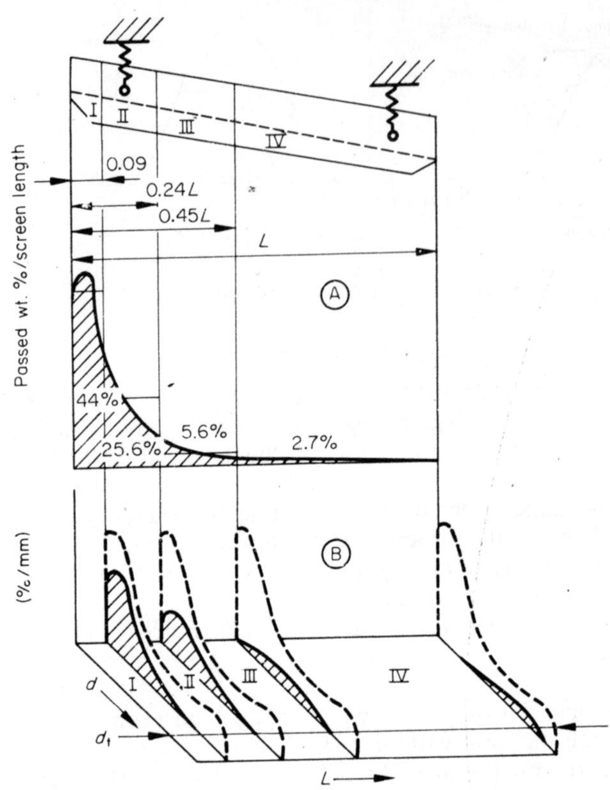

Figure 4.7 Schematic representation of a typical sieving process (starting material: dry limestone of particle size 0.05–1 mm). Mesh width $l = 0.6$ mm; flow rate $q = 2600$ kg h^{-1} m^{-2}; angle of sieve slope $\beta = 20°$; frequency f of vibration $= 100$ Hz; amplitude a of vibration $= 1$ mm; flow rate $q = 2600$ kg h^{-1} m^{-2}. Area A, fine material separated over the sieve length L; area B, particle size distribution in sections I, II, III and IV.

Figure 4.7 [4.6] shows a schematic representation of the course of continuous sieving on a vibrating sieve.

Although the finest granules fall through in a vigorous stream in the first part of the sieving process, the next sieving zone shows a much smaller throughflow, in this case already coarser grains. In this section the quantity of fine grains flowing through, depending on the mesh width, can often be described by an exponential function. The greater part of the sieving process after this merely serves to sieve off the particles of near-borderline size. The first stage of the sieving process takes an increasingly long time with increasing quantity of material with which the sieve is charged. If overloaded the zone of throughfall of finest particles goes right to the end of the sieve. In this case the coarse material will contain a considerable proportion of particles of $d/l < 0.8$ and the sieving efficiency is unsatisfactory.

The sieving effect deteriorates when, with increasing loading of the sieve, the fine material has insufficient time to fall to the bottom of the sieve and, as a result, eventually appears in the coarse material.

The efficiency of industrial sieving can easily be assessed by sieve analysis of the coarse and of the fine material by the method of Paul [4.9 and 4.10], as is shown by the following example [4.11] based on a 2 mm sieve. Table 4.11 shows the sieve analyses of loading, overflow and passage divided into series 2, 3 and 5. The products of these values with the pertinent total quantity found in the overflow or passage are given under series 4 and 6.

The numerical values of series 2, 4 and 6 are shown as total residue curves in Figs 4.8 and 4.9. The Figures only show the sector which is of interest. The curves for sieve overflow and throughflow pass on the one hand into the characteristic granulation loading curve and on the other hand into the separation curve, which separates the yield. This separation curve is intersected by the characteristic granulation loading curve at a point which is equidistant on the ordinate from the sieve overflow and throughflow curves. This distance is the measure for the content of 'bad grain', i.e. the same in the sieve overflow and throughflow. In the example of good sieving the discharges of bad grain amount to 1%, whereas in bad sieving the discharges amount to 4.8%, relative to the loading quantity. The actual grain separation or balanced grain is the point of intersection on the abscissa of the characteristic granulation loading curve with the separation curve. In good sieving the grain is separated at 1.9 mm, which is very close to the mesh width

Table 4.11 Examples of sieving efficiency

1	2	3	4	5	6	3	4	5	6
		Good sieving				Poor sieving			
Particle classification	Loading	Overflow		Throughflow		Overflow		Throughflow	
	%	%	0.785	%	0.215	%	0.895	%	0.105
+30	0.5	0.6	0.5	—	—	0.6	0.5	—	—
30−20	3.0	3.8	3.0	—	—	3.4	3.0	—	—
20−10	17.5	22.3	17.5	—	—	19.5	17.5	—	—
10−5	20.3	25.9	20.3	—	—	22.7	20.3	—	—
5−3.15	24.7	31.4	24.6	0.5	0.1	27.6	24.7	—	—
3.15−2	11.6	14.4	11.3	1.4	0.3	12.9	11.6	—	—
2−1	10.8	1.0	0.8	46.4	10.0	7.7	6.9	37.1	3.9
1−0.5	6.2	0.4	0.3	27.4	5.9	3.4	3.0	30.5	3.2
0.5−0.315	2.4	0.1	0.1	10.6	2.3	1.1	1.0	13.3	1.4
0.315−0.1	2.0	—	—	9.3	2.0	0.8	0.7	12.4	1.3
−0.1	1.0	0.1	0.1	4.4	0.9	0.3	0.3	6.7	0.7
	100.0	100.0	78.5	100.0	21.5	100.0	89.5	100.0	10.5

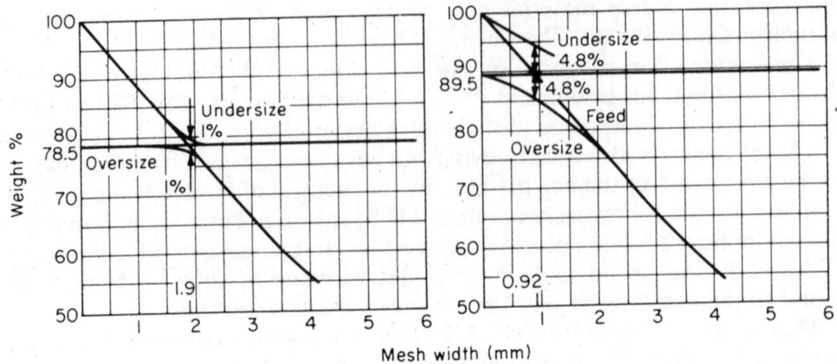

Figure 4.8 Sieve performance measured by Paul's method—efficient sieving.

Figure 4.9 Sieve performance measured by Paul's method—poor sieving.

of the sieve. In bad sieving, however, the grain separation is displaced further towards the fine range and lies at 0.92 mm, i.e. the sieve has worked as if its lining had had a mesh width of only 0.92 mm.

Blast sifting, also known as dry flow classification, is used in drug processing both for removal of dust and for classification by particle size. Blast sifting originated from the separation of the 'chaff from the wheat' by winnowing cereals in the wind. It is now also used for the separation of fine grain [4.12 and 4.13]. Blast sifting is a settling or sedimentation process using a gas, usually air, as the carrier medium. The simplest blast sifter consists of a tube through which air is blown from bottom to top. The material to be sifted is fed into the centre of the tube.

The terminal velocity of a particle in still air is determined by the Stokes equation

$$v = \frac{d^2 sg}{18\eta_A}$$

where v = terminal velocity of particle (m/s), s = density of solid material (kg/m³), g = acceleration due to gravity (m/s), and η_A = viscosity of air (Pa s).

Although the formula does not apply in the turbulent flow in most blast sifters, it shows in principle the influence of particle size (squared) and density on the sedimentation velocity.

If the medium flows against the particle at a speed of v', the fall velocity is reduced and is then only $v_{eff} = v - v'$. Particles with $v_{eff} > 0$ descend, those with $v_{eff} < 0$ ascend and those with $v_{eff} = 0$ remain suspended. These conditions are disturbed by the following influences:

(i)   Varying particle shape.
(ii) Varying flow-speed of the air near the walls of the tube and in the centre of the tube.
(iii) Coating of material on the walls of the sifter.

As a consequence of these processes, and in a manner similar to that in sieving, a certain proportion of the fine material is found in the coarse material and a certain amount of coarse material appears in the fine material and hence a 'separation sharpness' can also be spoken of here.

4.1.4. COMMINUTION AND CLASSIFICATION OF DRUG PLANT MATERIALS

4.1.4.1. General procedure

The object of comminution and classification of drugs is to obtain certain fractions in high yield and as free of dust as possible. Clean, narrow, dust-free fractionation is of great importance not only for further processing to infusion mixtures but also for subsequent extraction. Excess dust, for example, results in clogging of percolators and discharged extracts which are turbid and difficult to clarify. Large differences in the particle size of the drug also results in long extraction times, as the menstruum takes longer for complete extraction of the coarser material. The way in which plant drug material is shredded and ground is therefore very important and generally consists of a multistage process.

Figure 4.10 shows a flow chart for the preparation of shredded infusion materials. Plant drug materials are never free of impurities (i.e. sand, dust, pieces of wood and metal, dead beetles, etc.) and a careful watch must be kept for constituents such as larger pieces of wood, which also seriously upset the milling process, when the bundles of plant drug material are being torn apart.

Pneumatic removal of sand, which is usually combined with magnetic removal of metals, is followed by a preliminary sieving which prevents the material which is already of the desired particle size from unnecessary processing, which would result, for example, in losses of ethereal oils. Dust components which would affect subsequent comminution processes are also removed.

In the shredding process which follows the drug plant materials are cut on shredders (see Section 5.2.1) to the correct size for the next stage of comminution. As dust is also produced in addition to the desired grain fractions, this shredding is followed by a second sieving before the drug material undergoes the actual preparation of the finely shredded material. The coarse material produced by grinding is recycled back to the grinder after the subsequent third sieving. The dust is usually burned. The drug

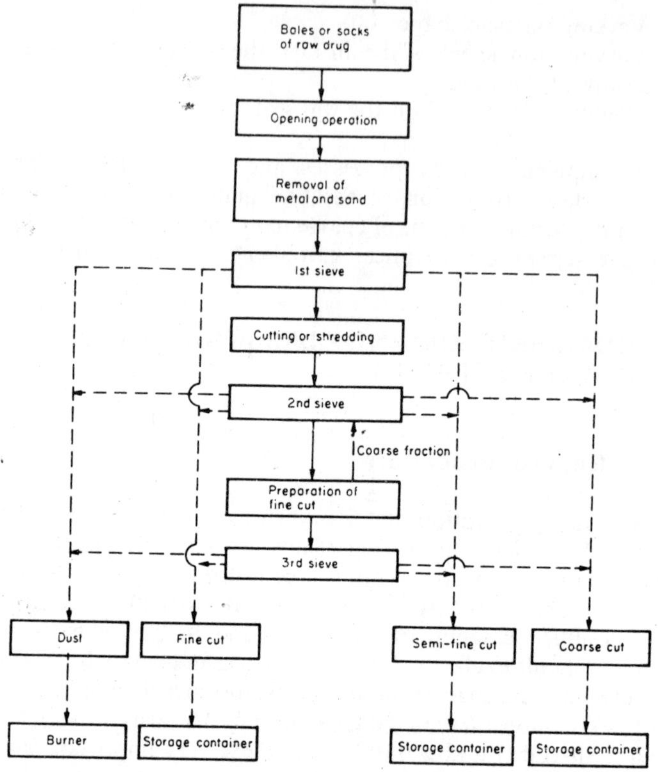

Figure 4.10   Stages in the preparation of shredded infusion materials.

fractions obtained are collected in storage bins, which are usually made of wood, as this helps to prevent condensation and the growth of moulds.

Comminution and classification plants for phytopharmaceuticals, depending on the requirements, are complete manufacturing plants (see Section 5.3.3) or equipment made up by the user according to his own needs.

The 'finely shredded', 'semi-finely shredded' and 'coarsely shredded' fractions obtained as per the above flow chart are used for various purposes. Finely shredded fractions are used for the preparation of teabags or phytopharmaceuticals in sachets and semi-finely shredded teas and drug plant materials are used for tea or infusion mixtures. Coarsely shredded material is put on the market as a ready-to-use drug or used for extraction. A coarsely precomminuted drug as free of dust as possible is required for extraction, particularly if this is to be done by percolation.

Figure 4.11 shows a diagram of a plant for grinding drug material to powder, in which the material is precomminuted on a hammer mill. Sand,

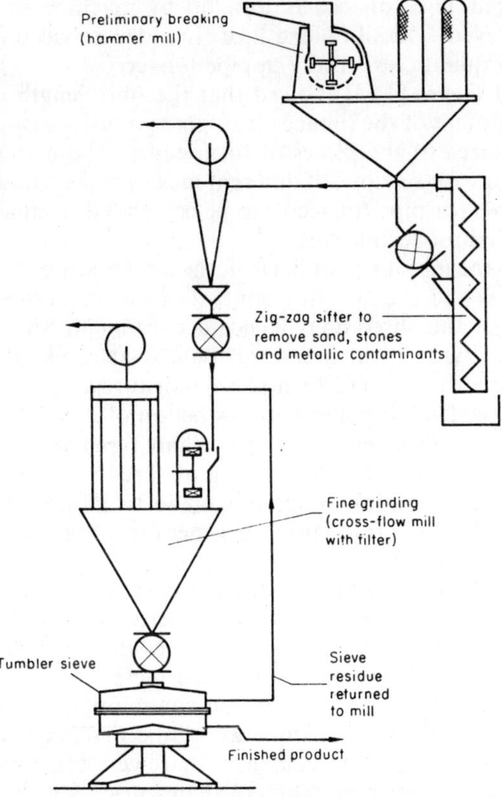

Figure 4.11 Mill for producing powdered drugs.

stones and metal fragments are then removed with a zig-zag sifter (see Section 5.3.1.2) and the material ground to powder on a cross-current mill, to which a tumbler sieve is attached for the classification.

4.1.4.2. Medicinal leaves and herbs

Leaf drugs are distinguished from herbal drugs by the absence of stalk and stem components. Both groups are treated in the same way; this is because leaf components are generally predominant in herbal drugs and—apart from a few exceptions—the same comminution machines are used.

The leaf drug grown in the largest quantity is certainly tobacco, which is a non-pharmaceutical product. The comminution and classification of tobacco therefore provides valuable hints for the comminution of phytopharmaceuticals. Before shredding, tobacco is torn up by machine or by hand, then shredded and sieved. The fibre length of shredded tobacco is an important characteristic of quality, especially in pipe tobacco.

Flesselles and Gayet [4.14] showed that the fibre length depends on the preliminary treatment of the tobacco leaf, on its moisture content and on the average surface area of the pieces of torn leaves. The material is sieved in large drum sieves with sections of different mesh widths. Quality characteristics and standards for pipe tobacco can be calculated from sieve analyses of torn and shredded tobacco leaves.

In addition to the initial size of the leaf, the temperature is important in the shredding process and the moisture content of the material is important for the final result of the shredding. Hence according to Smit [4.15] 'harder' cigarettes can be made from tobacco shredded at room temperature than from tobacco shredded at 40°C. Controlled moisturization prior to shredding reduces brittleness, friability and the production of dust.

Multistage vibrating machines (Engelsmann type) with horizontal sieve surfaces have proved useful in the sieving of shredded tobacco [Kruszinski and Larsen, 4.16]. Vibrating sieves are made to vibrate by means of an eccentric. The following fractions are generally obtained: > 15 mm, 15–10 mm, 10–6 mm, 6–2 mm.

Exhaustive investigations on sieve analysis of tobacco have been published by Hübschen [4.17]. These show that the water content of the tobacco, the sieving machine, the size of the sieve and the sieving time all affect the result. Reproducible results are obtained only when these values are fixed.

Many and varied problems arise in the comminution of pharmaceutical leaf and herb material as a consequence of the diversity of the medicinal plants which exist. *Cutting or shredding mills* are very often used for grinding herbs which contain a high proportion of stem and stalk material. The nature of the leaves of medicinal plants also varies enormously, and it can easily be seen that the hard and somewhat brittle leaves of the bearberry plant require comminution conditions different from those for peppermint or coltsfoot.

The following list gives the various leaf and herbal drugs from a technical viewpoint of comminution. Only drugs currently on the market are considered.

1. Easily friable medicinal leaves
   Folia Belladonnae
   Folia Digitalis
   Folia Melissae
   Folia Myrtilli
   Folia Orthosiphonis
   Folia Rubi fruticosi
   Folia Sennae
   Folia Stramonii

2. *Strong but friable medicinal leaves*
   Folia Uvae Ursi
   Folia Lauri
   Folia Eucalypti

3. *Soft, fibrous medicinal leaves*
   Folia Althaeae
   Folia Farfarae
   Folia Malvae
   Folia Salviae
   Herba Fumariae
   Herba Pulmonariae

4. *Medicinal herbs with high stem/stalk content*
   Herba Chelidonii
   Herba Hyperici
   Herba Majoranae
   Herba Serpylli
   Herba Thymi

5. *Medicinal herbs containing fat/resin*
   Foila Betulae
   Folia Boldo

6. *Very hard and very brittle medicinal herbs*
   Herba Ephedrae
   Herba Equiseti
   Folia Laminariae
   Folia Maté
   Folia Rosmarini
   Folia Visci albi

As already mentioned, *shredding mills* are used for grinding many medicinal leaves and herbs; these are particularly effective for material with a high content of stem and stalk, though they are also useful for soft, fibrous material. They produce only small amounts of unwanted fine material and are mainly used for coarse comminution.

Hammer mills can be used for easily friable and for resinous drugs. *Pin mills*, usually in the form of wide-chamber mills, are advantageous for drugs with a high content of fats or ethereal oils as they can grind very finely. When being ground in hammer and pin mills drug plant material can become heated; attention must be given to this, as it can, for example, reduce the content of ethereal oils and water. For these reasons *fluted roller grinders* are used for drugs with a high content of ethereal oils. Rollers fluted in one direction running against rollers fluted in a different direction and running at a different speed (causing friction) tear the material with their 'cutting edges' by the principle of shearing. The uniformity of the resultant product is often better than that obtained by impact comminution. There is also less heating of the product with fluted roller mills. The finenesses obtained are 100–500 µm.

The possibility of cold grinding also exists, though this is more useful for seeds and fruits.

4.1.4.3. Roots and barks

Roots and barks are moderately hard or hard, but sometimes brittle and friable plant parts. They vary greatly in their nature and condition after harvesting. Fine, thin barks (Cortex Cinnamomi, Cortex Frangulae) and coarse, thick types (Cortex Condurango, Cortex Quercus) exist. Medicinal roots include thin, sometimes clustered forms (Radix Ipecacuanhae, Radix Valerianae) as well as woody types (Radix Levistici, Radix Ratanhiae, Rhizoma Rhei, Rhizoma Tormentillae). There are also forms (Rhizoma Iridis, Rhizoma Zingiberis) on the market as stem or roots from which the bark has been peeled off and which have a smooth appearance and smooth fracture surfaces.

The following general scheme can be given for the comminution of roots and barks:

$$
\begin{array}{c}
\text{Purification of starting materials} \\
\downarrow \\
\text{Shredding (breaking)} \rightarrow \text{Shredded drugs} \\
\downarrow \\
\text{Grinding} \rightarrow \text{Powdered drugs}
\end{array}
$$

Purification is carried out by blast sifting and removal of iron fragments with a magnetic separator.

Drug plant cutters or shredders, which often have blades for achieving a so-called square cut, or *shredder mills* are used for *cutting and shredding*. The shredded drug is obtained after sieve classification.

Grinding of powdered drugs is accomplished with *hammer mills* and *fine mills with sieves*.

4.1.4.4. Seeds and fruits

The comminution of seeds and fruits often proves to be particularly difficult because of their content of fats and ethereal oils. Both *fine mills with sieve*

Table 4.12 Percentage residue after grinding fruits and seeds using fluted rollers

Ground material	Sieve				
	2000 µm	1000 µm	500 µm	200 µm	100 µm
Roasted hazel nut kernels	0	0.5	5.0	26.8	47.0
Coriander	0	13.6	81.0	97.4	99.5
Almonds (peeled)	0	0.1	5.9	28.8	51.0
Nutmeg	2.8	19.0	61.6	84.4	93.5
White pepper	4.8	76.1	90.6	97.0	98.5

attachments and *shredder* and *pin mills* are used. Roller and fluted roller mills are also used, particularly for grinding coffee and cocoa.

The values given in Table 4.12 are total residue percentages. It should be noted that sieves with a mesh-width < 0.75 mm cannot as a rule be used with fine grinders with sieve-attachments as they quickly become clogged. The results of milling of pepper with a Perplex universal mill B 315 [4.18], which can be fitted with a percussion disc or an oscillating beater, are given by way of example in Table 4.13.

Table 4.13 Typical percentage residue after grinding pepper

Sieve (µm)	Total residue (%)
500	1–2
300	~ 10
200	20–30
100	50–60

Similar results can also be obtained with a Contraplex wide chamber mill in two passages. The choice of mill depends on the desired fineness. Shredder mills yield larger product units than pin mills, particularly when the latter are wide chamber pin mills.

Cold grinding, particularly of seeds and fruits, has gained favour in the last decade [4.19]. Here the products, before grinding, are rendered brittle and friable by intense cooling with liquid nitrogen and then comminuted in a similarly cooled grinder. Cold grinding not only protects the ethereal oils but also enables an increased throughput, as the mill does not heat up during grinding. Too much heating results in sticking of the grinder parts and sieve attachments. Table 4.14 demonstrates its advantage for drugs containing ethereal oils [4.19].

Table 4.14   Ethereal oil content of spices before and after various types of grinding

Spice	Original		After conventional grinding		After cold grinding	
	(v/g)	(%)	(v/g)	(%)	(v/g)	(%)
Black pepper	3.37	100	2.21	65.7	3.09	92.0
Pimento	3.19	100	2.71	85.0	3.08	97.0
Cardamom with peel	3.61	100	2.20	60.0	3.15	87.0
Mace	16.10	100	9.10	56.5	14.50	90.0
Clove	17.30	100	11.50	66.0	16.50	95.0

$(v/g) = (mL/100\ g)$

Components of fruit which have to be removed during preparation give rise to special problems, e.g. in the case of rose hips (rose-bush pseudocarp), the hairs and seeds must be separated from the fruit pulp. The preparation proceeds as shown in Table 4.15.

Table 4.15   Procedures for the preparation of rose hips

Procedure	Machines	Product
Coarse shredding of fruit	a) Oscillating beater mill with 8 mm sieve attached b) Shredder mill c) Toothed disc mill	Crushed fruit
Separation of hairs	Blast sifting, e.g. with a) Multiplex zig-zag sifter b) Pneumatic conveyance with cyclone	Fine material: hairs Coarse material: peel, core and seeds
Separation of core and seeds from fruit pulp	Blast sifting, see above	Fine material: pulp Coarse material: core, seeds, parts of pulp
Grinding of fruit pulp	Shredder mills Oscillating beater mills	Ground fruit pulp Grain sizes: 0.5–1.5 mm, 1.0–3.0 mm

4.1.4.5.   Other drug plant materials

These include flowers and parts of flowers, which have not yet been discussed, and products such as alginates, agar and gelatins.

Shredder mills are used for the comminution of flowers and flower parts, where grinding must be followed by blast sifting for the removal of dust. The amount of dust produced by the grinding depends on the moisture content of the material being ground and hence the material is sometimes moisturized to a certain extent before it is ground.

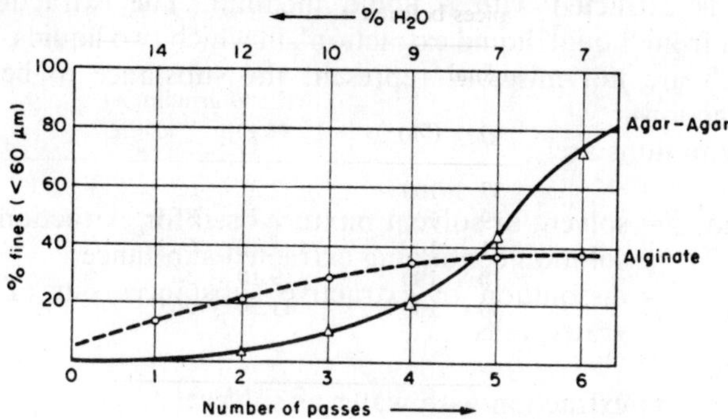

Figure 4.11a Varying behaviour of agar-agar and alginate, two structurally related tough elastic substances, during impact comminution in a pin mill (Contraplex mill 250 CW: 12 000 and 5400 rev/min, equivalent to 17 400 rev/min).

Beushausen [4.4] has observed that in the comminution of tough elastic substances such as agar and alginates the fineness can be improved by repeated grinding on a pin mill. The ultimate fineness of alginate is practically attained after four passages through the pin mill; it is not possible to grind it any finer because of the structure of this substance. The proportion of fine material < 60 μm increases only a little in the case of agar, significant increases in fineness occurring only after the fourth passage. This can be attributed to the fact that the starting material contains too much moisture, but that after four passages so much of this has been expelled that a certain degree of brittleness and friability has been attained. Beushausen's results are reproduced in Fig. 4.11a. According to Lauer [4.20] the effect of moisture on the result of grinding can generally be characterized as follows: in coarse grinding (~ 0.5–2 mm) the product should contain a certain amount of moisture (e.g. 15%), otherwise an undesirably high proportion of dust will be produced by the comminution process. In fine grinding, as low a moisture content as possible is advantageous, as particularly high fineness is produced with low energy expenditure. Small differences in moisture content, say between 10–16%, have had no effect either on coarse comminution or on fine grinding of, for example, senna leaves.

4.2 Extraction of drugs

4.2.1. TERMS AND DEFINITIONS

The extraction of drugs (solid extraction) represents a 'solid from solid' separation, as solid components must be extracted from a solid substance. This type of extraction is generally known as 'solid–liquid extraction', as the 'solid' drug is extracted with a liquid medium. The extraction is thus distinguished from 'liquid–liquid extraction', in which two liquids, the carrier phases, which are not miscible, represent the substance to be extracted (solvent extraction).

Further definitions are:

Menstruum	= solvent or solvent mixture used for extraction
Miscella	= solution containing extracted substance
Rinsing	= dissolution of extractive substances out of disintegrated cells
Lixiviation or leaching	= extraction with water as solvent

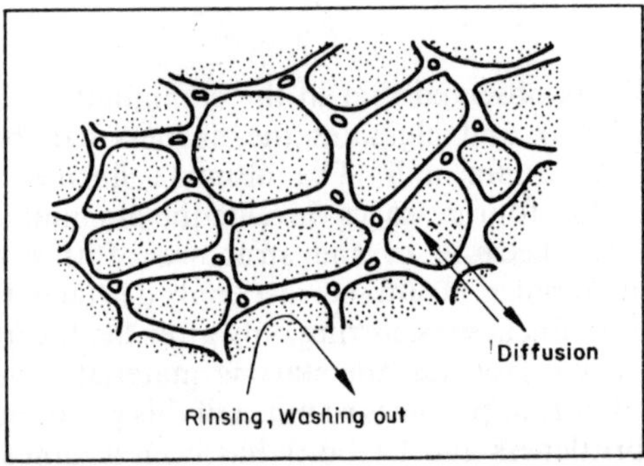

Figure 4.12   Processes of extraction (List, 4.7.21).

Two processes run parallel with each other in the extraction of drugs: (1) the rinsing of extractive substances out of disintegrated plant cells, and (2) the dissolution of extractive substances out of intact plant cells by diffusion. The latter requires the prior steeping and swelling of the drug plant material in order to increase the permeability of the cell walls (Fig. 4.12). Muravev *et al.* [4.22] have attempted to separately analyse the three stages:

Penetration of the solvent into the plant cells and swelling of the cells.
Dissolution of the extractive substances.
Diffusion of the dissolved extractive substances out of the plant cell.

The investigations were carried out with lumps of liquorice root (Radix Liquiritiae). The authors found that:

The solvent (0.25% ammonia solution) penetrates axially into the roots more rapidly than radially and this process is accelerated by raising the temperature.

The steeping and swelling process is strongly influenced by particle size and is more evident radially.

Upon penetration into the drug plant material the solvent becomes enriched with extractive substances and hence the highest content of extractive substances is found on the solvent front.

These investigations on liquorice root are not relevant to all drugs. They are applicable only to certain root and rhizome drugs but not to leaves, barks, etc.

The greatest value of the work lies in the demonstration of differing speeds of transport during extraction.

Extraction processes are still today carried out with a great variety of methods and apparatus. The reasons for this diversity are (a) large differences in the required quantity of extract, (b) great differences in starting material, and (c) the use of various menstrua.

In addition to the actual extraction, further preparative stages such as the removal of ballast materials, evaporation of the solvent in order to thicken and concentrate the extract, and drying are frequently necessary. Further purification and crystallization stages are also required to obtain pure substances from crude plant substances.

Although the isolation of pure substances from plants is not the subject of this book, we shall show by the example of the recovery of strychnine and brucine from the seeds of Nux vomica that extraction is but one of the many process stages in the overall process of isolation. The flow chart (Fig. 4.13) for the isolation of strychnine and brucine from the seeds of Nux vomica was taken from the paper by Srinivasulu *et al.* [4.23].

Extraction processes for drugs can be divided into two major groups:

Processes which result in the establishment of a concentration equilibrium between solution and solid residue.

Processes in which the drug is extracted exhaustively.

4.2.2. METHODS ESTABLISHING A CONCENTRATION EQUILIBRIUM

4.2.2.1. Principles

Common to all processes resulting in a concentration equilibrium is the fact that the extraction process comes to a halt when the distribution of the extractive substances between miscella and drug residue reaches the value K, i.e. when the concentration gradient between miscella and residue has become zero.

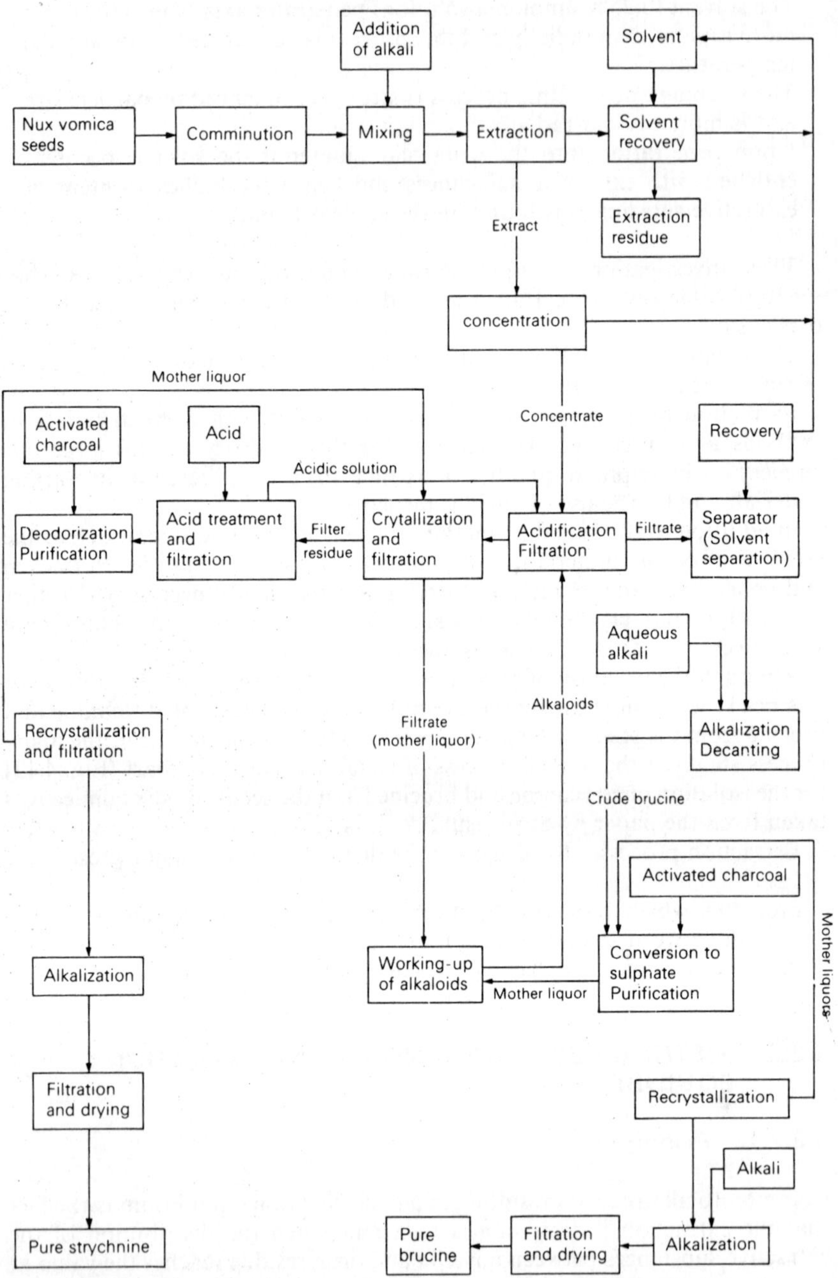

Figure 4.13   Isolation of strychnine and brucine from Nux vomica seeds (Srinivasulu *et al.*, 4.7.23).

$$K = \frac{\text{concentration of extractive substances in the miscella}}{\text{concentration of extractive substances in drug residue}}$$

Numerous factors affect the equilibrium, as shown by the flow chart in Fig. 4.14 for the simplest case of extraction, namely simple maceration.

The position of the equilibrium here is determined both by the properties of the drug, such as type, quantity, moisture content and degree of comminution, and by those of the solvent, such as selectivity and quantity. Although according to Melichar [4.24] the degree of comminution of the drug does not affect the position of the concentration equilibrium, it is essential for the speed of establishment of the equilibrium.

The parameters listed below affect the establishment and position of the equilibrium as follows:

Mixture ratio. The yield of extract decreases with constant quantity of solvent and increasing proportion of drug material [4.25–4.27].

Dissolution from disintegrated cells. Substances are dissolved out of disintegrated cells more rapidly than from intact cells. A greater degree of comminution of the drug plant material results in a larger proportion of disintegrated cells and hence in more rapid establishment of the equilibrium. Finely divided drug material would therefore in principle be preferred, so

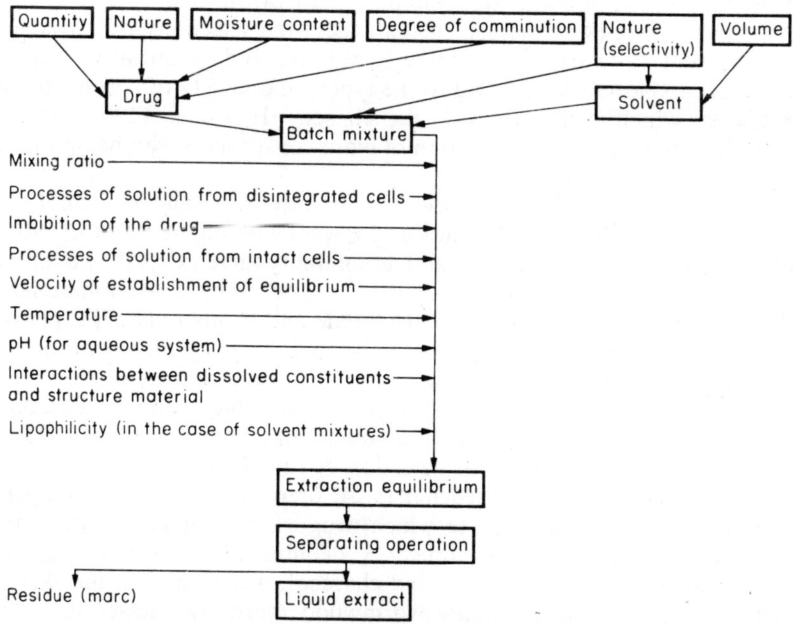

Figure 4.14 Factors that affect maceration.

long as it does not make the subsequently necessary separating processes more difficult.

Steeping and swelling of drug plant material. This is very important for extraction, as it dilates the cell capillaries and hence increases diffusion. Extraction can be hindered or prevented altogether in medicinal plants containing a large amount of mucilage. Ideally, therefore, the swelling should be only of an extent such as will facilitate diffusion of the extractive substances without hindering extraction.

Diffusion from intact cells. The solvent must first diffuse into the intact cells before it can dissolve the extractive substances out of them. The substances to be extracted must be sufficiently soluble in the solvent to produce an increase in the concentration of the solution inside the cell so that equalization of the concentration with the surrounding solvent can take place by diffusion.

Rate of establishment of equilibrium. This is decisive for the time taken to complete the extraction. In addition to the influences of particle size and degree of swelling of the drug plant material, the extraction temperature, the properties of the solvent such as viscosity and polarity, the type and intensity of movement of drug and solvent and any additives such as salts or surfactants all play a part here.

Temperature. Raising the temperature generally has the effect of shifting the extraction equilibrium towards dissolution. We have already mentioned the effect on speed of extraction running parallel with this.

pH value. The pH of the solvent influences the selectivity of extraction. The selectivity is related both to the qualitative and quantitative yield of extractable active substances and to the type and quantity of accompanying substances, which can vary with the chosen pH value. The selectivity, especially of the analytical extraction of plant constituents, can be optimized by altering the pH of the solvent.

Interaction of dissolved constituents with insoluble support material of plants. Even with the optimum choice of solvent with regard to its selectivity for the substances to be extracted and its solubility there exists the possibility of adsorption of already dissolved constituents on to the support material of the drug plant. Hence as early as 1916 Palme and Winberg [4.27] explained errors in the content determination of alkaloid drugs by the adsorption of active substances on to the support material. In addition to adsorption of quinine on to cinchona bark and of strychnine on to Nux vomica seeds, these authors even detected adsorption of atropine on to the skeletal material of alkaloid-free liquorice root. This detection was made in acidic aqueous and alcoholic extraction media. Melichar [4.24] prepared a leached drug plant skeleton by Soxhlet extraction to which drug extracts of known extractive substance content were then added. A definite adsorption, which was, however, inferior to that on activated charcoal, was found on the skeletal material of the medicinal plants wormwood (vermuth), lesser centaury, thyme and also on the leaves of henbane (*Hyoscyamus*). This adsorption was considered responsible by other authors [4.28–4.30] for the relatively low extract yield with increasing proportion of drug plant material in maceration. Attempts at the quantitative description of the conditions have led to Muller's maceration isotherms [4.26] and to Melichar's theory of colloidal dissolution [4.31].

Degree of lipophilicity. The chosen degree of lipophilicity is of great importance in the utilization of organic solvents or solvent mixtures. Any alteration will alter both the quantitative ratio of the extracted substances and the qualitative composition of an extract. It is therefore necessary to check the composition of the extract again if the extraction agent is changed. This also applies to the interchange of similar solvents such as ethanol/methanol.

We have so far discussed only the influences on the process of maceration arising from the drug plant material itself, the solvent and the surroundings. We have not considered here effects which can be obtained by special pH adjustment and additions of surfactants, salts and cosolvents.

The numerous influences on extraction show that even with simple maceration the processes cannot be fully described in detail. The various attempts which have been made at the mathematical description of solid–liquid extraction all represent merely summary statements.

Voeste [4.32] lists the 'diffusion theory', the 'soaking theory' and the 'capillary velocity theory'.

The 'diffusion theory' was described by Schoenemann [4.33] with the aid of the McCabe–Thiele graph commonly used in distillation. The theory states that the rate of extraction depends on the rate of diffusion. The theoretical deduction was given for absolute counterflow extraction and shall therefore be described as a continuous process.

The 'soaking theory' was developed by Boucher *et al.* [4.34]. The authors conclude that not only the rate of diffusion but also the rate of dissolution of the extractive substances in the solvent critically affect the rate of extraction.

The 'capillary velocity theory' originated from Karnowsky [4.35]: it represents the rate of extraction as a function of the rate of flow in the capillaries.

Schultz and Klotz [4.36] have worked out an equation for maceration in the pharmaceutical industry:

$$G = \frac{(L_M - x)a}{a(L_M - x) + 1} \times 100 \tag{1}$$

where G = percentage active substance content in the miscella obtained without expression

L_M = quantity of solvent (menstruum) used, measured in parts of drug, e.g. when 1000 mL solvent is used for 200 g drug, then $L_M = 1000/200 = 5$

x = quantity of solvent absorbed by 1 part of drug, measured in parts of drug, e.g. if 200 g drug retains 40 g solvent, then $x = 40/200 = 0.2$

a = maceration constant.

The values x and a were determined as follows:

The active substance content is determined from the quantity of drug to be processed. A small sample containing known quantities of drug and solvent is macerated for 10 days, with occasional shaking. This sample is then filtered quantitatively under suction at slightly reduced pressure and the filtrate accurately weighed. The quantity of solvent retained is determined from the quantity of filtrate obtained. Hence x is known.

The active substance content in the filtrate is accurately determined and expressed as a percentage of the total active substance content of the drug, for example:

Total active substance content in 200 g drug: 20.0 g
Total active substance content in 960 g filtrate: 15.0 g

$$\frac{20.0}{15.0} = \frac{100}{W}, \quad W = 75\% = \text{percentage content of extracted active substance to be determined}$$

When equation (1) is solved for a, then

$$a = \frac{W}{(L_M - x) - W(L_M - x)} \tag{2}$$

W, which is first divided by 100 (thereby making the total active substance content of the drug equal to 1), can be used in this equation and hence a can be calculated:

$$a = \frac{0.75}{4.8 - 0.75 \times 4.8} = 0.625$$

Hence a and x are known.

If some of the menstruum absorbed by the drug is recovered by expression, the formula is modified as follows:

$$G = \frac{a(L_M - x + y)}{a(L_M - x) + 1} \tag{3}$$

where y = menstruum recovered by expression.

The results obtained by Schultz and Klotz with cinchona bark agree well with the theory. On the basis of these results, Müller [4.26] gives the following equation:

$$c = am^q,$$

where c = concentration in the macerate (kg/m³), m = weight of solvent used per unit quantity of drug (kg/kg), and a and q are constants.

The constants must be determined from the measured values of at least two samples. The resultant concentration in the macerate can be calculated from these for any quantity of solvent (menstruum).

After Muller [4.37] had shown that the maceration can be described by a Freundlich adsorption isotherm only when the ratio of drug to solvent in the sample is greater than 10, Muller and Mielck [4.38], on the basis of Langmuir's considerations, succeeded in producing graphical and mathematical methods of determining the maceration parameters and their variations by applying the law of mass action.

Melichar [4.31] contrasts these 'adsorption theories' with his theory of 'colloidal dissolution', in which the extraction mixture, i.e. drug plant skeleton + extractive substance + extraction agent, is regarded as a colloidal system in which, according to the Ostwald–Buzagh solid substance rule, the yield of macerate depends on the quantity of solid drug material present at the bottom of the system.

The theoretical deductions were tested and confirmed in laboratory experiments with *Calamus* and gentian roots, henbane leaves and buckbean, wormwood, lesser centaury and thyme [4.39]. Goncharenko [4.40] has recently investigated the effect of time, temperature and particle size on the extraction of tannin from Turkish nuts and of isovalerianic acid from valerian root. He concluded that temperature and particle size have a decisive influence on the result of extraction, whereas the extraction time is not influenced by the solvent/drug ratio. An exponential function was used as a mathematical model for this.

4.2.2.2. Maceration, kinetic maceration, remaceration and digestion

4.2.2.2.1. Definitions

Maceration (also known as *simple maceration*) is defined as the extraction of a drug with a solvent with several daily shakings or stirrings at room temperature. Compared with other methods of extraction the intensity of movement is so low that we use the term stationary conditions.

Kinetic maceration is carried out at room temperature, like simple maceration, the difference being that the material is kept in constant motion.

In *remaceration* some of the solvent is added to the drug. After filtration the residue is extracted a second time with the remainder of the solvent and the drug residue squeezed out to express as much solvent as possible.

Digestion is maceration at higher temperature, normally at 40–50°C.

4.2.2.2.2. Pharmacopoeia specifications

In the more important Pharmacopoeias maceration has various meanings. *DAB 8* includes macerates, among other things, under the heading 'aqueous drug extracts'.

Macerates. The drug plant material, comminuted as prescribed, has the stated quantity of water at room temperature poured on to it and is then left to stand for 30 min at room temperature, with occasional stirring. The extract is then filtered and made up to the prescribed weight with the washings.

Maceration (simple maceration) is used for preparing aqueous extracts of mucilaginous drugs such as marsh mallow root or linseed. *DAB 8* gives equal preference to maceration for the preparation of extracts, fluid extracts and tinctures.

According to *DAB 8*, maceration is carried out as follows:

"The quantity of extraction fluid prescribed in the monograph is poured on to the comminuted drug materials, the mixtures then being kept for 5 days in tightly sealed vessels at room temperature, protected from sunlight and shaken several times daily. After decanting or straining, the residual liquid is expressed from the solid and the combined extract kept for 5 days at below 15°C, filtered and, if necessary, adjusted to the required concentration with the prescribed extraction liquid; losses due to evaporation are to be avoided during the preparation."

BP 80 permits maceration only for the following tinctures and extracts: senna liquid extract, squill tincture, compound benzoin tincture, catechu tincture and opium tincture.

The last three drugs are preparations which contain little support substance in the starting drug plant materials which upon percolation would clog the percolator. Drug plant material low in supporting substance must therefore be macerated in the same way as mucilaginous drugs (compare Section 1.2.2.3.2).

In addition to the preparations named above, *BP 80* permits all other extracts and tinctures to be prepared by percolation. In contrast to *DAB 8* it thus strictly prescribes the method of preparation for the individual drugs.

USP XX permits maceration for a few tinctures, but prescribes percolation for the preparation of the majority of tinctures and for extracts and fluid extracts, though with these last two the percolation can be preceded by maceration phases of varying length. Table 4.16 summarizes the conditions of extraction methods.

Mother tinctures of the *Homoeopathic Pharmacopoeia* are in most cases prepared by maceration and in a few cases by percolation with ethanol of varying concentration. The strength of alcohol to be used is stated in each monograph.

Table 4.16 Conditions of extraction as prescribed by major pharmacopoeias

	DAB 8	*USP XX*	*BP 80*
Extracts	P/M	P(M)	Liquid extracts
Fluid extracts	P/M	P(M)	P/M
Tinctures	P/M	P/M	P/M
Aqueous drug extracts	M		

P, percolation; M, maceration

4.2.2.2.3. Execution of maceration

Maceration is the most widely used of all methods of extraction. Its advantages over percolation and countercurrent extraction are that small samples, such as are made in the pharmacist's laboratory and in the preparation of homoeopathic mother tinctures, can be prepared in exactly the same way as technical and production batches.

For versatility maceration is thus unsurpassed. However, against this advantage is the disadvantage that the processes do not exhaustively extract the drugs. This fact becomes more important, the more expensive the drug plant material used and the extract recovered from it. Percolation or countercurrent extraction should therefore be used for such products wherever possible.

On the other hand there are a number of drugs which, because of their strong swelling properties or their high mucus content, can be processed only by maceration. Moreover, a considerable proportion of the extract is retained in these drug plant materials as the so called adsorbed fraction and hence in these cases the treatment of the drug residue (see Section 4.2.2.6) is of particular importance.

The influence of certain important parameters on the result of the extraction can be illustrated by a few examples.

Goncharenko *et al.* [4.41] investigated the influence of the *particle size of the drug* in the extraction of *Hypericum perforatum*. The content of tannin and of extractive substances were chosen as target values. The experiments were carried out on a laboratory scale with 20 g of drug plant material per sample, with kinetic maceration in a beaker with a glass stirrer. The results are shown in Table 4.17, which shows that the maceration equilibrium is reached much earlier with a high degree of comminution, though the actual yield is independent of particle size, as previously confirmed by the results of Melichar [4.24).

These and similar experimental results show that the degree of comminution for the extraction must be kept as fine as possible. This may be acceptable for small samples where the sediment can be quickly and efficiently separated by centrifugation, but for large batches it must be

Table 4.17   Influence of particle size on extraction

Fineness of drug	Target value (%)	Maceration time (min)					
		1	2	3	240	360	720
1–3 mm	Tannin				1.04	1.13	1.23
	Extractive substance				3.42	3.76	3.82
<0.25 mm	Tannin	1.23	1.24	1.25			
	Extractive substance	3.53	3.59	3.73			

remembered that fine, swollen sediments frequently cause difficulties in filtration. Centrifugation is therefore the only method left for removing finely divided drug residues, but even with this the fact remains that the separation becomes increasingly difficult the finer the particles. It is often not possible to extrapolate experimental results obtained with finely divided drugs to the industrial production scale.

The *type and intensity of the movement* used in kinetic maceration is an important factor affecting maceration. Isaac [4.42 and 4.43] showed that increasing the speed of the stirrer from 20–30 rev/min to 50–250 rev/min in a bucket mixer has a significant influence on the extraction result (Table 4.18).

Table 4.18   Effect of mixer rotation speed on the maceration of camomile flowers

Parameter	Yield with mixer rotation speed	
	66 rev/min	20 rev/min
Azulene (mg%)	10.5	7.2
Ethereal oil (mg%)	115	68
Extractive substances (%)	7.55	4.1–4.2

Khagi *et al.* [4.44] obtained similar results in their investigations on the content of saponins, sugars, total glycosides and k-strophanthin in relation to the type of movement used in the extraction process. In addition to simple maceration, a kinetic maceration and a kinetic maceration under pressure were carried out. Table 4.19 shows the results.

Table 4.19   Influence on yield of method of extraction of camomile flowers

Extraction method	Ethanol (l)			Saponins	Sugars	Total glycosides	k-Strophanthin
	Total	Extract	Residue in drug			Yield (g)	
Maceration	7.1	5.45	1.65	17.7	38.8	22.4	1.24
Kinetic maceration	10.0	8.0	2.0	29.5	18.8	22.7	2.25
Kinetic maceration under pressure	10.7	9.0	1.7	23.5	44.2	24.2	3.2

The experiments were carried out on a 1 kg scale, the experiment times being 72 h for simple maceration, 24 h for kinetic maceration and 9 h for kinetic maceration under pressure.

Generally speaking, kinetic maceration appears to be used more frequently than simple maceration. Various apparatus can be used for carrying out kinetic maceration, depending on the batch size and the type and intensity of the movement:

Apparatus without stirrers

Cube mixers
Gyro mixer
Ball extractor
Tumbler mixer
Siphon mixer
Turbomixer

Apparatus with stirrers

Ploughshare mixers
Planetary mixers
Bucket mixers
Intensive mixers
Masticators

The batch sizes here can be considerable. Madaus [4.45] for example describes batch sizes of 3000 L for the extraction of the seeds of milk thistle in an intensive mixer. Another important parameter is the temperature at which the maceration is carried out. In accordance with the laws of reaction kinetics, a rise in temperature results in a more rapid establishment of the concentration equilibrium and in certain cases also improves the yield. An increase in temperature is, however, often detrimental to the stability of the active substances which are to be extracted. Blaich [4.46] describes the maceration of senna leaves expressly with cold water at 15°C, whereas Madaus [4.45] used anhydrous ethyl acetate at 75°C for the maceration of milk thistle.

In addition to temperature, additives are also very significant.

The following can be considered as additives here:

Solvent additives for altering the polarity when water is being used, e.g. lower alcohols.
Acids or bases added to form salts or to restrain dissociation.
Surfactants added to improve wetting and to increase the permeability of cell membranes.

Horner *et al.* [4.47.] describe the use of methylene chloride as a principal solvent to which 0.5–5% of an alcohol with 1–4 C-atoms has been added for the extraction of morphine from poppy seeds. This alteration of the polarity is increased by the use of a base, usually ammonia, which shifts the distribution coefficient of the substances to be extracted towards the extract phase.

Lakoza *et al.* [4.48] showed that the yield of solasodine, a starting compound for the preparation of steroids in the extraction of *Solanum laciniatum*, can be increased by additions of sulphuric acid. The conditions are shown in Fig. 4.15. The results show that addition of 1% sulphuric acid is sufficient for an optimum yield and that more sulphuric acid has no additional effect. Blaich [4.46] converted the sennosides which he extracted from senna leaves into their calcium salts by addition of milk of lime; the calcium salts are sufficiently soluble in water to be extracted by large quantities of water.

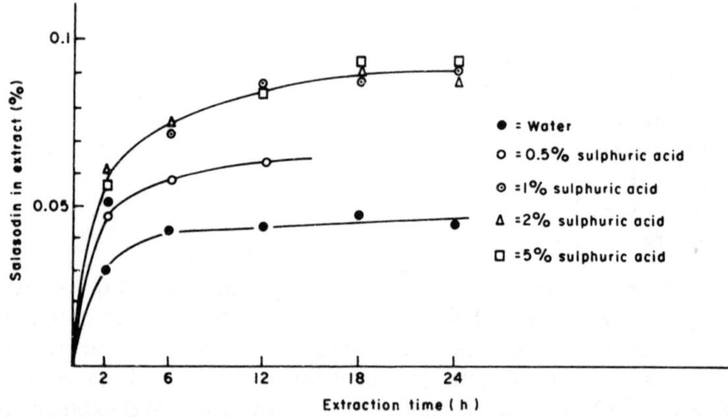

Figure 4.15 Effect on solasodin yield of addition of acid during its extraction from *Solanum laciniatum*.

4.2.2.3. Vortical (turbo) extraction

Simple maceration is a very slow extraction process. There has therefore been no lack of endeavours to reduce the length of time involved. In addition to kinetic maceration with shaking and stirring, which gives the same yield but attains the concentration equilibrium in a shorter time, Melichar *et al.* [4.49–4.51] have introduced vortical or turbo-extraction. Here the drug to be extracted is stirred in the menstruum with a high-speed mixer or homogenizer. The shredding and shearing forces break down the drug material to a particle size which is smaller than that of the material when it is first put in the mixer. The cells become highly disintegrated. The diffusion of extractive substances through the cell membranes is largely replaced by washing out from the destroyed cellular tissues (see Fig. 4.12). This results in substantially faster establishment of the maceration equilibrium and hence in a considerable saving of time.

The energy supplied for the high speed stirring and comminution of the drug material raises the temperature during the extraction, which is undesirable because of the risk of decomposition of thermolabile constituents. The temperature-rise must therefore be kept as small as possible. This is achieved either by stopping the process from time to time or by cooling the vessel. The further comminution of the drug, which favours rapid establishment of the equilibrium (see above) brings with it the disadvantage of making separation of drug residues from the miscella more difficult. The separation can be carried out by filtration, sedimentation or centrifugation.

Bogs [4.52] showed that with valerian root and belladonna leaves the extract and active substance levels of a tincture prepared by vortical extraction are higher than those obtained by simple maceration.

More exhaustive investigations of turboextraction of belladonna leaves by Bogs and Damm [4.53] revealed that with vortical extraction the alkaloid yield can be $\sim 30\%$ greater than that obtained with maceration. The authors recommend vortical extraction for 10 min for the preparation of small quantities of belladonna tincture. It was shown in storage trials that the alkaloid content of the tincture prepared in this way remained practically unaltered for 6 months.

These results were supplemented by a further study by the same authors [4.54] which compared the various methods of extraction with one another and investigated the effect of the degree of comminution and of storage of the comminuted drug material on the result of extraction. Tables 4.20–4.22 are taken from this study.

The results confirm earlier findings, according to which vortical extraction is superior to maceration and kinetic maceration. The alkaloid yield in belladonna tincture is comparable with that obtained by percolation (Table 4.20). The yield of ethereal oils in bitter orange tincture is higher the coarser the drug material used (Table 4.21), which shows the loss due to comminution prior to extraction and hence the positive effect of vortical extraction.

However, these results cannot be generalized, as is shown by Table 4.22. The hard roots of bloodroot produce the highest yields of dry residue and tannins after pulverization. High yields are obtained from shredded material by first steeping the roots and then subjecting them to vortical extraction, which also has advantages over maceration in this case.

The vortical extraction of *Cinchona* bark was investigated by Walter [4.55]. He compared it with maceration and percolation and found it to be superior to maceration and about as good as percolation over six stages. On the basis of the results of the investigation, two or three vortical extractions of the material, each extraction being done with half or a third of the menstruum, is suggested for an efficient and rapid extraction. The extract liquid is centrifuged off between the individual extractions.

In addition to vortical extraction with high-speed stirrers with a comminuting action, vortical extraction in a fluidized bed has also been described. Ainshtein *et al.* [4.56] investigated the extraction of solanin from potato seedlings in a conical vessel into which acidified water was forced from below to vigorously stir the drug.

Table 4.20 Effect of extraction method

Method	Time	Alkaloid content (mg/100 ml)
Percolation	5 days	52
Maceration	10 days	45
Laboratory stirrer (slow)	15 min	43
Mixing-beaker with fast blade stirrer (10 000 rev/min)	10 min	54
Elektroquirl (14 000–16 000 rev/min)	5 min	55

Table 4.21 Effect of degree of comminution and of type of extraction on the dry residue and ethereal oil content of Tinct. Aurantii

Extraction method	Degree of comminution	Content immediately after comminution		Content 3 months after comminution	
		Dry residue	Ethereal oils	Dry residue	Ethereal oils
Maceration	large particles	5.85%	612 mg/100 ml	5.82%	610 mg/100 ml
Maceration	concis.	5.85%	580 mg/100 ml	5.80%	300 mg/100 ml
Maceration	minut. consis.	6.14%	571 mg/100 ml	6.18%	300 mg/100 ml
Maceration	pulvis gross.	6.52%	530 mg/100 ml	6.57%	290 mg/100 ml
Vortical extraction	large particles	6.45%	621 mg/100 ml	6.45%	610 mg/100 ml
Vortical extraction	concis.	6.77%	590 mg/100 ml	6.70%	305 mg/100 ml
Vortical extraction	minut. concis.	6.95%	567 mg/100 ml	6.95%	300 mg/100 ml
Vortical extraction	pulvis gross.	7.00%	535 mg/100 ml	6.95%	285 mg/100 ml

Table 4.22 Effect of degree of comminution and of type of extraction on the dry residue and tannins content of Tinct. Tormentillae

Extraction method	Degree of comminution	Dry residue	Tannins
Maceration	concis.	4.40%	2.15
Maceration	minut. concis.	4.75%	2.28
Maceration	pulv. gross.	5.05%	2.45
Vortical extraction	concis.	4.72%	2.35
Vortical extraction	minut. concis.	5.00%	2.45
Vortical extraction	pulv. gross.	5.12%	2.60

Whereas maceration is a time consuming but energy saving process, vortical extraction is a time saving, but intensely energy consuming process. It is therefore mostly used for producing small quantities of extract in the laboratory. The method is rapid and gives a high yield and the expected difficulties in the separation of the drug residue are not very great in small batches.

A combination of vortex extractors in the form of batch wise or continuously operating Rotorstat dispersion machines with decanters or centrifuges for separating off the drug residue is necessary for vortical extraction on a large scale. The process is discussed further in Section 5.4.

4.2.2.4. Ultrasound extraction

Whereas the yield in vortical extraction is improved by powerful mechanical stressing of the drug material, sound waves are the force used to accelerate the extraction in ultrasonic extraction. Ultrasound is defined as frequencies above 20 000 Hz.

In pharmaceutical practice, according to List [4.57], ultrasound is usually produced with magnetostrictive or piezoelectric ultrasonic transmitters. In the magnetostrictive transmitter use is made of the change of length undergone by ferromagnetic substances upon magnetization (magnetostriction). In practice, nickel steels, which are composed of numerous nickel foil discs isolated from each other to prevent eddy currents, are used. The rod begins to vibrate in an alternating magnetic field produced by a high-frequency alternating current. The amplitude is at its greatest when the frequency of the alternating current is the same as a natural frequency of the vibrating rod. Frequencies of up to 200 kHz can thus be produced.

Higher frequencies are obtained with piezoelectric ultrasound transmitters, which make use of the so-called reciprocal piezoelectric effect produced when a quartz crystal undergoes a change of length when an electric current is applied. (Conversely, an electric current, the piezoelectric effect, is produced upon compression or expansion of a quartz crystal.) When such a crystal is placed between the plates of a condenser to which an alternating electric current is being applied it begins to vibrate at the frequency of the applied current. The amplitude is also at its greatest here when the frequency accords with one of the natural frequencies of quartz.

The principal effects of ultrasound in extraction are: (1) to increase the permeability of the cell walls, (2) to produce cavitations (i.e. the spontaneous formation of bubbles in a liquid below its boiling-point resulting from strong dynamic stressing), and (3) to increase mechanical stressing of the cells (so-called interface friction).

Haul *et al.* [4.58] showed in 1952 that ultrasound accelerates the distribution of phenanthrene between a light petrol as the lighter phase and methanol as the heavier phase according to the Nernst law of distribution. This discovery proved that ultrasound can accelerate liquid–liquid extraction.

Experiments on solid–liquid extraction were first carried out by Specht [4.59] on hops. Quartz vibrators with a frequency of 1000 kHz and a capacity of 35 W for laboratory experiments and 6 × 35 W for large scale technical investigations were used. It was found that the extraction of the active substances of hops proceeds more economically and productively in the ultrasound field at 50°C than at boiling temperature without ultrasound, which is important for the easily oxidizable bitter acids of hops.

The first pharmaceutical paper on improvement of extraction yields by ultrasound treatment was published by Schultz and Klotz [4.60]. These authors succeeded in significantly increasing the alkaloid yield and the total extract content in the extraction of *Cinchona* bark by using an electric siren (i.e. in the normal sound range), with water as the sound-transmitting medium. Surprisingly, the extraction experiments in the ultrasound range carried out simultaneously with an ultrasound apparatus at a frequency of 2400 kHz were unsuccessful; neither the yield nor the alkaloid content of the extract could be increased.

In another paper [4.61] the frequency of 2400 kHz used by Schultz and Klotz was shown to be too high. Most authors work with frequencies of 25–1000 kHz. The effect of ultrasound in extraction processes depends on the frequency and capacity of the apparatus and the length of time for which it is applied. Specht [4.59], using hops, showed that all three parameters affect the result of extraction. Often only the frequencies used, but not the capacity (output), are given in the literature, which makes comparative evaluation of these studies difficult.

When ultrasound is used as an aid to extraction, a possible alteration of the constituents must also be taken into consideration and carefully investigated. As early as 1928 Schmitt *et al.* [4.62] reported on the oxidation of halogenides and hydrogen sulphide. The mechanism of oxidation was first investigated in 1934 by Liu and Wu [4.63]. The authors are of the opinion that cavitation is responsible for the activation of oxygen. Henglein [4.64 and 4.65] discusses radical mechanisms in the formation of hydrogen peroxide from water. The primary chemical reaction is the radical decomposition of water:

$$H_2O \rightarrow \dot{H} + O\dot{H} \tag{1}$$

In solutions containing oxygen the H radicals are caught by oxygen:

$$\dot{H} + O_2 \rightarrow H\dot{O}_2 \tag{2}$$

Both the O$\dot{\text{H}}$ radicals formed in (1) and the H$\dot{\text{O}}_2$ radicals formed in (2) are capable of forming hydrogen peroxide:

$$O\dot{H} + O\dot{H} \rightarrow H_2O_2 \tag{3}$$

$$H\dot{O}_2 + H\dot{O}_2 \rightarrow H_2O_2 + O_2 \tag{4}$$

1 mole of hydrogen peroxide is produced from 1 mole of water.

Suss and Hanke [4.66] have investigated the influence of ultrasound on demineralized water in relation to the nature of the gases present. They found varying quantities of hydrogen peroxide, nitrous acid and nitric acid, depending on the intensity and duration of the ultrasound treatment. The maximum formation of hydrogen peroxide was found when a frequency of 800 kHz and an output of 2.9 W/cm^2 were used. Figure 4.16 shows the course of hydrogen peroxide production in relation to output. In a further paper the same authors [4.67] describe the investigation of the significance of cavitation and interface friction in the dispersion of suspended solids by ultrasound. According to this paper, interface friction is usually more significant for dispersion than cavitation, as was demonstrated on talc suspensions.

Interfaces are defined here as the contact surfaces between different media touching each other; in suspensions these interfaces would be those between particles and liquid. Relative movements are produced in front of and behind an interface, thus producing friction. Table 4.23 surveys the studies carried out so far on extraction of drugs by ultrasound. All authors confirm that ultrasound shortens the extraction time. In most cases 5–15 min is sufficient to obtain the same results as after at least several hours of maceration or percolation. The process must be optimized individually for each drug.

Decomposition of the alkaloids in jaborandi leaves was observed after only 30 s ultrasound treatment on the laboratory scale at 20 kHz [4.71]. In the case of foxglove leaves the content of digitalis glycosides fell when an ultrasound output representing the optimum formation of hydrogen peroxide during the extraction was used.

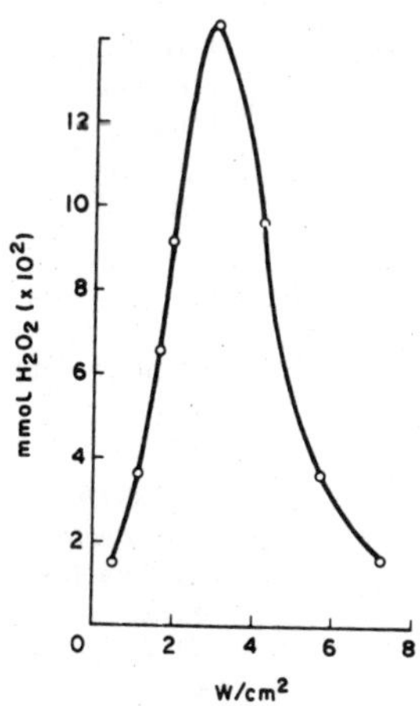

Figure 4.16 Variation of hydrogen peroxide formation with ultrasound power (treatment time 10 min, vol 100 ml).

Table 4.23 Examples of drugs extracted using ultrasound

Author and year	Reference	Ultrasound frequency (kHz)	Extracted drugs	Remarks
Schultz, O. E. & Klotz, J. (1954)	4.60	2400	*Cinchona* bark	No improvement of yield due to ultrasound frequency being too high
Thompson, D. & Sutherland, D. G. (1955)	4.76	400	Peanut oil from peanuts	n-Hexane as solvent, increased oil yield
Wray, P. E. & Small, L. D. (1958)	4.68	500	Belladonna leaves	Extraction results comparable with maceration; no decomposition of alkaloids
Bose, P. C. *et al.* (1961)	4.69	25	*Rauwolfia* root	
DeMaggio, A. E. & Lott, J. A. (1964)	4.70	20–40	Thorn apple (*Datura stramonium*) leaves	Comparison of simple and ultrasound maceration
Ovadia, M. E and Skauen, D. M. (1965)	4.71	20	*Cinchona* bark, ipecacuanha	Comparison with Soxhlet extraction
			Jaborandi leaves	Alkaloids from leaves tend to decompose
Srinivasula, C. Scrivastava, S. C. & Mahapatra, S. N. (1970)	4.72	1000	Nux vomica seeds	Comparison with kinetic maceration
Suss, W. (1972)	4.73	800	Foxglove (*Digitalis*) leaves	Comparison of maceration/ percolation/ turboextraction/ infusion preparation/ ultrasound extraction. Decrease of glycoside content at maximum H_2O_2 formation in ultrasound extraction
Shehepilov, N. S. *et al.* (1972)	4.74		Fruits of *Ammi visnaga* (khelline)	Ultrasound intensity, temperature, time, solvent concentration
Thakkar, V. J. *et al.* (1974)	4.75	43	Yams (*Dioscorea batatas*) Nux vomica seeds, *D. stramonium* leaves	Various extraction times according to type of drug

The two examples show the importance of a careful choice of the extraction conditions and method of detection of decomposition products. Particular attention must be given not only to the risk of oxidative decomposition of the active substances but also to the possibility of traces of metal being given off by the ultrasound transmitter; traces of metal can accelerate the decomposition of active substances.

Ultrasound extraction has so far not been widely used on a large scale because of the high energy costs.

4.2.2.5. Extraction by electrical energy

Electrical energy has been used in the form of an electric field, an electromagnetic field and as electric discharges to accelerate extraction and improve its yields. Bozhko [4.77] extracted scopolamine from the seeds and capsules of Indian thorn apple with the aid of a steel plate as a cathode at the bottom of the extraction vessel and several carbon electrodes as anodes at the top. He was able to show that the alkaloid yield could be significantly increased by application of a current. Rakhman-Zade *et al.* [4.78] surrounded an extraction column with an electric coil producing an alternating electromagnetic field of 50 Hz. It was shown with the example of extraction of valerianic acid from valerian root that extraction in an electromagnetic field is faster and more complete than with simple maceration.

Extraction by means of electric discharge is described more frequently in the literature. Issaev and Mitev [4.79 and 4.80] first suggested this process in 1968. The design of such an apparatus is shown in principle in Fig. 4.17. The electrical source V charges the capacitor F to the threshold tension. Upon reaching this tension the capacitor is discharged via the spark-gap and the two electrodes in the extraction vessel. The discharge frequency can be altered by varying the distance between the poles of the discharge gap and the strength of the discharge is given by the expression

$$W = \frac{CV^2}{2}$$

where W = energy of impulse, C = capacity of condenser, and V = voltage.

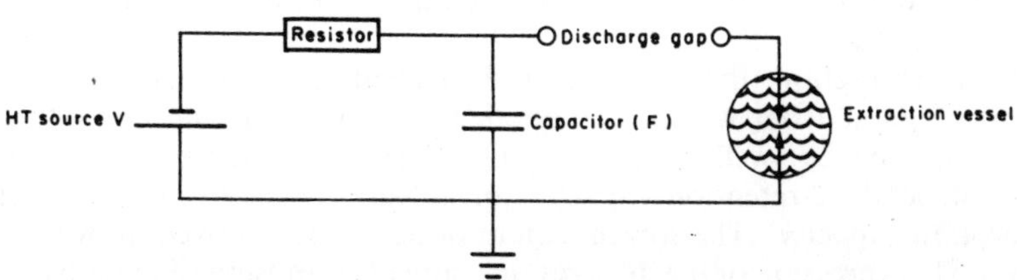

Figure 4.17   Schematic design of electrical discharge extraction apparatus.

The electrical discharge can be regarded as an ultrasound source with a broad frequency band resulting in the production of cavitation phenomena. Both the frequency and the intensity of the impulses affect the extraction by partially disintegrating the plant cells. The mechanical effects of the electrical discharges are attributed to: (1) the production of cavitations, and (2) the spreading of pressure waves produced by the electrical discharge at ultrasonic velocity.

The following process parameters are generally recommended:

Capacitor size: up to $2\,\mu F$
Discharge frequency: 500–600/min
Degree of comminution of the drug: 1.5–2 mm.
Ratio of drug to solvent: 1:4–1:5
Extraction time: 5–6 min.

Issaev and Mitev [4.80] investigated the extraction of alkaloids from *Cytisus laburnum* L. with this method. Boiko and Mizineko found a 25–20% increase in the alkaloid yield in the extraction of *Rauwolfia* with electrical discharge.

Dimov, sometimes with the co-author Bojdshieva [4.82–4.85], has published a series of papers on extraction with electrical discharges. Belladonna tinctures and extracts were prepared more quickly and with higher alkaloid yield with electrical discharges, which also eliminated decomposition of the alkaloids [4.82]. Factorial experiment planning made it possible to optimize the yield of glaucine, an alkaloid from *Glaucium flavum* [4.83]. Extraction with electrical discharge was also used successfully in the recovery of alkaloids from *Erysimum repandum* [4.84], as well as of santonin from *Artemisia maritima* and of the alkaloids [4.84] from *Delphinium orientale* [4.85].

4.2.2.6. Treatment of the drug residue

Further treatment of the drug residue may be necessary for several reasons: (1) the drug residue contains considerable quantities of absorbed solution with valuable extractive substances, (2) the solvents used should be recovered, and (3) there will be further uses of the drug residue, which necessitate removal of the solvents (which must not be equated with recovery of the solvents).

During extraction, the drug material, depending on its nature and the species of plant from which it is obtained, absorbs varying quantities of solvent and hence swells up to varying degrees. The drug residue has a species-dependent retention capacity for solvents. which is known as the 'absorption capacity'. The solvent can in principle be removed in two ways, namely: (1) expression of the drug residue, and (2) expulsion of the solvent by warming with or without pressure reduction.

Expression is used not only for the further treatment of the drug residue, but also for the recovery of juices from fresh plants. It is a basic operation of processing technology which has a very ancient tradition (e.g. preparation of wine), though it has so far been described very little. The individual processes are complex and vary from substance to substance. Although attempts have occasionally been made to put the process of expression on a theoretical basis, it still remains largely an empirical process. Even the apparatus used is devised on an empirical basis. The oldest and simplest presses are *basket presses* or *wine presses*, which are used in the preparation of wine and fruit juices. They consist of a sieve basket into which the material to be expressed is put. A punch presses the material down from above and the expressed juice flows out through the sieve basket and is collected and led away through a channel in the bottom.

Basket presses operate on the batch principle and are closely related to *filter presses* and *pack presses*, which likewise operate batchwise. These consist of a pressure resistant casing in which perforated plates are placed to form a cubical press-chamber. In pack presses the material is wrapped in cloth and many of these packs are laid one on top of the other between the perforated plates.

The *Willmes press* represents a special development which makes possible repeated pressing with loosening of the material at intervals.

Helical presses and *sieve belt presses* are used for continuous expression. Gromova *et al.* [4.86] described the extraction of plant material with an extraction press operating on the batch principle in which, among other materials, valerian root was investigated. The apparatus used, like the Willmes press, operates by compressed air. A pressure of up to 30 atmospheres is suggested, depending on the type of drug material.

The machinery used for expression is discussed in Section 5.4.5.

The complete removal of solvents is not possible with expression apparatus; this can be achieved only by following the expression process by a special step for the removal of solvent by heating with or without reduction of pressure. The simplest way of doing this is by batch wise maceration in sealed containers. In such cases the drug residue, after draining, is heated, if necessary under vacuum, to expel and draw off the solvent vapours. In percolation and continuous extraction the drug residue is usually removed and freed of solvent in separate apparatus.

4.2.3. METHODS FOR EXHAUSTIVE EXTRACTION OF THE DRUG

4.2.3.1. Principles

'Exhaustive extraction' is defined as the complete removal of the desired extractive substances from the drug material. The skeletal material of the drug plant remains behind. The objective is a quantitative extraction, which can be achieved in various ways (Fig. 4.18).

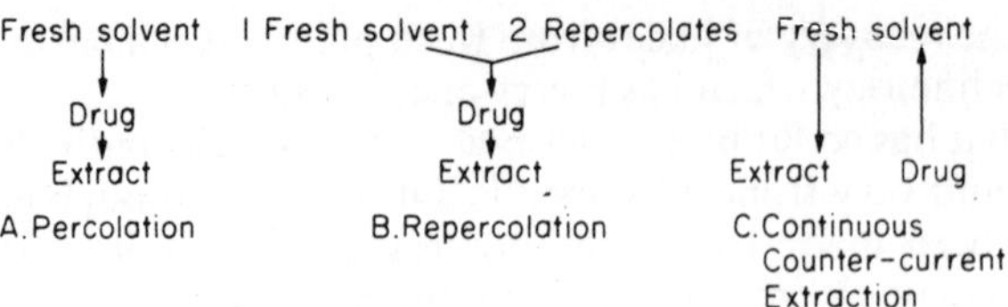

Figure 4.18   Exhaustive extraction procedure.

In *percolation* the drug plant material is exhaustively extracted by fresh solvent. Only fresh solvent is used and the extraction consumes a large quantity of it and takes a long time. We speak of *repercolation* when the drug is first extracted with fresh solvent and then some of the percolate is used for exhaustive extraction by stagewise concentration in another percolator. *Continuous countercurrent extraction* is a process in which fresh drug plant material is brought into contact with loaded/charged solvent at the same time as fresh solvent is being brought into contact with already pre-extracted drug. The first runnings are known as the *extraction head*. The subsequent fractions are the *part percolates*.

4.2.3.2. Percolation and repercolation

4.2.3.2.1. Principles

The first attempts at the mathematical description of the percolation process were made by Schultz and Klotz [4.87]. These authors worked out separate theories on the wash-out process (i.e. washing the extractive substances out of disintegrated cells of the drug plant) and the diffusion process which takes place during percolation, and combined the resultant equations to calculate the total process. Their results may be summarized as follows:

The diffusion process is dependent on the parameters of percolation rate, quantity of menstruum, diffusion constant of drug into menstruum and diffusion constant of menstruum into drug.

The wash-out process is completed more rapidly than the diffusion process and hence after a certain time pure diffusion prevails. Conclusions about the extent of the wash-out process can be drawn from the course of this pure diffusion.

The wash-out process is practically completed after 3 part percolates have been taken (where 1 part of percolate $\rightleftharpoons$ 1 part of drug).

Under these conditions 4/7 of the total yield of 3 part percolates appear in the first and 6/7 in the first plus the second part percolate.

The calculations agree well with the results of percolation experiments with *Cinchona* bark.

The theory of percolation is discussed in detail in a paper by Muller *et al.* [4.88]. These authors used a system of differential equations to describe the concentration curve in relation to elution quantity (percolation isotherms). The concentration curves for slow and fast dropping rates are shown.

4.2.3.2.2. Pharmacopoeia specifications

DAB 8 generally permits maceration and percolation for the preparation of extracts. *USP XX* and *BP 80* prescribe the method to be used in individual monographs. All three Pharmacopoeias specify the degree of comminution of the drug. The various percolation conditions are listed in Table 4.24, which shows that considerable differences exist between the three cited Pharmacopoeias. According to *BP 80* preswelling takes twice as long as per *DAB 8*. The American Pharmacopoeia gives only individual regulations here. Both *DAB 8* and the *BP 80* prescribe 24 h for the subsequent intermediate

Table 4.24 Pharmacopoeia specifications for percolation

Process	Pharmacopoeia		
	DAB 8	*USP XX*	*BP 80*
Preswelling of drug	At least 2 h with weight of prescribed liquid equal to 30% of weight of drug	Individually per monograph	Moistening of drug with sufficient quantity of menstruum leave for 4 h
Intermediate maceration	24 h	2–48 h, per monograph	24 h
Percolation rate	4–6 drops/min per 100 g drug	per 100 g drug: percolate slowly (0.1 mL/min). Percolate at a moderate rate (0.1–0.3 mL/min). Percolate rapidly (0.3–0.1 mL/min)	Not specified
Expression of drug residue	+	−	+
Total quantity of menstruum	Dry extracts: 3–4 parts percolate per 1 part drug Fluid extracts: 1–2 parts percolate per 1 part of drug	Sometimes 'exhaustive extraction', varying with monograph	Varies with monograph, in which $\frac{3}{4}$ of total quantity is recovered by percolation, rest by expression and making up with solvent

maceration, whereas the American Pharmacopoeia prescribes individual regulations. Differences also exist as regards percolation rate and subsequent expression of the drug. The total quantity of solvent to be used is specified consistently only in *DAB 8*, whereas *BP 80* and *USP XX* anticipate variable quantities. Overall, it can be stated that the *USP XX* has the most divergent provisions on percolation.

4.2.3.2.3. Percolation procedure

Next to maceration, percolation is the most widely used process in drug extraction. Its widespread use is due not least to the fact that it can be carried out both in the laboratory and on an industrial scale.

Prior to actual percolation, the drug material must be rendered into a suitable form. This is usually done by comminution which, as in maceration, must be carried out in such a way as to produce as little fine material as possible, and hence *shredding mills* have also proved useful here. Too much finely ground material in percolation has worse consequences than in maceration, as there is not only the problem of removing the fine material after completion of the extraction, but difficulties due to clogging of the percolator can occur even during the extraction. It must also be remembered that many percolation processes are carried out in such a way that the miscella runs clear and fulfils the conditions, for example, of a fluid extract directly after completion of percolation. These extracts are not subjected to any further clarification. An expensive and time-consuming clarification must on the other hand be carried out if fine particles get into the extract as a result of too much finely ground material in the drug. Comminution of the drug is followed by its preswelling. The Pharmacopoeias recommend pre-swelling *outside* the percolator because of the risk of glass percolators bursting. The specification of a uniform preswelling time, such as is for example given by *DAB 8* and *BP 80*, appears hardly logical here because of the varying steeping and swelling properties of drug plant materials, even if the preswelling times were so long that even the most obstinate drug plant materials were steeped. The process of presteeping and preswelling on a laboratory scale can be mastered without great difficulty. Preswelling outside the percolator on an industrial production scale requires a separate presteeping vessel and transport equipment and hence if the preswelling cannot be carried out during transport of the dry drug material to the percolator this separate process usually has to be abandoned. Preswelling in the percolator is possible with a suitably designed percolator with drug materials which do not swell too much, though these must not produce such a dense cake of drug material that fluid can no longer percolate through it. For these reasons maceration has been able to maintain its favoured position in the extraction of a number of drugs.

In percolation as per the Pharmacopoeias preswelling is followed by intermediate maceration. For intermediate maceration the presteeped, preswollen drug material is poured into the percolator and solvent is added through an opened stopcock so that it just begins to percolate dropwise out of the bottom of the percolator. The stopcock is then closed immediately and the material left to stand with the solvent in the percolator for 24 h. Percolation is not done when the dry drug material has been poured in (and compacted slightly) and preswelling and intermediate maceration take place

Table 4.25 Percolation optimization studies

Author and reference	Year	Drug/plant	Solvent	Investigated parameters	Batch size
Astakhova Minina [4.90]	1977	*Scopolia tangutica*	Chloroform	Intermediate maceration time, percolation rate, degree of comminution of drug material, solvent-drug material ratio, packing density of drug material	10 g drug
Ferrada, J. *et al.* [4.91]	1977	*Solanum tomatillo*	5% Acetic acid in water	Temperature, solvent-drug material ratio, initial content of drug material	Laboratory scale
Lipkovskii, A. [4.92]	1975	*Strophanthus* seeds	Ethanol Acetone/ water	Influence of acetone	1 kg
Osmanov, U. [4.93]	1973	*Vinca erecta*			
Muravev, I. [4.94]	1972	*Glycyrrhiza uralensis*	1% aqueous ammonia solution	Inter-maceration time. Number of percolators. Solvent-drug material ratio	22 g
Zinko, M. [4.95]	1977	Garlic	70% aqueous ethanol	Detailed study of battery of 6 percolators	240 g

simultaneously. This is permitted by *DAB 8* as 'another preparative method'. The actual percolation process is influenced by:

(1) Selectivity of the solvent.
(2) Quantity flow (dropping-rate) of the solvent.
(3) Temperature.

The selectivity of the solvent is important not only for the yield of one or more principal substances, but also for the qualitative and quantitative composition of the accompanying substances.

The flow rate of the solvent is governed by the fixed dropping rate. This therefore determines the contact time between solvent and drug. The drug/ solvent ratio is the ratio of quantity of drug used to the total quantity of solvent consumed. Although temperature is important in percolation, it is in practice only rarely used as a controlling factor.

The numerous variables have led a number of authors to look for ways of improving the percolation process. Table 4.25 gives a selection of these studies.

Some of the studies were carried out as factorial experiments (see page 127 and Section 2.1.3.2.5). The Box–Wilson optimization method was also used. With the exception of the study by Zinko, all these studies were experiments performed in the laboratory. Extrapolation of laboratory results to extraction on a large scale is dubious and hence the results have only a limited use for production purposes.

On the other hand the study by Zinko [4.95] is an example of the accurate investigation of an industrial percolator battery of six percolators each with a capacity of 240 kg of drug material. Garlic was used for the experiments. The battery operated on a countercurrent system, 240 kg of drug material first being steeped and preswollen in each percolator with 480 L of 70% ethanol for 24 h. Five percolators were in operation and the sixth was simply charged. Table 4.26 shows the precise course of the extraction in all six percolators.

When interpreting Table 4.26 it should be noted that a rigid system was chosen for numbering the percolators, as shown by the scheme corresponding to Table 4.26.

Extraction in a countercurrent-operated percolator battery can be termed continuous relative countercurrent extraction. The potential variations in the size of percolators make it a flexible and, as regards equipment, simple solution for extraction. Other continuous processes have however gained favour in the last few years. This is attributable to four difficulties which can arise in the operation of batteries of percolators:

Clumps of dry drug material which cannot be reached by the solvent throughout the entire extraction process can form in a percolator, which means that the drug is not completely extracted.
Too much swelling can cause increased pressure and clogging of the percolator.

Drug materials containing mucilage and pectin are particularly liable to cause difficulties in this respect.

If the drug material is filled unevenly into the percolator there is also the risk of channel formation, i.e. the solvent seeks preferred routes for flowing through the cake of drug material.

Table 4.26 Course of extraction reported by Zinko [4.95]

Parameter	Percolator number					
	1	2	3	4	5	6
Temperature (°C)	22	21	23	26	23	
Time material is in percolator	3	9	6	12	18	
Dry material in extract (%)	0.88	3.5	1.6	6.5	8.6	
Content of extractable material (%)	27.5	18.4	22.3	16.5	12.8	
Density of ethanol + water mixture	0.892	0.910	0.908	0.932	0.940	
Ethanol content (%)	67.36	59.6	60.5	49.1	44.2	

Factor experiments (Principles of Factorial Design, see Section 2.1.3.2.5.2)

The design of factorial experiments is illustrated here by an example of drugs extraction.

Let us consider the investigation of the influences of temperature, alcohol content and preswelling or steeping time in an aqueous alcoholic menstruum on the alkaloid yield of an extraction. Here the influence of temperature with constant content and constant time of exposure to solvent, the influence of the alcohol content with constant temperature and steeping time and the influence of steeping time with constant temperature and alcohol content would normally be investigated in three separate series of experiments.

The resultant optima of the series of experiments are then combined as optimum extraction conditions. Optimization of extraction thus consists of three stages:

Stage 1:

Constant alcohol content and steeping time, variable temperature

Alcohol content	50%	50%	50%
Steeping time	1 h	1 h	1 h
Temperature	40°C	60°C	80°C
Alkaloid yield	2%	2.3%	1.4%[a]

[a] Active substances begin to decompose because extraction temperature is too high

Stage 2:

On the basis of the above result the extraction temperature of 60°C is chosen for investigation of the influence of alcohol content and the following experiments are carried out:

Alcohol content	30%	50%	70%
Steeping time	1 h	1 h	1 h
Temperature	60°C	60°C	60°C
Alkaloid yield	1.7%	2.3%	2.9%

Stage 3:

Constant alcohol content (70%) and temperature (60°C), variable steeping time

Alcohol content	70%	70%	70%
Steeping time	0.5 h	1 h	2 h
Temperature	60°C	60°C	60°C
Alkaloid yield	1.6%	2.9%	2.9%

On the basis of the results of the three series of experiments, 60°C, 70% alcohol and 1 h preswelling time are chosen as the conditions for the extraction. Nine experiments, two of which were repeated, were performed to determine these data. However, the arrangement of the experiments does not permit any statement on the influence of temperature when, for example, 60% alcohol is used, or how the alkaloid yield at 40°C or 50°C would be changed in relation to the alcohol content of the menstruum. In other words, the influence of one factor is investigated while the other two factors are kept constant (which eliminates the need to detect any possible interactions between the factors). This deficiency in the experimental procedure is remedied by the use of factorial experiments in which each factor is investigated at each stage, and this will also detect any interactions between the individual factors. To avoid performing too many experiments it is necessary to select the number and order of the stages carefully. A problem is generally first tackled with a simple 2^3-factorial experiment, i.e. three factors are investigated on two levels. The factors and their levels can be shown in a simple scheme.

	Level	
Factor	Lower	Upper
Alcohol content	50%	70%
Steeping time	0.5 h	1 h
Temperature	40°C	60°C

The factors can be visualized as being arranged in a three-dimensional space in which the higher level always lies in the direction of the arrow, away from the origin.

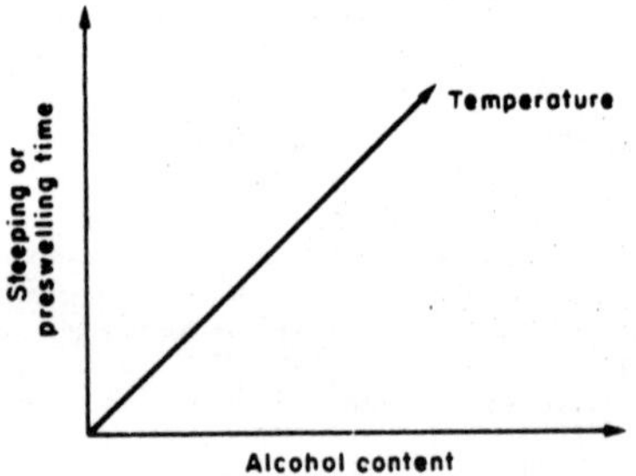

If all the lower levels are placed at the origin of the co-ordinate system a cube is produced in which the experiment combinations are situated at the corners:

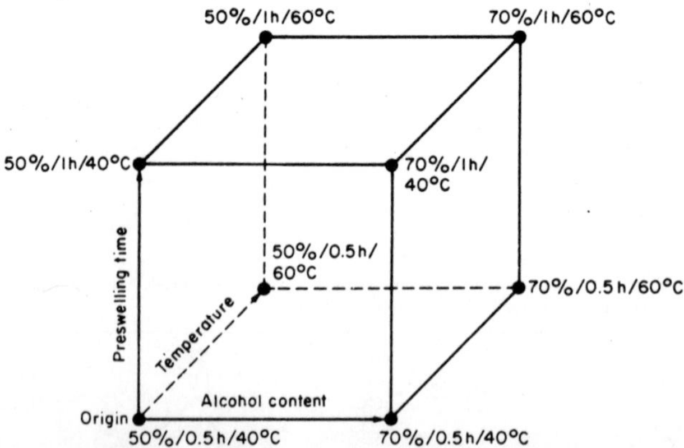

Figure 4.19 Experimental parameters used in a 2^3 factorial design.

The system covers all possible experiment combinations of three factors at two levels with a total of only eight experiments. The evaluation (see, for example, O. L. Davies, *The Design and Analysis of Industrial Experiments*. Oliver and Boyd, Edinburgh, 2nd Ed, 1967), makes possible the recognition of the influence of each individual factor and of the interactions of two or more factors: for example of alcohol content and temperature.

The performance of the first factorial experiment can be followed by a second and third experiment by extending the experimental conditions, thus making their range of validity and applicability more clearly and accurately defined. The validity of the results obtained depends on two conditions. ● They are valid only within the investigated range (i.e. within the 'cube'). ● They require linearity of the variables between the individual test-points of the experiment. This latter condition in particular represents a certain disadvantage of factorial experiments, as is shown by the curve in Fig. 4.20:

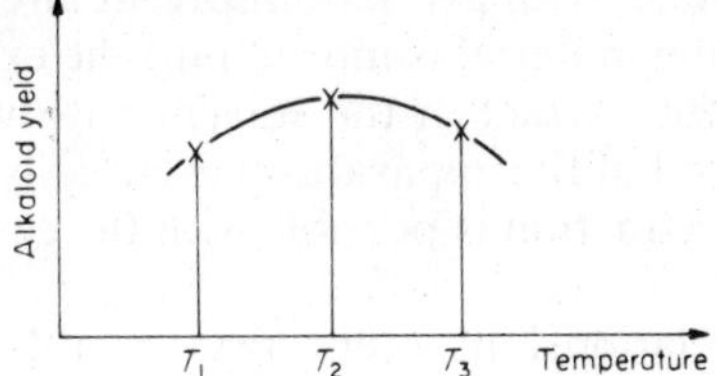

Figure 4.20 Variation of alkaloid yield with temperature.

It could be stated from the above procedure using temperatures T_1 and T_2 that temperature does not affect alkaloid yield. Only the performance of a third experiment at T_3 can reveal the error of this statement. It is therefore vitally necessary when carrying out factorial experiments to select the experimental conditions very carefully and in doubtful cases to alter the spacing of the graduations.

The actual optimization strategies, freed from the rigid scheme of factorial experiments, come into operation here by attempting to find the optimum by machematical formulation. For this we refer to the specialist literature, e.g. B. G. Bandermann, in *Ullmann's Enzyklopädie der Technischen Chemie,* Vol. 1, p. 347 [2.20].

The charging and then the emptying of the percolator after completion of an extraction are additional disadvantages to those mentioned above. The drug residue emptied out of the percolator still contains 'bound solution', i.e. it is not free of solvent and requires further treatment (see Section 2.2.6).

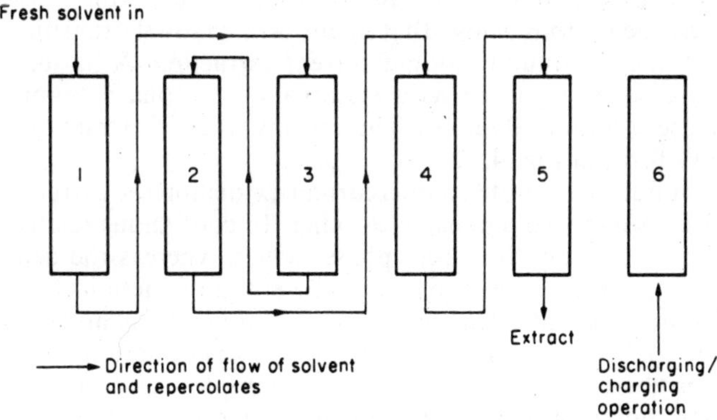

Figure 4.21 Schematic of the percolator used by Zinko (4.95).

4.2.3.3. Countercurrent extraction

4.2.3.3.1. Principles

In continuous countercurrent extraction a moving solution, emulsion, suspension or solid mass is extracted by a liquid phase flowing against it. The processes may be represented schematically as follows:

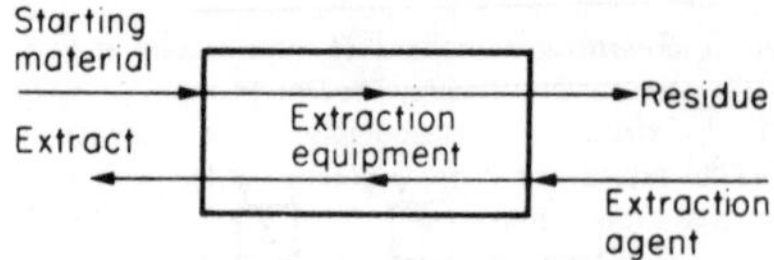

The starting material for the drug is put in the extraction apparatus, where it first comes into contact with extraction solvent already containing extract. The further the starting material is moved into the extraction apparatus, the less concentrated is the extract in the solvent with which it is coming into contact, until at the end of the apparatus it eventually meets fresh solvent. In this way a complete extraction is possible with the correct choice of quantity and velocity of flow.

The theoretical relationships were first established for liquid–liquid countercurrent extraction and subsequently applied to solid–liquid countercurrent extraction. Liquid–liquid countercurrent extraction plays only a minor role after the actual extraction of the drug and we shall not discuss its theory.

The scheme given above illustrates a continuous countercurrent extraction in the strictest sense, in which both streams of material continuously move against each other. This type of extraction is also called *absolute* countercurrent extraction.

In *relative* countercurrent extraction, on the other hand, only one phase (as a rule the extraction solvent) is in motion, the other phase (usually the solid) remains stationary. Such conditions apply, for example, in carousel extractors. Strictly speaking, these do not really operate on true continuous countercurrent extraction, although, in practice, this term is also used for them. These definitions show that there is a gradual transition between percolation and continuous countercurrent extraction. A battery of percolators may also be regarded as a stationary solid phase against which a solvent phase continuously flows. The various types of extraction are shown in summary form in Fig. 4.22.

In discontinuous absolute countercurrent extraction the extraction solvent and the drug are moved against each other. In continuous relative countercurrent extraction only the liquid phase moves, whereas the drug material remains in the same vessel throughout the entire extraction. In continuous absolute countercurrent extraction both extraction solvent and drug material are in continual motion. A percolator battery operating on countercurrent therefore produces a continuous relative counterflow extraction.

The quantitative principles of solid–liquid extraction were described for the Schoenemann and Voeste [4.33] and the Figlmüller [4.96 and 4.97] countercurrent extractions; we refer to these papers as theoretical models. The following processes are described here:

Mathematical solution to the stagewise extraction in absolute countercurrent (plotting of working-curve).
Calculation of a theoretical stage in continuous extraction.

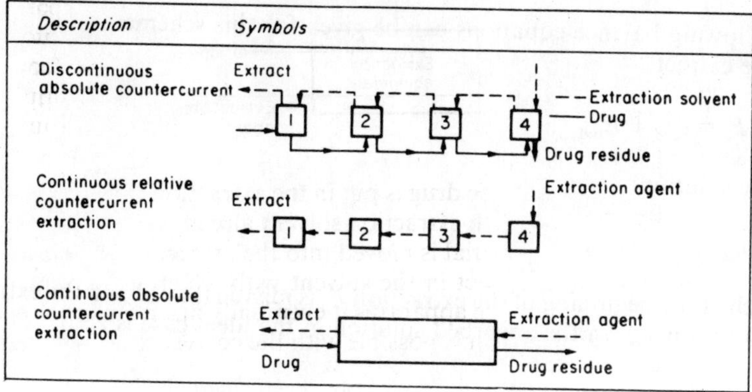

Figure 4.22 Countercurrent extraction procedures.

Graphical solution by means of equilateral triangle coordinates.
Graphical solution by means of right-angled triangle coordinates.

In the quantitative treatment of solid–liquid extraction it must be remembered that some of the extraction solvent and the partially enriched extract is absorbed by the drug material, i.e. part of the extract solution is free, whereas another part exists in bound or 'absorbed' form. In the symbols in Ref. 4.33 the lower case letters represent the 'absorbed solution' and the upper case letters the 'free solution'.

Mathematical solution to stagewise extraction in absolute countercurrent (plotting of working curve). Figure 4.23 contains a schematic representation of stagewise extraction in absolute countercurrent

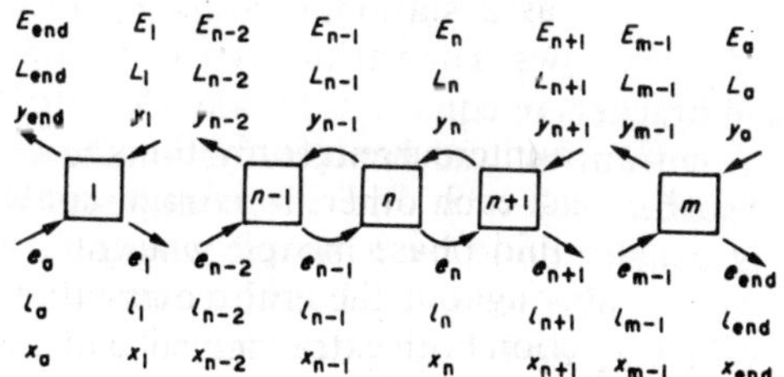

Figure 4.23 Absolute countercurrent stepwise extraction.

Parameters		Subscripts	
L_0, l_0	Pure solvent	a	Substances entering the system
E	Extract in free solution	end	Substance leaving the system
e	Extract in absorbed solution	1, 2	Extration stage: stage 1 is at highest concentrations (note flow betweeen stages identified by subscript index of lower stage
$L = L_0 + E$	Free solution		
$l = l_0 + e$	Absorbed solution		
$y = E/L$	Extract concentration in free solution	n	Any intermediate stage
$x = e/l$	Extract concentration in absorbed solution	m	Final stage

The following balance equations can be given for this scheme:
For the extract:

$$e_a + E_a = e_{end} + E_{end} \qquad (1)$$

For the solution:

$$l_a + L_a = l_{end} + L_{end} \qquad (2)$$

in which at the beginning of the extraction L_a is identical to L_0 and the extract concentration x_{end} in the absorbed solution in the ideal case is zero.

Furthermore,

$$e_a + E_n = e_n + E_{end} \qquad (3)$$

applies to any stage n for the extract and

$$l_a + L_n = l_n + L_{end} \qquad (4)$$

for the solution.

Conversion and division of (3) by (4) gives

$$\frac{E_n}{L_n} = \frac{(E_{end} - e_a) + e_n}{(L_{end} - l_a) + l_n} = y_n \qquad (5)$$

and substitution as per (1) and (2) gives

$$y_n = \frac{(E_a - e_{end}) + e_n}{(L_a - l_{end}) + l_n} \qquad (5a)$$

The expressions in brackets in equations (5) and (5a) are the known initial or required end concentrations, and hence only the still unknown terms e_n and l_n, i.e. the absorbed solution and its extract content, need to be determined. This is usually done by a simple maceration experiment, in which the conditions resemble those of the subsequent extraction as closely as possible. The values e_n and l_n are determined after establishment of equilibrium and percolation of the solution.

As $x_n = \frac{e_n}{l_n}$ and y_n can be calculated from equation (5a), y_n can be plotted against x_n in a so-called working curve by determination of y_n and x_n over the whole working range concerned. It represents a function:

$$y_n = f(x_n) \qquad (6)$$

As free solution with a lower extract content is continually coming into contact with absorbed solution under the conditions of real extraction, an

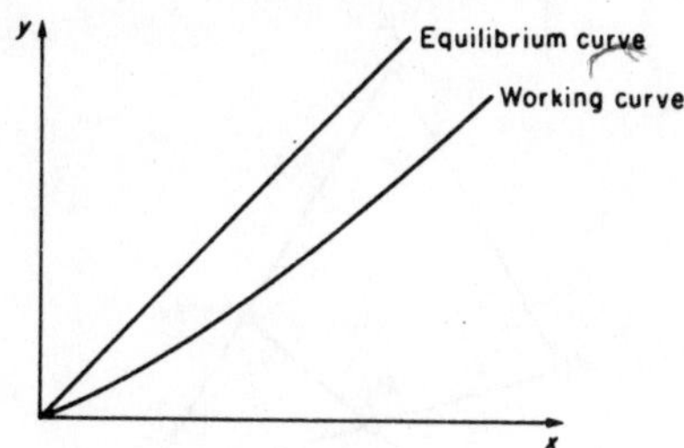

Figure 4.24 Comparison of equilibrium curve and working curve.

equilibrium, for which $y_n = x_n$, is not established, and so this equilibrium curve can be drawn into the y_n vs x_n graph as an auxiliary line with a slope of 45° (Fig. 4.24). The theoretical base value is then determined from Fig. 4.25. The drug material saturated with solution enters the first extraction stage (Point 1) with an initial concentration a, corresponding to equilibrium concentration b (Point 2). Assuming complete establishment of equilibrium, the extraction material with this equilibrium concentration goes into the next stage (Point 3), corresponding to another point on the equilibrium line, etc. This produces a series of right-angled steps which begins at the initial concentration a on the working curve and continues to the desired end concentration on the working curve. The number of points on the equilibrium curve here gives the number of required theoretical separation stages.

Calculation of a theoretical stage in continuous extraction. As an equilibrium curve is not obtained in continuous extraction, the introduction of a theoretical separation stage point is necessary. This is done in practice by determining the extract concentration at the beginning and end of the extraction in the free solution in a small extractor in which the conditions approximate as closely as possible those anticipated in the subsequent

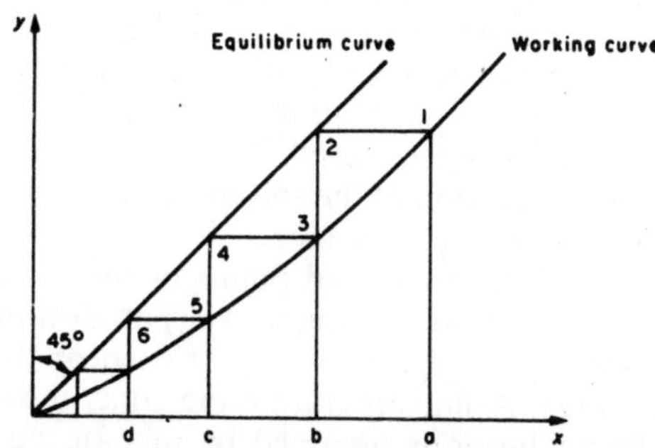

Figure 4.25 Graphical determination of the required theoretical extraction stages.

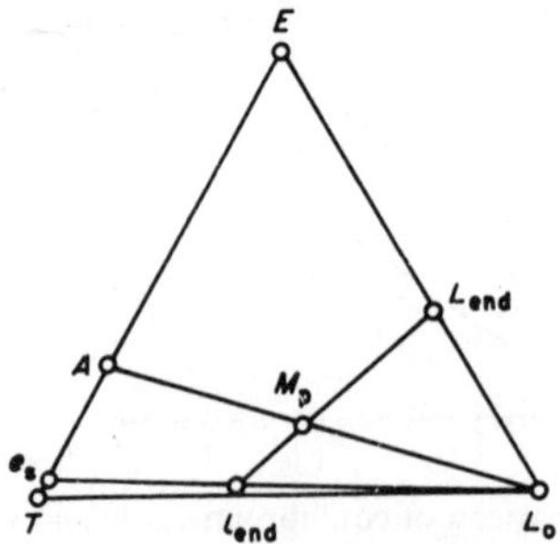

Figure 4.26 Determination of the number of separation stages. E = Extract; T = carrier substance (insoluble part of drug material); L_0 = solvent; A = starting-mixture (carrier substance + extract = drug material, considering the drug material as a two-component mixture); M_p = mixture point; L_{end} = free solution with extract content of end stage; l_{end} = drug residue with the residue content of extract and residual solvent (this stage point is situated close to the $\overline{TL_0}$ line because of the low extract content); e_s = residual extract content in carrier substance (drug plant skeleton) after separation of solvent from the sediment l_{end}.

extraction. A series of steps is drawn between the found concentrations plotted on the working curve, and the number of theoretical separation stages is determined. The distance in metres between the two measurement points divided by the number of separation stages gives the theoretical number of stages.

Graphical solution by equilateral triangle coordinates. The determination of the number of separation stages by equilateral triangle coordinates can be done on a phase diagram (Fig. 4.26):

The corners of the triangle represent the carrier substance, the solvent and the extract. The drug material is regarded here as a two-component mixture of extract and carrier substance. The starting-point A of the extraction accordingly lies on the line $\overline{TE}$. The higher the content of extract in the drug material, the closer A lies to E. A mixture is produced by addition of solvent L_0. The mixture point M_{p1} must lie on the direct connection $\overline{AL_0}$. The position of the point M_{p1} is determined solely by the ratio of solvent to drug material. The more solvent is added, the closer M_{p1} moves towards L_0. The mixture M_{p1} is resolved into free solution L_{end} and solution l_{end} (absorbed solution) adhering to the drug material. L_{end} is the effluent extract with extract content E. After expulsion of the solvent from l_{end} the residual extract content e_s remains in the carrier substance.

The points L_{end} and l_{end} represent end points of the extraction. They are reached via a series of intermediate stages. A separation into free solution and absorbed solution L_n and l_n takes place at each of these intermediate stages up to the nth stage. A line separating the zone of absorbed solution from the zone of free solution is obtained by joining the points l_n of the various stages. This line is required to determine the number of theoretical

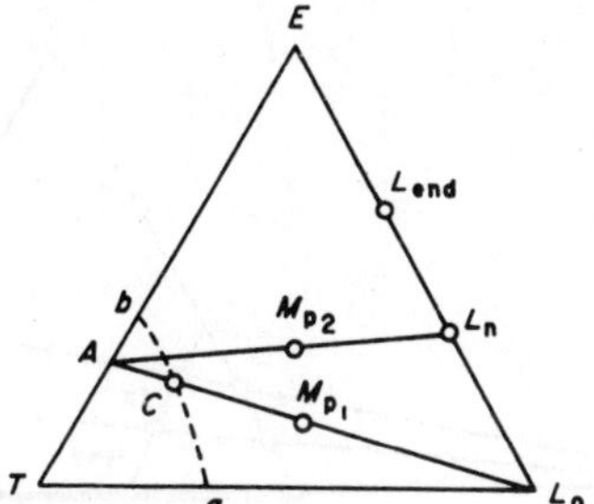

Figure 4.27   Determination of the boundary curve of the sediment phase.

stages. It is determined by adding a certain quantity of solvent to the starting mixture A, whereupon the mixture point again lies at M_{p1}. Point C, which is a point on the boundary line between free and absorbed solution, is obtained by determining the quantity of absorbed solution in this mixture (e.g. by cautious removal of the free solution by suction filtration through a Büchner funnel) (see Fig. 4.27). Further points on this boundary line could be obtained by treating drug materials of different extract contents in the same way, hence by shifting point A on the line $\overline{TE}$ and delineating this zone as a slightly curved line after determination of the absorbed solution in each of the mixtures.

As drug materials of varying content are usually not available, one can use partially extracted drug materials for constructing the points of lower extract content, thus shifting A towards T. Solvent L_n containing extract, instead of solvent L_0, is mixed with the drug material for plotting the points above A. The mixture point M_{p2} then no longer lies on the connecting line $\overline{AL_0}$ but on the line $\overline{AL_n}$. The pertinent point on ab is in turn obtained by determination of the absorbed solution by careful suction-filtration.

The pole P and the conodes (lines of equal extract concentration) are also still needed for determination of the theoretical number of stages. The pole P (Fig. 4.28) is obtained by lengthening $\overline{L_0}$ via l_{end} and e_s up to the point at which this line intersects the extension of $\overline{AL_{end}}$. This point is the pole P (Fig. 4.28).

The conodes radiate from the triangle point $T (= 100\%$ carrier substance). They connect this point with the side $\overline{EL_0}$ of the triangle (Fig. 4.29). The ratio E/L_0 is the same at each point on such a connecting line, only the content of carrier substance being altered, i.e. a transition takes place from the zone of absorbed solution into the zone of free solution (Fig. 4.29). The theoretical number of stages can then be determined with these auxiliary lines (Fig. 4.30). The solution L_{end} is in equilibrium with point l_1 via a conode (l_1 lies on the boundary line for the absorbed solution). The extension of the line $\overline{Pl_1}$ via l_1 leads to the point L_1, which represents the solution flowing against and

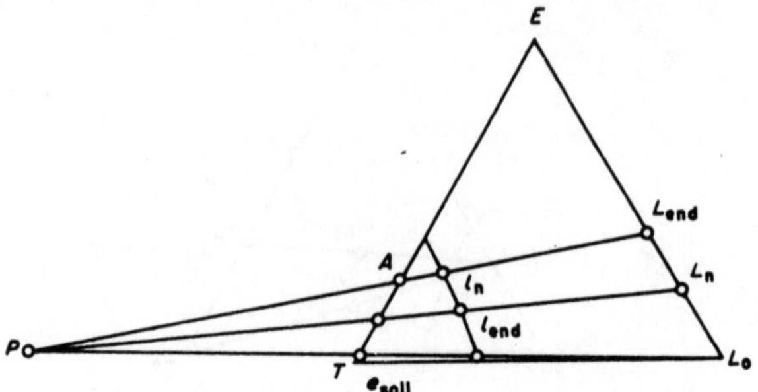

Figure 4.28　The pole.

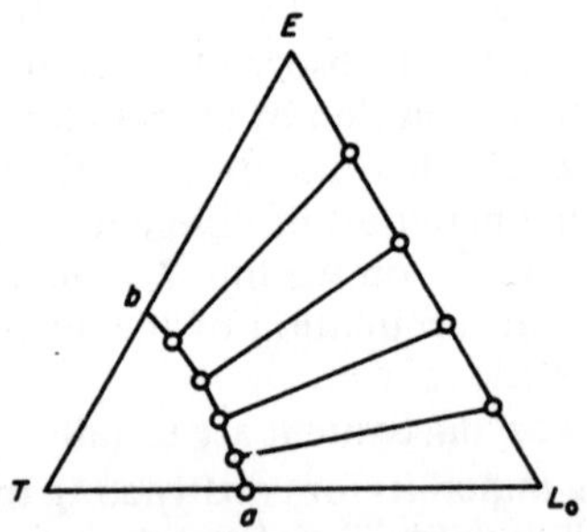

Figure 4.29　The conodes.

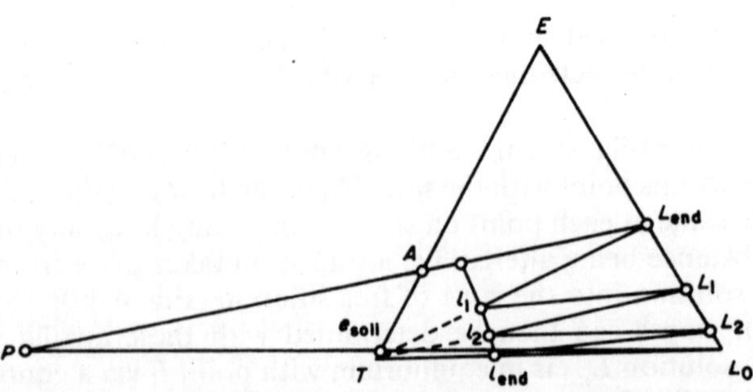

Figure 4.30　Determination of the theoretical number of stages.

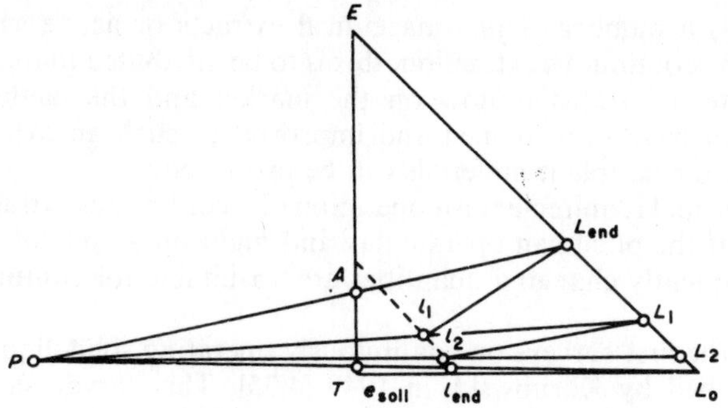

Figure 4.31   Right-angled triangle coordinates.

hence not in equilibrium with l_1, and in turn standing in equilibrium with l_2 via a conode, etc.

Starting from L_{end}, a series of steps is then set up with conodes and pole lines until the point l_{end} is reached. The number of conodes used here is the number of theoretical bases.

Graphical solution by right-angled triangle coordinates. A right-angled system can also be used instead of an equiangular coordinate system. The conditions are shown in Fig. 4.31. The carrier substance T then stands in the right-angle of the triangle. The line $\overline{EL_0}$ has zero carrier substance concentration. The other conditions correspond to those of the equilateral triangle. The advantage of the right-angled triangle is that the values of extract content E and solvent content L are easier to read off.

4.2.3.3.2.   Countercurrent extraction procedure

Continuous countercurrent extraction is used to advantage when large quantities of material have to be processed. For this reason it originated in the recovery of edible oils from oil seeds. Some of the plant designed for this has considerable daily output. Thus Kehse [4.98] reported in 1970 on a plant designed as a carousel extractor which, with a diameter of 8 m and two storeys each 2 m high, was capable of extracting 2000 tons of soybeans per day. The oil content of these beans was 18% and they were extracted down to a residual content of 0.8%.

These and similar plants have proved useful in the food, beverages and condiments industry for the extraction of, for example, paprika, tea, orange peel, coffee and hops, which are processed in large quantities. The problem with pharmaceutical products is that often the required quantities are not sufficient for operating a continuous extraction plant and moreover the products are frequently changed and have greatly differing properties. If despite this a number of pharmaceutical extracts or active substances are obtained by continuous extraction, this is to be attributed to the fact that the appropriate apparatus is now on the market and the methods of drug preparation have been refined and improved to such an extent that even difficult medicinal plant materials can be processed.

The principal requirement for operation of a continuous extraction plant is always that the plant can operate day and night on a cycle of at least 24 h. Small, frequently changing quantities are unsuitable for continuous extraction plants.

One of the first surveys on continuously operating solid–liquid extractors was published by Karnowsky in 1949 [4.35]. This survey shows that the apparatus now commonly used were at that time already in use, at any rate as forerunners of the modern apparatus. Continuous extraction apparatus can basically be divided into four major groups:

> *Helical countercurrent extractors*, i.e. apparatus in which both the drug material and the solvent are in continuous motion against one another.
> *Carousel extractors*, in which the material to be extracted is moved against the solvent stream in chambers as stationary subunits. They also include related apparatus such as *rotary extractors* and *basket* or *belt vat extractors*.
> *Extraction centrifuges*, which represent a combination of a high-speed mixer and a centrifuge or a decanter.
> *Column extractors*, in which the particles of drug material are extracted in an excess of solvent in a counterflow apparatus with built-in columns.

Helical countercurrent flow and column extractors operate on the principle of continuous absolute countercurrent, the carousel extractor on the principle of continuous relative countercurrent and extraction centrifuges operating in combination with high-speed vortical mixers represent the discontinuous countercurrent extraction type, although this type of extraction is also regarded as continuous. Table 4.27 shows the essential differences between the four types.

The way in which the medicinal plant material is prepared can be crucial for continuous extraction. Apart from the case of extraction centrifuges, in which very fine medicinal plant particles are desirable, since in these even the finest particles are cleanly separated by the centrifugal forces operating, a high proportion of powder is best avoided in continuous extraction, as it raises additional problems in separation and clarification. In this connection Kehse [4.98] defines fine material as particles of a diameter of less than 0.5 mm and of a proportional content in the whole material of less than 5% and certainly of at most 10%. On the other hand, more rapid solubilization and extraction of the medicinal plant material and sometimes even a higher yield of extract are possible with small particles.

Table 4.27 Extractor types for countercurrent extraction

Extractor type	Drug preparation	Mode of operation	Examples
Helical countercurrent extractors	Shredder mill	Drug material and solvent in permanent opposite motion	Continuous extraction coil (NIRO Company)[a]
Carousel extractor	Shredder mill	Drug material stationary in chambers which are moved against a solvent stream	Carousel extractor Extechnik[b]
Extraction centrifuge/mixer	Hammer mill	Spatially separated zones of intensive mixing (disintegration of drug material) and of extraction and separation centrifugation) in several successive stages	Solid–liquid extraction plant Westfalia Co.[c]
Column extractors	Shredder mill	Drug material and solvent in permanent opposite motion	Pulsation column Eries Co.[d]

[a] Niro Atomizer A/S, 305 Gladsaxevej, DK-2860 Soeborg, Denmark
[b] Extraktionstechnik, Gesellschaft für Anlagenbau m.b.H., Postfach 760147, D-2000 Hamburg 76, W. Germany.
[c] Westfalia Separator AG, Postfach 3720, D-4740 Oelde 1, W. Germany
[d] Eries, Rue de Genève, B.P. 20, F-69740 Genas/France.

The determination of the best/optimum extraction parameter and its transfer from the laboratory to the industrial production scale represent major problems in continuous extraction. Industrial pharmaceutical extractors are seldom designed and operated for just one medicinal plant material; rather, an extractor perhaps originally designed for a certain medicinal plant material may also subsequently be used for other extraction problems. In such cases a production plant is suggested for the problem, which must then be optimally adapted to the plant material. The method of determining the required number of separation stages in the triangle diagram is generally used for this. Such investigations must be carried out for each extraction problem. As an investigation on the industrial production scale is not possible because of the large amount of material required, the required number of stages is determined in laboratory experiments with simple percolators.

The following investigations must be carried out:

(1) Percolation of a cake of drug material of a height corresponding to that which would be used under actual production conditions by cross-flow extraction, in which the drug is extracted in stages with fresh solvent. The solvent ratio shall correspond as closely as possible to the subse-

quent production conditions. The extract content is determined in each extraction stage.

(2)   Determination of the 'bound solution', i.e. the quantity of liquid retained by the cake of drug material after addition of liquid has stopped.

(3)   Plotting of the so-called 'compensation curve'. This is done by pumping fresh solvent (in the ratio adapted to the subsequent production) through the cake of drug material and determining the decrease in the extract concentration in the solid with time. The difference from the initial content of the drug gives the total content in the extract. Figure 4.32 [4.98] shows typical curves of various materials. These curves also show that the extraction of oil from oil seeds, for which continuous extraction was originally developed, is considerably faster and more complete than is the case with drugs.

The required number of stages for the cross-flow extraction is then determined graphically using the results obtained. The theoretical number of stages for the cross-flow extraction determined here will not agree with the previously experimentally determined number. The reason for this is that the graphical determination always assumes establishment of equilibrium, which in actual practice is not always the case. Cross-flow extraction is likewise converted to countercurrent extraction using the triangle diagrams.

The yields of the individual stages of the percolation experiments can be represented graphically as yield curves, which permit comparison with the theoretically determined stages. Figure 4.33 [4.98] gives an example of this.

Continuous solid–liquid extraction is best investigated in the recovery of edible oils from oil seeds. Detailed papers have been published on the choice of suitable solvents [4.99] and on the theory of batch-wise and continuous extraction [4.100] in this field. Complete extraction in a few stages, including rapid establishment of the extraction equilibrium within one stage, is the basis of the process now being used worldwide for the extraction of vegetable oils. A further advantage is seen in the use of lipophilic solvents, which

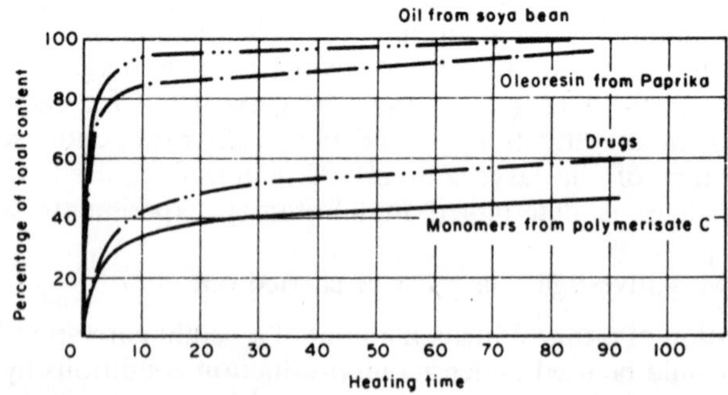

Figure 4.32   Adjustment curves in continuous extraction.

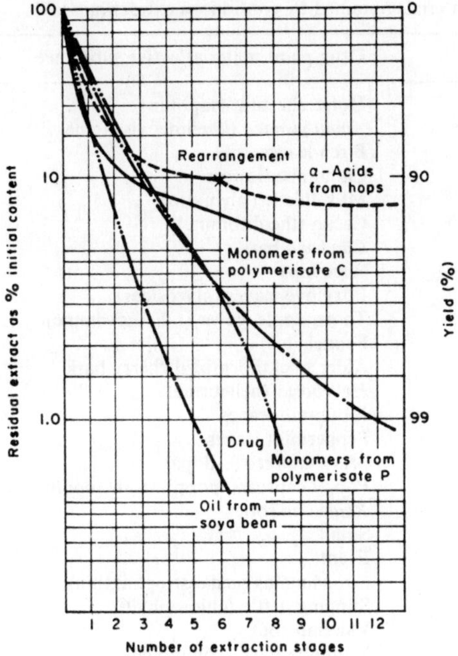

Figure 4.33 Variation of yield with number of extraction stages.

prevent swelling of the oil seed during extraction, though the use of such solvents is now becoming less frequent.

These advantages, which do not apply solely to oil seeds, are offset to a certain extent by the following disadvantages:

Smaller quantities of drug material cannot be processed very efficiently.

Frequent changes in the product to be processed result in long periods when the plant is standing idle.

Drug materials containing steeping solvents can cause difficulties as a result of swelling during extraction

Swelling is aided by water and hence the drug materials will undergo preswelling when solvents with a high water content are used.

In addition to extraction of oil seeds, continuous countercurrent extraction is also used for the preparation of fruit juices. Dousse and Ugstad [4.101] describe the extraction of apple juice with tap water at 60°C.

Masters [4.102] describes the extraction of liquorice root with a helical countercurrent extractor, where 750 kg of the comminuted roots with a moisture content of 8–10% were processed hourly in an extractor 11 m long.

Table 4.28   Plant materials obtained by continuous extraction

Apparatus	Drug plant material/active substance
Carousel extractor	Horse chestnut (saponin)
	Ammi visnaga (khelline, visnagine)
	Birch leaves
	Camomile flowers
	Milk thistle (silymarine)
	Cacao (theobromine)
	Cinchona bark
	Rose hips
	Foxglove leaves (glycosides)
	Thorn apple leaves (L-hyoscyamine)
	Fennel
	Alder buckthorn/bird cherry bark
	Helleborus (hellebrine)
	Jaborandi leaves
	Peppermint leaves
	Mucuna preta (L-dopa)
	Poppy capsules, poppy straw (opium alkaloids)
	Rauwolfia root
	Squill
	Ergot
	Strophanthus seeds (strophanthin)
	Stinging nettle (chlorophyll)
	Valerian root
	Yohimba bark (yohimbine)
Helical countercurrent extractor	Liquorice root
	Ginger
	Alder buckthorn/bird cherry bark
	Eucalyptus
Centrifugal extractor	Poppy straw

The extraction of poppy straw (poppy heads and stems), ephedra herb and the buds of liana vine was investigated in the pharamaceutical sector [4.103]. The extraction of caffeine and morphine from plant material has also been described [4.104].

In addition to these references, the makers of extraction production plant have reported on drug plant materials which they have investigated in extraction trials. Table 4.28 summarizes these.

4.2.4.   EXTRACTION WITH SUPERCRITICAL GASES

4.2.4.1.   Principles

4.2.4.1.1.   Ideal and real gases

Ideal gases, i.e. gases in which the forces of attraction between the molecules

are so small that they are imperceptible, obey the general Boyle and Mariotte
gas equation:

$$pV = nRT$$

where p = pressure (bars), V = volume (m^3), n = molar number of the gas,
T = absolute temperature (K), and R = general gas constant (J K^{-1} mol^{-1}).

Graphical representation of the product pV against p for each temperature
produces the so-called pV,p isotherms and for ideal gases produces a line
parallel to the p-axis as shown in Fig. 4.34.

Figure 4.35 shows the pV,p-isotherms of carbon dioxide at various
temperatures. This shows that carbon dioxide by no means obeys the general
gas equation in the range shown here.

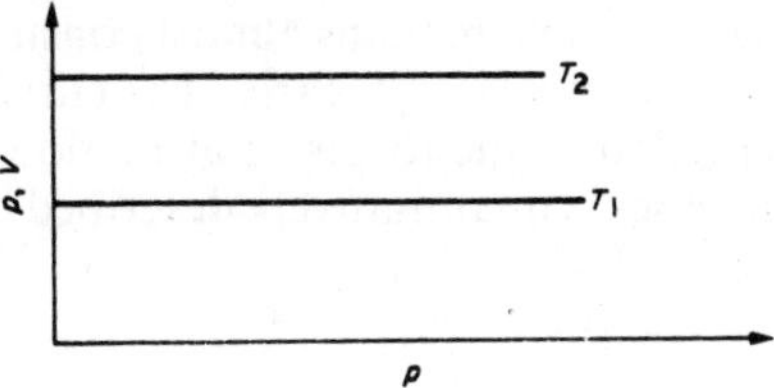

Figure 4.34 pV, p-Isotherms of an ideal gas.

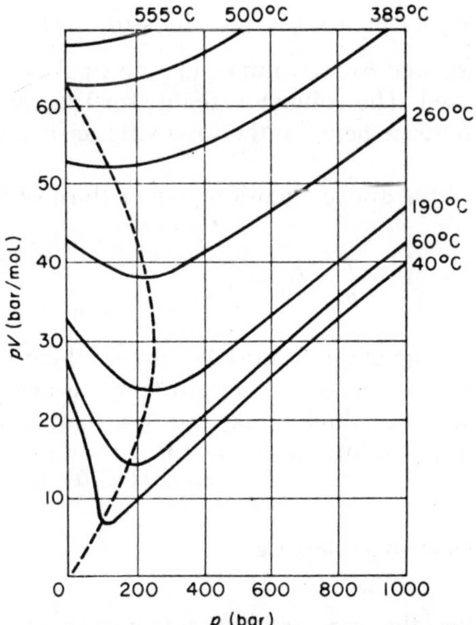

Figure 4.35 pV, p-Isotherms of carbon dioxide.

Ideal behaviour of a gas is possible only when the number of molecules per unit volume is small and their velocity sufficiently high. However, this means that with falling temperature (decreasing velocity) and increasing pressure (increasing concentration) the gases pass from the ideal into the real state. A gas shows ideal behaviour at high temperature and low pressure.

The pV, p isotherms of carbon dioxide clearly show this behaviour. In this graph there is a temperature at which the isotherm begins horizontally and rises slowly with increasing pressure. At this temperature (555°C) and at low pressures the gas behaves as an ideal gas. Below this temperature the isotherms have a minimum, which becomes more pronounced as the temperature decreases. The reason for this is that the forces of attraction increase with falling temperature and hence the product pV becomes smaller. If the external pressure is increased further the molecules can be pushed closer to one another and their motion becomes less random. This decrease in volume acts against the forces of repulsion and, of necessity, produces an increase in pressure, as the volume remains almost constant. The product pV thus becomes larger and now increases linearly. The curve of the isotherms is determined by the interaction of the forces of attraction and repulsion.

The behaviour of real gases is quantitatively described by the van der Waal equation of state:

$$(p + \frac{a}{V^2})(V - b) = RT$$

where $\frac{a}{V^2}$ = internal pressure (or cohesion pressure), which must be added to the external pressure, and b = covolume, i.e. the smallest volume to which a gas can be compressed. The volume available for free movement of the gas molecules is the difference between the total volume V and the covolume b.

Rearrangement of the above equation gives a third order equation:

$$V^3 - (b + \frac{RT}{p})V^2 + \frac{a}{p}V - \frac{ab}{p} = 0$$

The equation states that there must be one or three real values for the volume for a certain pressure and a certain temperature, namely one value for high temperatures and three values for low temperatures. Figure 4.36 [4.21] shows the corresponding isotherms for carbon dioxide.

4.2.4.1.2. Phase diagram of a substance

Figure 4.37 shows the pressure (P)/temperature (T)-projection of the phase diagram of a pure substance in general form. The graph shows the

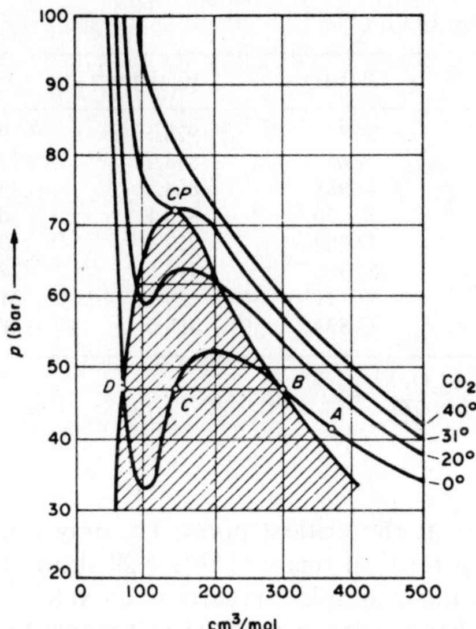

Figure 4.36 Isotherms for 1 mol ($= 44$ g) carbon dioxide.

sublimation curve, which separates the solid and the gaseous phases; the melting pressure curve, which separates the solid and liquid phases and also the vapour pressure curve, which separates the liquid and gas phases. All three curves intersect at a point, the triple-point, T_r, at which all three phases coexist, which means that three different molar volumes are also present side by side. The vapour pressure curve starts from the triple point and ends at the critical point, K_p. As the vapour pressure curve moves towards the critical point the values approach the densities of the liquid and gaseous phases and

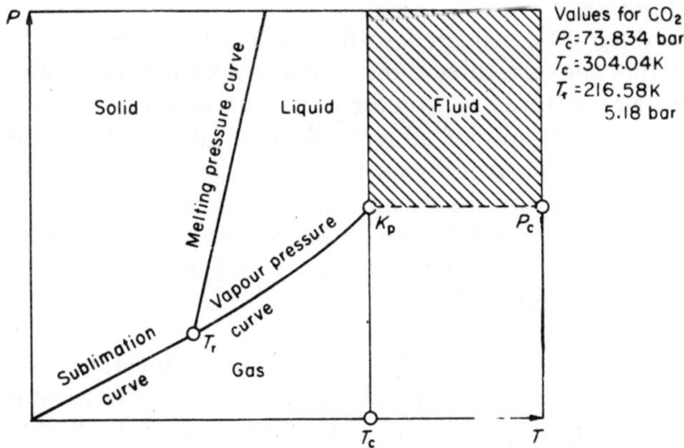

Figure 4.37 Phase diagram of a pure substance.

Table 4.29 Density of CO_2

$T(°C)$	P (bar)	$\rho_{(l)}$ (kg/m³)	$\rho_{(g)}$ (kg/m³)
0	34.817	928.5	96.26
5	39.657	898.2	113.78
10	44.988	864.2	134.59
15	50.850	825.1	160.50
20	57.289	777.7	193.90
25	64.356	713.8	242.4
30	72.111	629.9	312.5
31.04	73.834	468.0	468.0

$\rho_{(l)}$ = Density of liquid CO_2
$\rho_{(g)}$ = Density of gaseous CO_2

assume the same value at the critical point, i.e. only one phase remains, namely the fluid or supercritical phase. Table 4.29 shows how the densities approach equality in the example of carbon dioxide, and contains the densities of CO_2 in the vapour pressure curve as it moves towards the critical point K_p as a function of pressure and temperature. It can be seen that the densities of the two phases approach one another and become equal at the critical point, where $T_c = 31.04°C$, $P_c = 73.83$ bar and the density is 468 kg/m³.

Above the critical point K_p the density of the fluid continues to increase with increasing pressure and can attain values similar to those of liquids.

The critical point of a pure substance is thus characterized by its critical data P_c = critical pressure, T_c = critical temperature and ρ_c = critical density.

The 'fluid' or 'supercritical' phase region, which is shown shaded in Fig. 4.37, lies above P_c and T_c and starts at the critical point. The 'fluid' phase region is bounded by the 'gas' and 'liquid' phases and, at very high pressures, by the 'solid' phase.

The phase diagram considered so far has been two-dimensional. Figure 4.38 [from 4.106] shows a three-dimensional representation which considers the pressure, temperature and volume of carbon dioxide.

Isotherms are alteration of state in the spatial CO_2 p,V,T graph. The changes of state are described by means of the 'CO_2 cube', into which seven lines of equal temperature (isotherms) are drawn. These are T_3 (the critical temperature), T_6 (the temperature of the triple point at which the solid, liquid and gaseous phases can occur simultaneously) and T_7 (the sublimation temperature of solid CO_2). Also:

T_1: This temperature lies far above $T_3 = T_c$. At point A_1 CO_2 is gaseous and reduction of the volume merely increases the pressure, with simultaneous production of heat, up to line B at the top of the diagram. The gaseous state is maintained throughout this isothermic process.

T_2: Starting from point A_2 the same process proceeds as under T_1. The gaseous state is maintained.

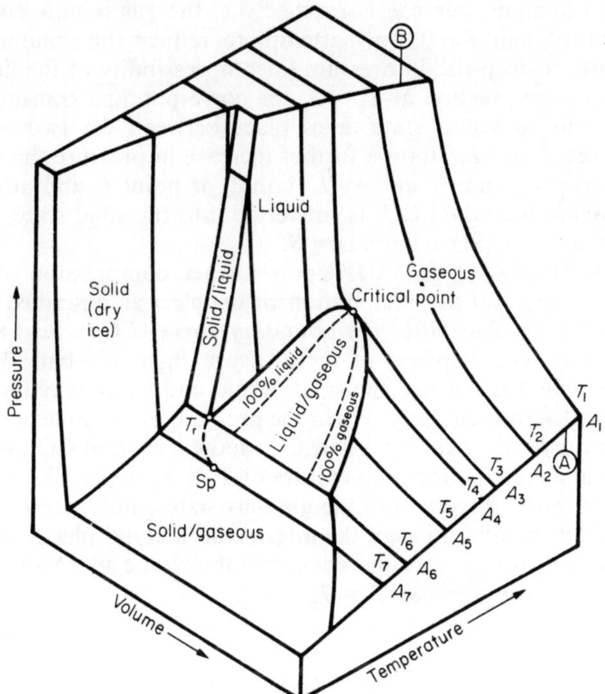

Figure 4.38 p,v,T Diagram for carbon dioxide.

Critical point	Triple point	Sublimation point
$T_c = 304.04$ K	$T(\mathrm{Tr}) = 216.4$ K	$T(\mathrm{Sp}) = 194.5$ K[†]
$= 31.04\,°C$	$= -56.6\,°C$	$= -78.5\,°C$
$P_c = 73.81$ bar	$P(\mathrm{Tr}) = 5.28$ bar	$P(\mathrm{Sp}) = 1$ bar
$V_c = 2.156\,\mathrm{dm^3\,kg^{-1}}$		[†]In 100% CO_2 atmosphere

(T, temperature; P, pressure; V, volume; Sp, sublimation point; Tr, triple point)

$T_3 = T_c = 31.04\,°C$ (304.04 K). Beginning from point A_3, the pressure increases, with a reduction of volume and release of heat, and at the critical point reaches the value $P_c = 73.8$ bar. At this point the molecules of the CO_2 gas are as densely packed as in liquid CO_2 and hence the medium is in a state which can simply be said to be both gas and liquid. This is expressed in the meeting of the boundary curves for the liquid F and for the gaseous state G. The pressure increases upon further compression.

T_4: Smaller than T_3, hence subcritical. Starting from point A_4, the pressure in the CO_2 gas increases with decreasing volume up to the point of intersection with the boundary curve G, at which the CO_2 medium is still 100% gaseous. ($x = 1$ signifies gaseous state). If the volume is then further reduced, CO_2 mist (CO_2 droplets) begins to form, its proportion continually increasing with reduction in the volume (e.g. $x = 0.8$ means 80% gas and 20% mist droplets) while the pressure P remains constant until all the gas is liquefied and the

$\quad$ left boundary curve F is reached, i.e. the gas is now completely liquified and $x = 0$. Any attempt to reduce the volume further causes a steep rise in pressure (incompressibility of the liquid).

T_5: $\quad$ Upon compression at $T_5 < T_4$ the corresponding transition from gaseous to liquid state takes place between the two boundary curves G and F. Upon a further increase in pressure the isotherm meets the phase boundary L (liquid) at point C and at constant pressure the liquid CO_2 is converted into the solid state, which is 100% at the phase boundary S.

$T_6 = T_{Tr} = 216.4\,K$ $(-56.6°C)$. Upon further compression at T_6 the same processes of precipitation of droplets as described under 4 and 5 take place after the boundary curve G (gaseous) has been reached. This happens at the pressure $P_{Tr} = 5.28\,bar$. When the triple point is reached the solid, liquid and gaseous phases are all present simultaneously. When the pressure of the state at the triple point is reduced to $P = 1\,bar$, CO_2 snow, i.e. solid CO_2, is formed and at a sublimation temperature of $T_7 = T_S = 194.5\,K$ $(-78.5°C)$ is converted directly into the gaseous state (sublimation point S_p) without passing through the intermediate liquid phase. Dry ice is obtained after a certain reduction of volume when the snow is further compressed at $T_7 = T_S$.

4.2.4.1.3. Properties of supercritical gases

The physical properties of a gas are altered when the conditions go beyond the critical point. Bruner and Peter [4.107] state that densities similar to those of true liquids can be attained even in gases compressed at relatively low pressures in the region of 50–300 bar (atmospheres). These properties can be varied within wide limits by a suitable choice of pressure and temperature. The lower the density of the super-critical gas, the more it behaves like a gas, but when its density is increased it increasingly assumes the properties of a liquid. Other properties, e.g. the dielectric constant, are also altered as well as the density. The variation of density and of the dielectric constant D with pressure at 50°C is shown in Fig. 4.39 [4.108].

These changes in the properties increase the ability of carbon dioxide to dissolve numerous substances, e.g. naphthalene (Fig. 4.40) and diphenylamine (Fig. 4.41) [4.109]. Only those gases which can be converted into the supercritical state at attainable pressures and temperatures can be considered for extraction use. Here it must be pointed out that one great advantage of advantage of extraction with supercritical gases lies in the use of low temperatures, and hence gases with a low critical temperature are particularly suitable for this. Table 4.30 [4.110] gives a survey of suitable gases.

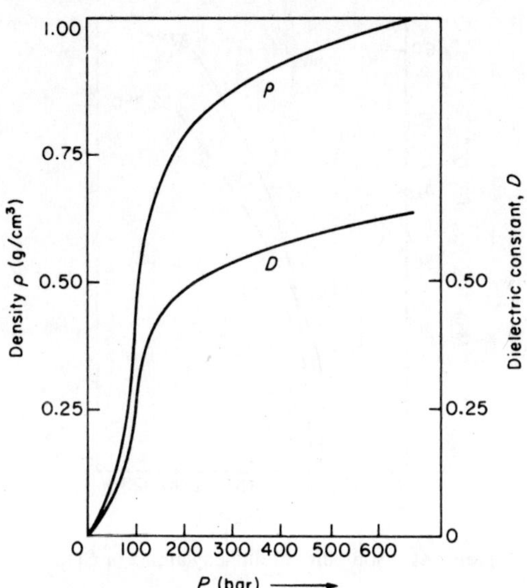

Figure 4.39 Variation of density and dielectric constant win pressure of carbon dioxide at 50°C.

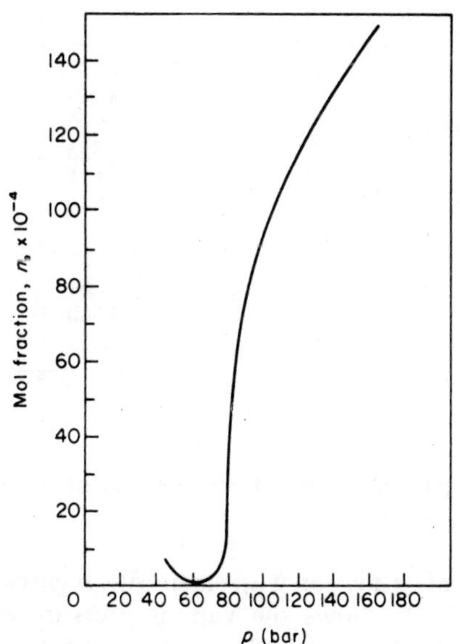

Figure 4.40 Solubility of naphthalene in CO_2 at 34.6°C.

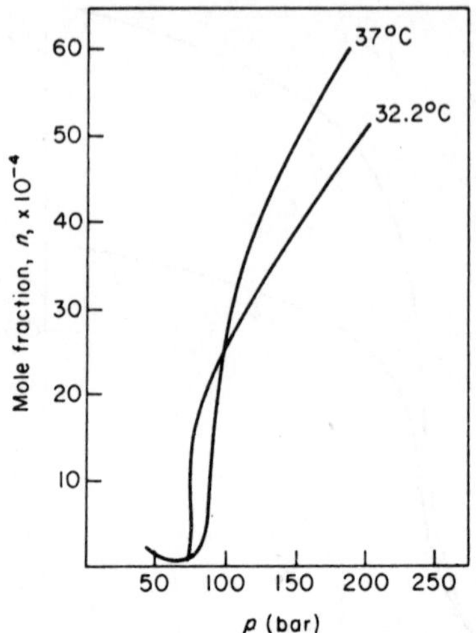

Figure 4.41   Solubility of diphenylamine in CO_2.

Table 4.30   Gases suitable for 'supercritical' extraction

Gas	T_c (°C)	P_c (bar)
Nitrogen	-147.00	33.934
Methane	-82.49	46.407
Carbon dioxide	31.00	73.836
Ethane	32.25	48.839
Ethylene	9.21	50.313
Nitrous oxide	36.45	72.549
Sulphur dioxide	157.50	79.841
Propane	96.85	42.557
Propylene	91.60	46.103
Ammonia	132.40	112.998
Sulphur hexafluoride	45.56	37.602

4.2.4.1.4. Fundamental possibilities of extraction and substance separation with supercritical gases

The principal features of extraction with supercritical gases are shown in Fig. 4.42 [from Ref. 4.111]. It shows the vapour pressure curves of the low boiling-point extraction solvent and of the volatile substance to be extracted, with their respective critical points. The two critical points are connected by the first order critical curve. This curve is dependent on the system and differs for each system. It can be plotted clearly only in two-substance systems.

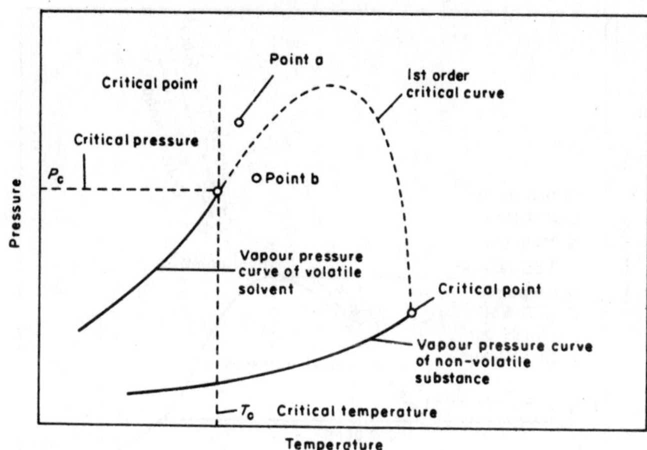

Figure 4.42   Schematic pressure–temperature diagram of a binary mixture (normal type).

Outside the first order critical curve (at point *a*) both components are in the supercritical state and totally miscible with each other (so-called monophasic region). Inside the curve (point *b*) loaded supercritical gas is present together with a liquid phase (biphasic region). A distinction must now be made between extraction and separation. Extraction, i.e. loading, can take place both in the monophasic and in the biphasic region. In the latter, the process is described as extraction with separate loaded gas phase. Separation cannot be carried out in the monophasic region.

The various possibilities of separation can be shown most clearly with a temperature/entropy graph. Figure 4.43 [from Ref. 4.112] shows the corresponding graph for carbon dioxide with the existence ranges of the various phases. The graph shows the various ranges of existence of solid, liquid and gaseous CO_2 in relation to temperature and entropy. The triple point, the critical point and three different isobars are given.

The critical isobar passes through the critical point and delineates the supercritical region (fluid) from the solid, liquid, solid–liquid, gaseous and liquid–gaseous states. The subcritical isobar (low pressure) illustrates the fact that the supercritical phase cannot be attained below the critical point. The supercritical isobar shows that transitions to the solid–liquid phase take place from the fluid region.

Figure 4.44 [4.113] shows the sum of all possibilities of separation of a substance from the supercritical gas phase.

For the following illustrations [from Ref. 4.113] P_e and P_a shall represent the pressures in the extraction vessel and separation vessel respectively. T_e and T_a the corresponding temperatures and P_c and T_c the critical pressure and the critical temperature of the solvent.

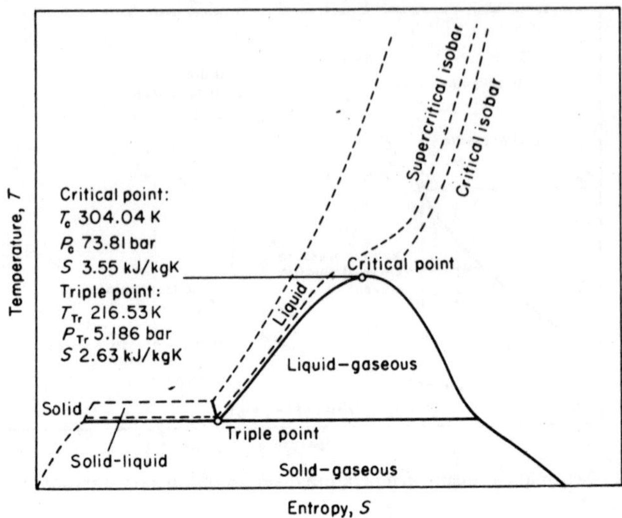

Figure 4.43   Temperature–entropy diagram for carbon dioxide.

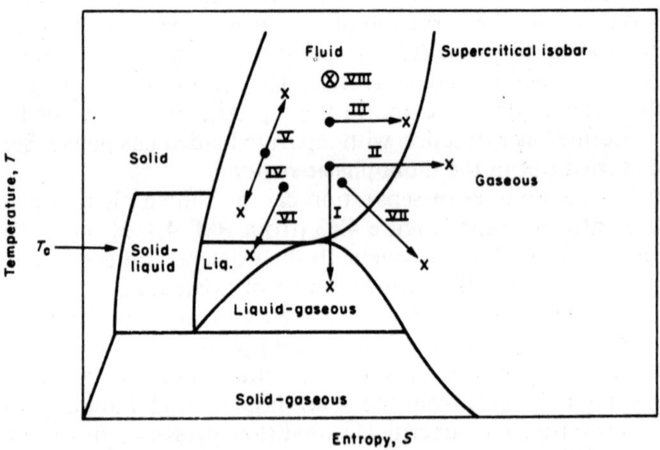

Figure 4.44   General temperature-entropy diagram. Separation can take place by any of the eight routes from starting position (●) to final phase (×).

Process I

$$P_e > P_c \qquad P_a < P_c < P_e$$
$$T_e > T_c \qquad T_a < T_c < T_e$$

Extraction is carried out in the supercritical state and a gas/liquid state is attained in the separation vessel by reduction of pressure and cooling. Here the liquid phase of the extraction agent in the separation vessel assumes the role of a trap for the extracted substances.

Process II

$$P_e > P_c \qquad P_a < P_c < P_e$$
$$T_e > T_c \qquad T_a = T_e > T_c$$

Extraction is carried out supercritically, the pressure is reduced isothermally and separation is carried out in the gaseous phase of the extraction agent.

Process III

$$P_e > P_c \qquad P_a < P_e \qquad P_a > P_c$$
$$T_e > T_c \qquad T_a = T_e > T_c$$

Extraction is carried out in the supercritical region, the pressure is reduced isothermally and separation is carried out at reduced pressure but still supercritically. The density of the extraction agent in the separating vessel and, hence, its dissolution power are thus reduced, resulting in a separation of the extracted substances.

Process IV

$$P_e > P_c \qquad P_a = P_e > P_c$$
$$T_e > T_c \qquad T_a < T_e \qquad T_a > T_c$$

4.2.4.2. Process principles

All the processes described in Section 4.2.4.1.4 may be attributed to alterations of pressure and temperature or to one of these. An extraction plant which exploits all possibilities must therefore be equipped with pressure and temperature adjustments both for extraction and for reduction of pressure. Figure 4.45 shows such a plant diagramatically [4.114]. The extraction agent (8), after passage through a heat exchanger and a filter (4 and 7), is brought to the required pressure by a pump or compressor (3). It then passes through the extraction vessel containing the drug material. The loaded supercritical gas leaves the extraction vessel under pressure and passes via a throttle valve into the separating vessel (2). The extract is drawn off through a valve (6a) at the bottom of the separating vessel, the gas is removed under pressure and recycled to the compressor and then to the extraction vessel.

The process can be considerably extended by fractionated extraction and separation. Figure 4.46 shows a scheme for fractionated extraction [4.113] in which the material to be extracted is successively compacted under varying pressures and the extract is separated into fractions corresponding to each compaction pressure.

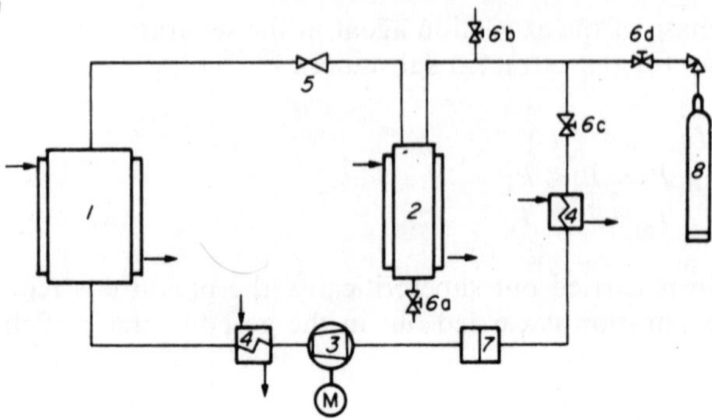

Figure 4.45 Schematic diagram of the high-pressure extraction process. 1, Extraction vessel; 2, separating vessel; 3, pump/compressor; 4, heat exchanger; 5, throttle; 6, valves; 7, filter; 8, extraction agent.

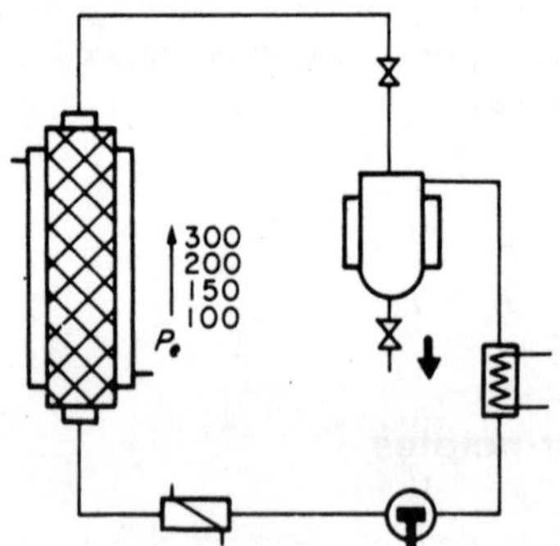

Figure 4.46 Fractionated extraction.

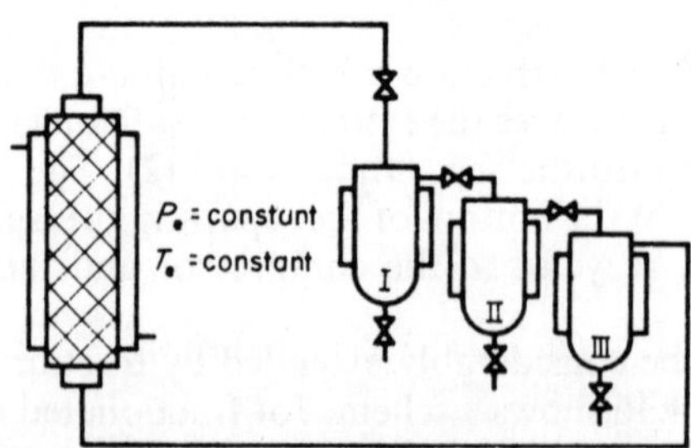

Figure 4.47 Fractionated separation.

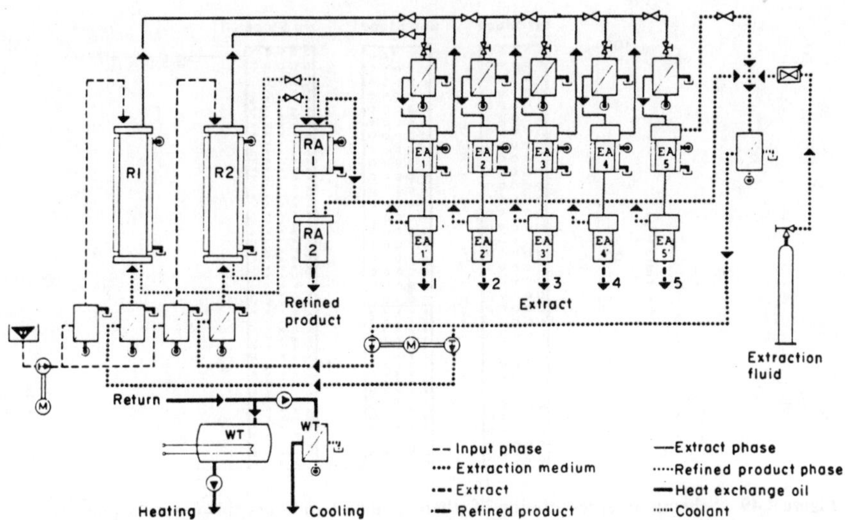

Figure 4.48   Extraction plant operating on fractionated separation.

The process of fractionated separation shown in Fig. 4.47 [4.113] is more refined. Here an extract is led through several successive separation stages and thus fractionated.

The production capacity of extraction plants can be greatly increased by a combination of several extractors with several separators. Figure 4.48 shows such a plant in principle [4.114]. These and similar plants can be operated on a relatively continuous countercurrent basis when they are of a suitable design, i.e. similar to that of percolator batteries.

Stahl [4.116] gives a very clear example for fractionated separation. In the extraction of the ethereal oil from caraway seeds with liquid carbon dioxide the fatty oil is separated from the ethereal oil component with a three-stage separator unit as illustrated in Fig. 4.49.

The first separation stage mainly produces fatty oil. In the second separation stage, which quantitatively constitutes only a small fraction of the total yield, the proportion of ethereal oil is higher. The bulk of the ethereal oil is found in the third stage. Although the operation here is carried out in the subcritical region, the example shows the possibilities of fractionated separation.

4.2.4.3. Applications

A symposium on the principles and application of extraction with super-critical gases, in which the state of the art was summarised [4.115], first took

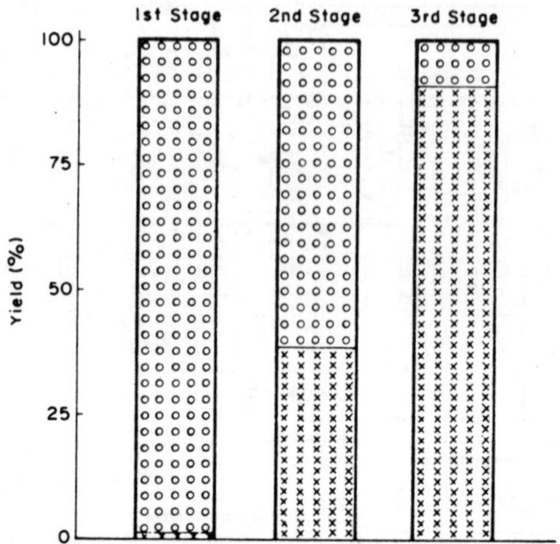

Figure 4.49   Schematic representation of ethereal oil ($\times$) and fatty oil ($\bigcirc$) in various fractions during the high-pressure extraction of caraway seeds with liquid CO_2. Extraction conditions: $P = 90$ bar; $T = 20°C$.

	Separation conditions	Extract yield
Stage 1	90 bar/40°C	3.8%
Stage 2	75 bar/40°C	0.8%
Stage 3	30 bar/0°C	3.7%

place in 1978. The papers read there were subsequently published in an expanded form as a book [4.116]. Brunner and Peter [4.107] have more recently published a survey paper. Further information is given by Coenen [4.111] and Zosel [4.117]. The applications known so far can be summarized in four groups:

> Purification of petroleum and used oil, recovery of bitumen from tar sand and recovery of grinding oil.
> Extraction of coal and preparation of coal tars.
> Processing of edible oils and fats, separation of fatty acids and triglycerides.
> Extraction of active substances from plants (hops, coffee).

Brunner and Peter [4.107] give a survey on the applications, especially on technical applications.

In the pharmaceutical field, Stahl especially has used extraction with supercritical gases, first analytically [4.119] and then preparatively [4.118].

The following were also treated:

> Camomile [4.120 and 4.121], extraction of matricin with carbon dioxide.

Pyrethrum flowers [4.122], pyrethrins.
Valerian root [4.123].
Papaver bracteatum Lindl, extraction of thebaine with fluoroform [4.124].
Various drug plants containing alkaloids [4.125].
Wool fat, extraction of lanolin with supercritical propane/propylene [4.126].
Orange peel, extraction of carotene with supercritical carbon dioxide with addition of 8% acetone as cosolvent [4.127].
Aniseed, star anise, caraway, cloves and Ceylon cinnamon with carbon dioxide [4.118].

Only the pharmaceutically relevant examples in this list were considered. The cited reviews [4.107–4.116] should be referred to for examples from the foodstuffs industry. Stahl [4.128] gives a more recent summarized description of the recovery of natural substances.

Despite the intensive research studies by Stahl, extraction with supercritical gases in the pharmaceutical industry is still in its infancy. The wide range of extractable substances, with their greatly varying polarities and the

Table 4.31 Drug constituents determined by Stahl [4.119]

Drug, air dried	Main constituents found by thin layer chromatography	Molecular weight	Chemical formula	MP (°C)	BP (°C)	Extract-able at (bar)
Caraway seeds	Limonene	136	$C_{10}H_{16}$	−74	178	70
	Carvone	160	$C_{10}H_{14}O$	—	231	70
	Triglycerides	>600	—	—	—	90
Peppermint leaves	Menthol	158	$C_{10}H_{20}O$	44	216	70
	Menthone	156	$C_{10}H_{18}O$	—	209	70
Camomile flowers	Herniarin	176	$C_{10}H_8O_3$	117	—	70
	α-Bisabolol	222	$C_{15}H_{26}O$	—	155	70
	"En-In-Dicycloether"	200	$C_{13}H_{12}O_2$	—	—	80
	Matricin	306	$C_{17}H_{22}O_5$	—	—	85
Coffee	Caffeine	194	$C_8H_{10}N_4O_2$	178 subl.	—	80
	Triglycerides	>600	—	—	—	90
Cannabis	Cannabidiol	314	$C_{21}H_{29}O_2$	66	—	85
	Cannabinol	310	$C_{21}H_{26}O_2$	76	—	90
	Tetrahydrocannabinol	315	$C_{24}H_{30}O_2$	76	—	90
Sunflower seeds	Squalene	410	$C_{30}H_{50}$	−20	280	80
	Cholesterine	366	$C_{27}H_{28}O$	148	360	85
	Triglycerides	600	—	—	—	90
Sesame seeds	Sesamine	354	$C_{20}H_{18}O_2$	123	—	85
	Triglycerides	>600	—	—	—	90

numerous potential variations of the extraction process, offers a large number of possibilities for future research. A number of advantages of extraction with supercritical gases in this field can be mentioned [4.118]:

Low thermal stress on the constituent substances.
Selective capacity for dissolution governed by the adjustable parameters pressure and temperature.
Guaranteed freedom of the resultant extracts from solvent.
Complete physiological harmlessness.
Carbon dioxide is a gas which, wherever it can be used, is a very cheap extraction agent which is relatively harmless.

4.2.5. SUMMARIZED EVALUATION OF EXTRACTION PROCESSES

The quantity of drug material to be extracted is the most important factor in the choice of an extraction process. This applies particularly to continous extraction, which is worth using only if the extractor can be in operation 24 h a day and the amount of drug material is so large that a daily change can be avoided. The nature of the drug material has an effect here in that it must be ensured that no interruptions of the continuous operation will occur, say, through clogging.

Whereas the quantity problem plays a dominant rôle in the choice between continuous and discontinous extraction processes, the choice between maceration and percolation is especially influenced by the properties of the drug material. Drug materials containing mucilage and those which swell strongly can often be treated only with maceration. Barks and roots are well suited to percolation.

The multiplicity of apparatus always makes a clear distinction difficult between these two processes. Remaceration in a closed vessel, in which the entire part extract is drawn off between two additions of solvent, can for example also be regarded as cross-flow percolation. This generally means that the transitions from simple maceration via motion maceration and remaceration to percolation and continuous countercurrent extraction are blurred. One and the same extraction problem can be resolved in different ways.

4.3. Purification of miscellae

4.3.1. NECESSITY AND OBJECTIVES OF PURIFICATION OF MISCELLAE

The miscella obtained by extraction with a solvent or solvent mixture frequently contains unwanted components in addition to the desired extractive substances. These undesirable substances may for example be tannins, which cause turbidity or opacity upon storage or further processing of the extract. Pigments are more rarely undesirable. Substances which impair the stability of the active substances concentrated in the extract must on the other hand be removed. In certain circumstances the degree of microbial contamination must also be drastically reduced. Last but not least, residues of physiologically harmful solvents or solubilizers must be removed.

The object of any purification must be to remove the unwanted constituents as quantitatively as possible without damaging the extractive substances or causing alterations which would have detrimental effects such as reduced stability or yield.

A distinction must be made between purely physical and physico-chemical methods of purification. The physical methods include decanting, filtration, sedimentation, separation, centrifugation and heating, whereas the physico-chemical methods include adsorption, precipitation and ion-exchange.

The use of physico-chemical methods usually has to be followed by the application of a physical method for removal of the added extraction aids.

A strict separation of the individual basic principles is not always possible with the present state of chemical technology. As far as appears meaningful, the methods shall therefore be summarized in the following groups. A bacteria-reducing purification must also be carried out in addition to the physical and physico-chemical purification in cases where the number of bacteria in the extract is not kept low by the use of suitable solvents.

4.3.2. PHYSICAL METHODS AND THEIR APPLICATION

4.3.2.1. Sedimentation, separation and centrifugation

4.3.2.1.1. Definitions

The terms sedimentation, separation and centrifugation are used side by side in the literature, where the same process is frequently covered by different names.

Sedimentation is, strictly, the settling of solid particles from a liquid or a gas under the influence of gravity.

Centrifugation is sedimentation under a large multiple of the force of gravity.

Separation is a process in which liquids are clarified, emulsions are separated or solids are extracted in centrifuges. The extraction of solids in separators is described in the section on counterflow extraction. The detailed descriptions in this section are only concerned with the clarification of liquids i.e. solid/liquid separation, and also liquid/liquid separation where necessary.

Surveys in the field of separation are given by Ullmann [4.129] under the title 'Zentrifugen und Hydrozyklone' and also by Hemfort [4.130], who suggests the classification shown in Fig. 4.50.

| Separation of gases | | De-emulsification | | | | Sedimentation | | Filtration | |
| | | | Separation | | | | | | |
Separation of isotopes	Degassing	Extraction	Two-phase separation	Three-phase separation	Decanting	Combination sedimentation/filtration	Sieve centrifugation	Filter centrifugation
G/G	L/G	L/L	L/L	S/L/L	S/L	S/L	S/L	S/L
Gas centrifuge	Thin-layer centrifuge	Ring chamber extractor	Tube separator	Three-phase tube separator	Decanter	Sieve decanter	Sieve-screw centrifuge	Shearer centrifuge (one or multistage)
		Spiral chamber extractor	Plate separator	Plate separator	Overflow centrifuge	Discontinuous cone centrifuge	Gliding centrifuge	Scraper centrifuge (horizontal)
		Plate-drum extractor		Self-regulating separator	Double-cone centrifuge	Baffle ring centrifuge	Oscillating centrifuge	Syphon-scraper centrifuge
				Plate nozzle separator	Tube centrifuge		Tumbler centrifuge	Scraper centrifuge (vertical)
					Ring-chamber centrifuge			Hanging pendulum centrifuge
					Self-regulating plate centrifuge			Three-column centrifuge
					Plate-nozzle centrifuge			Freely oscillating centrifuge
					Bucket centrifuge (laboratory)			
					Hydro-cyclone			

Figure 4.50   Classification of centrifuges.

In the purification of plant extracts, separators and decanters play the greatest role in the separation of solids.

Liquid/liquid extraction is of importance when substances dissolved in the extract are to be concentrated, as is the case in the isolation and preparation of natural substances in their pure state.

4.3.2.1.2.   Principles of gravitational sedimentation

Sedimentation under gravity can be described by Stokes's law, as the solid substances which are to be separated have small particle diameters and the prevailing flow under the operational conditions is laminar [4.129].

$$v = \frac{2r^2(\rho_s - \rho_l)}{9\eta} bf_\varepsilon$$

where v = sedimentation speed of particles (m s^{-1}), r = particle radius (m), ρ_s = density of solid substance (kg m^{-3}), ρ_l = density of liquid (kg m^{-3}), η = dynamic viscosity of liquid (Pa s), f_ε = semi-empirical function of the relative intermediate volume of the granules, which takes into account the influence of hindrance to sedimentation by increasing concentrations of solid substance in the suspension, and b = centrifugal acceleration (m s^{-2}).

Centrifugal acceleration replaces acceleration due to gravity in the original Stokes's formula. Here:

$$b = r\omega^2 = \frac{1}{2}(d_e - d_i)\frac{\pi^2 n^2}{900}$$

where ω = angular velocity (s^{-1}), r = average centrifugal radius (m), d_e = external diameter of centrifuge gap (m), d_i = internal diameter of centrifuge gap (m), and n = number of revolutions (s^{-1}).

In the case of simple sedimentation b is replaced by g. The results of experiments in a standing cylindrical vessel can then be converted to the conditions in the centrifugal field by multiplication with the factor b/g, which is also called the acceleration factor Z.

$$Z = \frac{b}{g} = \frac{r\omega^2}{g}$$

Z is dimensionless. The acceleration factors for various centrifuges are:

Tubular centrifuges:	13 000–17 000
Disc centrifuges:	5000–13 000
Sieve and filter centrifuges:	300–1500

4.3.2.1.3. Choice and use of centrifuges

Two parameters are important for the choice of centrifuges: (1) the particle size of the solid substances to be separated, and (2) the solid substance content. Figures 4.51 [4.130] and 4.52 show the recommended types of centrifuge in relation to particle size and solid substance. It can be seen from Figs 4.51 and 4.52 that separators are used for clarifying liquids containing low concentrations of solids in the form of small particles. Decanters are

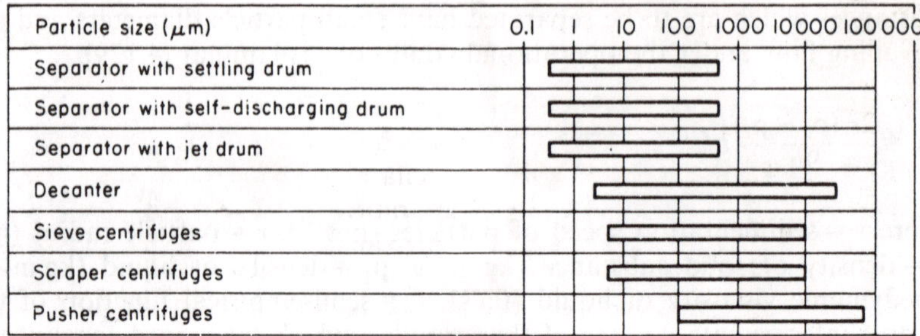

Figure 4.51 Centrifuges and the ranges of particle sizes for which they are used.

suitable for separating liquids from large-particle solids in moderate concentration, and sieve, shelling and shearing centrifuges are used for medium to large particle solids in high concentrations.

The type and method of extraction govern the solid content and particle size in the resultant drug extracts. Maceration extracts are as a rule roughly preclarified by decanting or filtration before being centrifuged. Separators with self-emptying drums are suitable for this. Miscellae obtained by percolation are normally very clean, clear extracts, though they do still need a further clarification. Separators with clarifying drums can be used for this. The same applies to extracts from continuous counterflow extraction when the latter is done on the percolation principle (e.g. carousel extraction).

A larger quantity of solid to be separated is involved in high-speed vortical extraction processes, in which comminution of the drug material very often takes place, and hence the liquid is first decanted. Separators are also used later in the process. The clarification process is thus carried out in several stages. Special investigations on the influence of the type of separator on the result of the separation have not been published for phytopharmaceuticals.

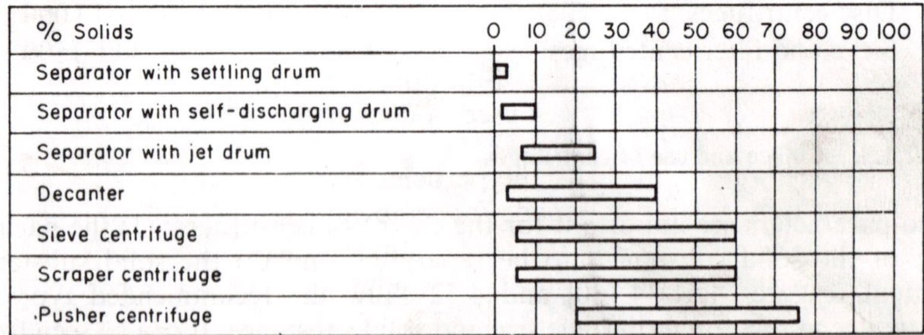

Figure 4.52 Selection of centrifuges in relation to the context of solids in the separation medium.

4.3.2.2. Decanting

Decanting is the oldest known process for clarifying liquids containing solids in suspension. Here the cloudy liquid, which in the pharmaceutical industry may be a phytopharmaceutical or miscella extract, is left to stand undisturbed and protected from light and evaporation until the solid particles have settled. As the sedimentation again approximately obeys Stokes's law, the sedimentation rate depends on the values described under Section 4.3.2.1.2. The sedimentation process can thus take several days or even weeks.

The clear supernatant liquid is then decanted off very carefully to avoid disturbing the sediment. When large volumes are being processed the supernatant liquid can be siphoned off. The liquid removed must often be left to stand again because it is still somewhat cloudy, and either decanted again or filtered. The sedimentation process, especially for the removal of the remaining cloudiness, can be accelerated with the aid of kieselguhr, cellulose flakes, etc. (see Section 4.3.2.3). The effective addition of anionic surfactants is rarely used for pharmaceuticals, as these auxiliary substances cannot be removed. The remaining sediment or drug residue must be further treated by washing and filtration or centrifugation. Decanting is a simple, basic operation requiring little apparatus in the 'solid from liquid' substance separation. Its one disadvantage is the large storage space required.

4.3.2.3. Filtration

Filtration is the separation of solids from a liquid by mechanical means with a filter. The suspension put on the filter may be a turbid liquid or a sludge; the residue left on the filter is called the filter cake and the liquid running out of the filter is the filtrate. Depending on the purpose of the filtration, we have

Clear filtration, when the clear filtrate is the product obtained.
Separation filtration, when the residue is the product obtained.
Analytical filtration, when filtrate and residue are the products obtained [4.21].

The filtration of extracts is clear filtration.
Depending on the type of filter used, a distinction is also made between:

(1) *Sieve filtration*, in which the particles are retained on the surface. However, provision is made for removal of the particles from the filter surface by some suitable means, so that a thick filter cake cannot build up. Sieve filtration includes, among others, membrane filtration and ultrafiltration.
(2) *Cake filtration*, in which the filter retains the particles on its surface like a sieve so that a filter cake is built up during the filtration process.

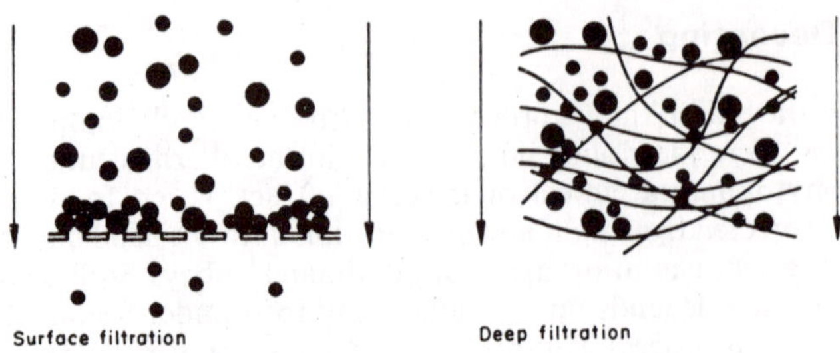

Figure 4.53 Surface filtration and deep filtration.

These types of filtration are also known as surface filtration. Figure 4.53 [4.21] illustrates the difference between surface and deep (or cake) filtration.

Ullmann [4.131] gives a detailed survey of the techniques of filtration.

Different types of filtration can also take place simultaneously on various types of filter. If the pores of a filter and of a filter cake are considered as narrow capillaries, the flow of the liquid being filtered can be described by the Hagen–Poiseuille equation. The rate at which the filtrate flows through the filter cake is then proportional to the pressure-drop Δp through the filter cake (and the filter) and inversely proportional to the thickness Δx of the filter cake (and of the filter);

$$\frac{V}{t} = \frac{r_0^4 \pi \Delta p}{8 \eta \Delta x}$$

where V = volume (m³), t = time (s), r_0 = pore radius (m), η = dynamic viscosity (Pa s), and $\Delta p/\Delta x$ = pressure drop in filter layer (bar m⁻¹).

The quantity of liquid passing through the filter per unit time is the greater, the lower the viscosity of the liquid being filtered and the thinner the filter layer. It also increases with increasing pore radius and with increasing filtration pressure. The Hagen–Poiseuille law cannot be applied quantitatively without restrictions to filtration, as too many parameters, some of which alter during the filtration, influence the process. The pore radius is not constant and the ratio $\Delta p/\Delta x$ alters with increasing filtration time and cake thickness.

These difficulties arise particularly in the filtration of extracts, e.g. muci-laginous substances, which require a greatly increased pressure even with thin filter cakes.

Filters made of cellulose fibres are mainly used for the retention of extremely small particles in the clear filtration of extracts. Filters with a loose, coarse structure and a low tendency to clogging must be used for cloudy substances. Normal clearing filters of a denser structure are also used. The sterilizing or bacteria-removing filtration of extracts is usually not possible as the filter rapidly becomes clogged.

Filtration aids (rigid granular or fibrous substances) are frequently used to counteract such difficulties, e.g. kieselguhr or aluminium silicates. They are used either as the preliminary coating of the filter, in which a suspension of the filter aid is put on the filter before the actual filtration, or stirred into the suspension which is to be filtered. Filter aids are supplied in various granule sizes and pouring densities.

The choice is governed by hourly output and filter cake thickness. Figures 4.54 and 4.55 [4.132] show a number of filter aids and their properties, and the dependence of the filtration capacity on the cake thickness.

Types, arranged in order of sharpness of filtration		Coarse		Medium	Fine		Coarse	Fine
	Seitz Ultra	Seitz Spezial	Seitz Super	Seitz Media	Seitz EF	Seitz Extra	Perlite Type B	Perlite Type A
Dry bulk density (g/l)	344	341	325	220	208	178	165	133
Specific filter cake density (g/l)	258	273	288	212	231	196	163	148
Specific filter cake thickness (mm m^{-2} kg^{-1})	3.9	3.7	3.5	4.7	4.3	5.1	6.1	6.8
Specific filter capacity at 10 mm cake thickness and 1 kp cm^{-2} (L m^{-2} h^{-1})	7900	27000	23000	65000	2200	2300	52000	24000
Mean sedimentation rate in water (cm/min)	3.4	1.2	2.9	0.5	0.4	0.2	2.0	1.2

Figure 4.54 Filtration aids.

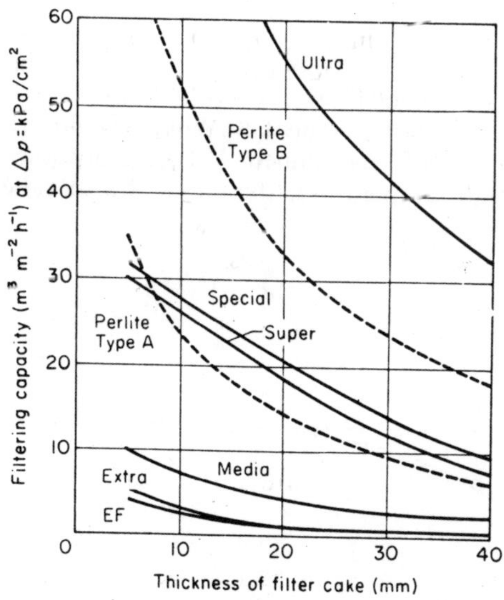

Figure 4.55 Filtering capacity of filtration aids for water at 20°C.

4.3.3. PHYSICO-CHEMICAL METHODS AND THEIR APPLICATION

4.3.3.1. Adsorption

Adsorption is the binding of a substance or a group of substances to a carrier. It can take place both on solid and liquid interfaces. In adsorption on solids a distinction must be made between adsorption of gases and that of liquids. Adsorption of liquids or, more precisely, adsorption of dissolved molecules from a solution, is important for the purification of extracts. This can be both the removal of impurities and the selective recovery of an active substance or of a concentrate from an extract.

The extent of the absorption depends on:

The chemical nature of the sorbant (the material which does the adsorbing).
The chemical nature of the adsorband (the material which is adsorbed).
The surface area of the sorbant.
The temperature.
The concentration of the dissolved substance.

The theoretical principle of the adsorption of dissolved substances produces the Langmuir adsorption equation:

$$a = a_0 \frac{c}{k + c}$$

where a = quantity adsorbed, a_0 = saturation quantity of a, c = concentration in the solution, and k = constant. When a is plotted against c a curve such as is shown in Fig. 4.56 is obtained. As the adsorption curve applies only to a certain constant temperature, the curve is also called an adsorption isotherm.

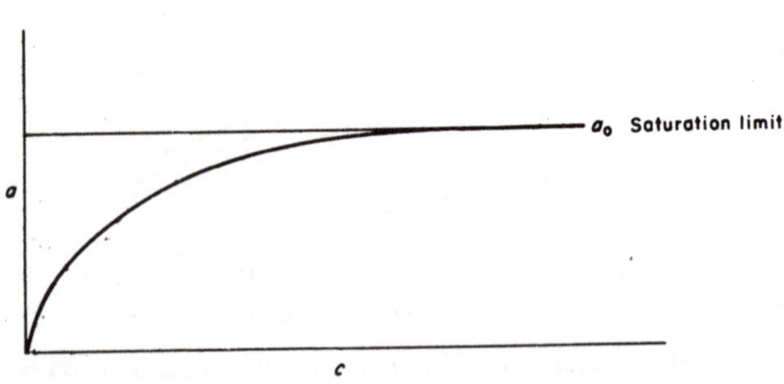

Figure 4.56 Langmuir adsorption isotherm.

The rate of adsorption is of great importance in the design of adsorption plants, the more so, the slower the adsorption process. Adsorption is generally slower from liquid than from gas. Small molecules are adsorbed more rapidly than large ones. The rate of adsorption also depends on the granule size of the sorbant.

The following substances are used as adsorbants:

Activated charcoal. This does not consist only of pure carbon. Depending on its origin (wood charcoal, brown coal (lignite), coal), it contains various impurities and ash (non-combustible components).
Silica gel.
Active aluminium oxide.
Fuller's earth.

In addition to these conventional adsorbants, other substances which are partly used for adsorption and partly for chromatographic purification are also used in the purification of extracts. In the last few years cross-linked polyvinylpyrrolidone has been used in the preparation of wine and beer because of its capacity for adsorption of accompanying impurities. In the pharmaceutical industry the use of this substance is still in its infancy. The use of Sephadex LH 20 for the chromatographic purification of hypericin has been described [4.133].

4.3.3.2. Ion exchange

Ion exchangers are used in the processing of miscellae mainly for the isolation of active substances. They are also used to a limited extent for the removal of impurities such as polyphenols [4.138].

Ion exchangers are capable of binding ionic constituents and hence removing them from a solution. There are two different types of ion exchanger: cation exchangers, which are capable of exchanging cations with H^+, and anion exchangers, which can exchange OH^-. The processes take place according to the following schemes:

$$R - H + C^+ = R - C + H^+ \text{ (cation exchanger)}$$
$$R - OH + A^- = R - A + OH^- \text{ (anion exchanger)}$$

The ion exchangers now commonly used in the pharamaceutical industry are based on polystyrene, which carries the exchanger groups on a three-dimensional framework.

The following are used as exchanger groups:

$- SO_3^- H^+$. Strongly acidic cation exchangers
$- COO^- H^+$. Weakly acidic cation exchangers
$- CH_2N^+(CH_3)_3Cl^-$. Strongly basic anion exchangers
$- CH_2NH_3^+Cl^-$. Weakly basic anion exchangers

Strongly acidic and strongly basic ion exchangers generally bind the substances to be exchanged more strongly.

Ion exchangers are particularly used in the preparation of plant medicines for the selective recovery of plant constituents. Their selectivity makes it possible to avoid other separation operations, as was shown by Winkler and Patt [4.134] with the example of saponins, which are isolated both from horse-chestnut seeds and from the leaves of *Hedera helix*. Kamphius *et al.* [4.135] describe a corresponding process for the recovery of aescin on the production scale.

Hietala and Sundman [4.136] isolated L-dopa from plant extracts using a strongly acidic cation-exchanger. Madaus [4.137] also used this ion-exchanger for the isolation of *Aesculus* saponins.

Yavich *et al.* [4.138] investigated the sorption kinetics of rutin, quercetin, tannin and gallic acid on ion-exchanger resins. These authors found that polyphenolic compounds have a high affinity for anion-exchanger resins and, therefore, can be used both for the isolation and for the purification of extracts of polyphenols. Bombardelli [4.139] suggests the use of weakly acidic ion exchanger resins for the isolation of saponins from *Panax ginseng*.

It must be stated overall that ion-exchangers are used predominantly for the isolation of pure substances, as their use just for the removal of impurities would be too expensive.

4.3.4. BACTERIA REDUCING PURIFICATION PROCESSES

The modern pharmacopoeias demand limitation of the total bacterial count as well as the elimination of pathogenic bacteria. These requirements are particularly difficult to fulfil in phytopharmaceutical preparations, as most medicinal plant materials are highly contaminated. Diding *et al.* [4.140] has published the bacterial counts on crude medicinal plant materials in a paper on the ethylene oxide fumigation of such materials. Table 4.33 shows some examples from this.

Table 4.33 shows that as well as highly contaminated drug plants there are also plants with moderate bacterial counts (Cortex frangulae) and even some with very low bacterial counts (Podophyllin and Cortex quillaiae).

It was also confirmed by serial investigations that certain groups of drug plants, e.g. the anthraquinone drug plants, are contaminated with bacteria to a lesser extent than others, as some of their constituents are converted into the bactericidally active anthranols.

The question of how the bacterial counts are altered during the preparation of phytopharmaceuticals is of special interest. Nikbacht [4.141] has attempted, with the example of the preparation of ready-to-drink active substance extracts, to monitor the variation in the bacterial counts during the preparation process and to suggest measures for reducing the bacterial contamination.

The individual processing stages have different effects on the bacterial

Table 4.33   Bacterial counts (per gram) in drug plant material

Drug plant	Incubation temperature	
	20°C	35°C
Cortex Frangulae	10.200	12.300
Cortex Quillaiae	675	1.200
Cortex Rhamni	1.000	1.950
Folia Belladonnae	825.000	640.000
Folia Digitalis	2.480.000	3.060.000
Folia Sennae	21.000	22.000
Herba Thymi	166.000	220.000
Podophyllinum	0	0
Radix Senegae	1.032.000	900.000
Rhizoma Valerianae	177.000	260.000

counts during the preparation of phytopharmaceuticals. Incorrect storage at too high humidity can result in bacterial multiplication and growth of fungal moulds. Spraying of the drug plant material with water to aid and improve its grinding also causes bacterial multiplication.

The use of water or of solvent mixtures containing water in the extraction always favours bacterial growth, particularly when the extraction is carried out at moderate temperatures or the extract is left to stand for a long time after or before further processing. The use of organic solvents and preparation stages such as concentration of a solution by evaporation reduce bacterial contamination, though this effect was regarded as insufficient for ready-to-drink active substances [4.141].

It can be seen that bacteria-reducing measures cannot usually be avoided in the preparation of phytopharmaceuticals.

Although, in principle, all the methods given in the Pharmacopoeias can be used, the possibilities are limited by the properties of the extracts.

Removal of bacteria by filtration on membrane filters, though undoubtedly the most economic method, fails because of immediate clogging of the filter. The use of layered filter-beds of the Seitz EKS type is also seldom possible in continuous operation because of the great fall-off in filtration capacity.

UV irradiation, which is used routinely for sterilizing water, has only a limited application in plant extracts, where the solid content is sometimes high and the penetration depth therefore small. Checks must also be carried out in every case to determine whether or not the high UV intensities damage the active substances. In most cases the process of pasteurization, acquired from the milk industry, is used. Ultra-high-temperature heating has also acquired a certain significance.

4.3.4.1. Pasteurization

Pasteurization is used wherever steam sterilization or tyndallization cannot be used because of instability of the preparation.

DAB 8 describes 'Pasteurization': 'The aqueous liquid to be treated is heated for at least 30 min at a temperature of at least 70°C.'

Only vegetative (multiplying) bacteria are, as a rule, killed by the process, which becomes increasingly effective with increasing temperature and heating time.

In the dairy industry various methods of heating milk are summarized under the term 'pasteurization':

> Prolonged heating, in which the milk is heated for 30 min at 62–65°C. This process now is no longer significant.
> Brief heating, in which the milk is heated for 40 s at 71–74°C. This has a 98% lethal effect on pathogenic bacteria. This method is now widely used.
> High-temperature heating, in which the milk is heated for 2–4 s at 85°C. The mortality rate for pathogenic bacteria is 99.5%.

These mortality rates in the pasteurization of milk apply only to pathogenic bacteria. Spores and heat-resistant microorganisms have low, and sometimes no noticeable, mortality under the given conditions.

4.3.4.2. Ultra-high-temperature heating

All processes which give a sterile product come under this heading. The term 'uperization' is also frequently used. The product is first preheated to ~ 80°C and then either heated at 135–140°C for 6–10 s in the 'indirect' method or at 145–150°C in the 'direct' method. In the indirect method the product does not come into direct contact with the steam used for the heating, whereas in the direct method the superheated steam is forced into the product and then drawn off again after the required exposure time.

Our own experiments on suspensions with *Bacillus stearothermophilus* as test bacterium have shown that the product was sterile after an uperization time of 2 s at 150°C in the direct method, whereas a temperature of 140°C was not sufficient to kill all these heat-resistant organisms.

When releasing the pressure *in vacuo* after completion of heating in the direct uperization process care must be taken to ensure that volatile components of the extract cannot be lost when the water added during the steam-sterilization evaporates again. This method is not suitable for extracts containing alcohol and/or ethereal oils.

4.4. Concentration of miscellae

4.4.1. DEFINITIONS

Concentration means increasing the proportion of a substance or solute in solution by evaporation or vaporization without converting the product to the dry state, though it might in the end have a very high viscosity. Evaporation is defined as the transition from liquid to gaseous state in the presence of other gases, e.g. air, in an evaporation chamber. In the case of vaporization, on the other hand, only molecules of the solvent are present in the chamber. Figure 4.57 shows the differences schematically.

Evaporation is a term little used when describing the concentration of extracts, 'boiling down' and 'vaporization' being the common descriptions. Boiling down is used when the object is the recovery of a solid or of a concentrate with a high solid content, whereas vaporization is used for the recovery of the solvent.

Billet [4.142] has made a detailed survey on the field of vaporization in his book *Verdampfung und ihre technische Anwendung* (*Vaporization and its Technical Application*) published in 1981.

Heat transfer in vaporization can take place either by conduction, convection or radiation.

Conduction plays a part only in the spreading of heat in solids. It takes place only in very thin films of liquids and gases (film vaporization).

Convection is the predominant means of heat transfer in liquids and gases. The heat is transported by the movement and mixing of layers of fluid of different temperatures.

Radiation of heat does not require any transfer medium and therefore has no connection with any material carrier medium. It plays only a minor rôle in vaporization. The separation of the components in vaporization takes place as a result of differences in vapour pressure. These must be large enough to make separation by vaporization possible. If this is not the case a modification must be carried out, i.e. a countercurrent distillation in which some of the condensate flows back against the vapour in order to produce better separation.

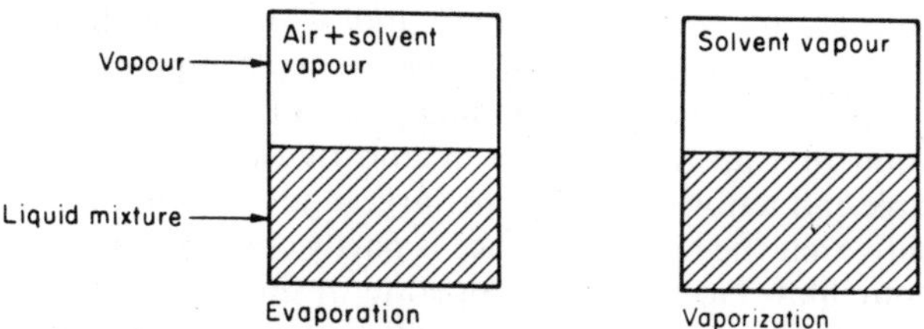

Figure 4.57 Difference between evaporation and vaporization.

4.4.2. PRINCIPLES OF VAPORIZATION

'It has so far been impossible to calculate the theoretical heat transfer from the solid wall of a vessel to a non-boiling liquid. It has been even less feasible to calculate, purely theoretically, the heat transfer to a boiling liquid—which is undergoing phase change—and to express this in generally valid equations' [4.143].

This statement explains why so many different types of vaporizer now exist.

All theoretical statements made so far produce equations of limited validity. The boiling-down of extracts is particularly problematical here, as their specific heats are as a rule unknown and also cannot be reliably determined because of the varying composition of the starting solution. The driving force of vaporization is the temperature difference between the heating vapour and the boiling point of the solution to be vaporized. They are separated from one another by the wall of the vaporizer. Heat is transferred from the heating vapour, the temperature of which must be above the boiling point of the liquid, through the wall to the solution. Vaporization of a liquid takes place only at the interface between liquid and vapour (gas). This interface can be a liquid surface or a vapour bubble. The frequency of production of vapour bubbles in a liquid increases with increasing energy supply. This can be explained by the activation of 'defective spots' on a solid wall brought about by a rise in temperature. Very smooth surfaces without defective spots do not induce bubble formation, with the result that the solution becomes overheated and its boiling is delayed.

There are four different regions of vaporization (Fig. 4.58). The Figure shows the course of the coefficient of heat transfer α and of the heating surface loading q^* in relation to the temperature difference between the heated vaporizer wall and the boiling-temperature of the liquid.

(1)   The region of free convection. This region occurs with small temperature differences between the wall and the boiling point of the liquid. No bubbles are formed, i.e. free convection prevails. The equations for free convection apply here. Vaporizers do not work in this region.

(2)   Region of bubble vaporization. Practically all vaporizers operate in this region. Many bubbles are formed on the heating surface. The coefficient of transfer of heat and with it the heating-surface loading, q^*, increases with increasing temperature difference between the wall and the boiling point of the liquid.

(3)   Region of transition. In the region of transition the number of bubble-induction points is so large and the bubbles develop so rapidly that the heating surface is partly covered with an insulating vapour film. No vaporizers operate in this region because of the greatly reduced heat transfer coefficients which result.

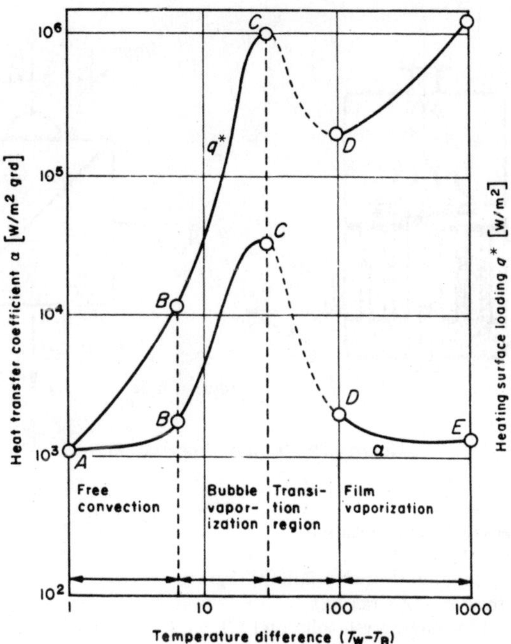

Figure 4.58 Variation of heat transfer coefficient and heating surface loading q* with temperature difference $T_W - T_B$, for boiling water at atmospheric pressure. T_W is the wall temperature and T_B is the boiling point (temperature) of the liquid (water, 100°C).

(4) Region of film vaporization. A stable, insulating vapour film is produced on the heating elements by further increases in the heating-surface temperature and hence in the difference from the boiling point of the liquid. This causes a decrease in the coefficient of heat transfer, α. The heating-surface loading in this region increases as a result of the greater difference in temperature between the vaporizer wall and the boiling-point of the liquid. Film vaporization plays no part in this technology because of its very small coefficients of heat-transfer. So called film vaporizers derive their name from the thin liquid (not vapour) film produced for example by fast-running rotors on the heated wall of the vaporizer.

4.4.2.1. Procedures in simple vaporization

The basic processes taking place in vaporization and the quantity and energy balances shall be examined more closely with the example of simple vaporization taken from the literature [4.142] and illustrated in Fig. 4.59. The

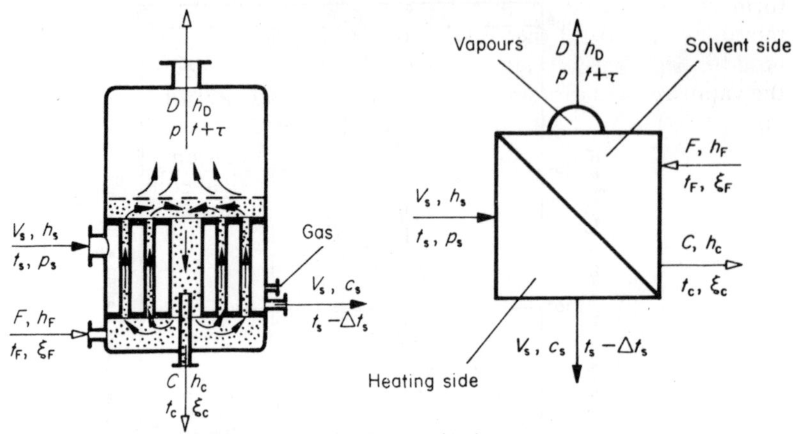

Figure 4.59   Simple vaporization.

V_s	steam input (kg/h)
h_s	enthalpy of steam (kJ/kg)
t_s	steam condensation temperature (°C)
p_s	steam pressure (bar)
F	supplied fresh solution (extract solution) (kg/h)
h_F	enthalpy of fresh solution (kJ/kg)
t_F	temperature of incoming fresh solution (°C)
ξ_F	solid content of fresh solution (% by weight)
D	weight of vaporized solvent (kg/h)
h_D	enthalpy of vaporized solvent (kJ/kg)
p	vapour pressure in the vapour space (bar)
$t+\tau(=t_c)$	boiling temperature measured in the vapour space. This is the sum of the boiling temperature t of the pure solvent and the boiling point rise τ caused by the solution
C_s	specific heat capacity of steam condensate (J/K)
$t_s-\Delta t_s$	exit temperature of condensed steam (°C). The temperature of the condensed steam must of necessity be lower than the temperature of the steam, as the steam has given up energy to the solution
C	concentrated/thickened solution leaving the vaporizer
h_c	enthalpy of thickened solution (kJ/kg)
t_c	temperature of thickened solution (°C)
ξ_c	solid content of thickened solution (wt %)

left-hand side of Fig. 4.59 shows the processes taking place in the vaporizer and the right-hand side the corresponding quantity and energy balances of the heated and the solution sides of the vaporizer wall, and of the liquid being evaporated (so-called vaporizer symbol). This type of vaporizer is a multi-tube vaporizer, which is one of the most frequently used types. It consists of a series of parallel heated tubes arranged vertically and up to 5 m in length. The diameter of a tube is normally about 1/25th of its length. In the centre is the wide fall-tube necessary for the automatic circulation of the liquid. The solution which is to be concentrated enters the tubes from the bottom and is heated above its boiling-point by hot vapour. The vapour bubbles which form rise in the tubes and carry liquid up with them. A liquid–vapour mixture issues from the tops of the tubes and separates into vapour and liquid in the space above the liquid being concentrated. The vapour leaves the vaporizer under pressure and is led into condensers or preheaters and the liquid flows down again through the fall-tube.

The object of the vaporization is normally to concentrate a solution F of an initial concentration ξ_F in % by weight to the final concentration ξ_C in % by weight at a rate of x kg/h. The quantity of solvent which is thus to be evaporated off is

$$D = F\left(1 - \frac{\xi_F}{\xi_C}\right) \tag{1}$$

The equation states that the more the solution is concentrated, i.e. the greater the concentration in the end-product, the smaller will be the fraction ξ_F/ξ_C and the larger will be the quantity of solvent which will have to be evaporated off.

The second factor involved is the quantity of heating-vapour V_s required to vaporize the pertinent quantity of solvent. This is calculated by the formula

$$V_H = \frac{D(h_D - h_C) + F(h_C - h_F) + Q_1}{h_s - C_s(t_s - \Delta t_s)} \tag{2}$$

Q_1 is the sum of all heat losses which occur.

The following quantities of heat go into the required quantity of heating vapour:

> The quantity of heat $D(h_D - h_C)$ required for the vaporization.
> The quantity of heat $F(h_C - h_F)$ required for heating the solution to boiling temperature.
> The heat losses Q_1.

We shall illustrate the calculation with an example [4.142].

Numerical example. $F = 20\,000$ kg/h of water containing dissolved substance at $\xi_F = 0.12$% by weight is to be boiled down in a single-stage vaporizer at atmospheric pressure to a final content of $\xi_C = 1.2$% by weight. Exhaust steam at $p_s = 3.5$ atmospheres is available as heating vapour. The values to be determined are the quantity of water boiled off and the consumption of heating steam for the two cases when, on the one hand, the original aqueous solution enters the apparatus at a temperature $t_F = 95°$C and, on the other hand, at only $t_F = 20°$C. The drop in temperature Δt_s occurring upon condensation in the vaporizer shall be ignored.

Solution. According to equation (1) the quantity of water evaporated

$$D = 20\,000\left(1 - \frac{0.12}{1.2}\right) = 18\,000 \text{ kg/h.}$$

When the original water flows into the apparatus preheated to $t_F = 95°C$, equation (2), using the enthalpy values which can be read off from the VDI steam tables [4.7.144] at the given pressures and temperatures, gives the heat consumption

$$V_s = \frac{18\,000\,(2675 - 418.7) + 20\,000\,(418.7 - 397.8)}{2730 - 4.187 \cdot 138} = 19\,000 \text{ kg/h},$$

which corresponds to a specific steam consumption of

$$d_s = \frac{V_s}{D} = \frac{19\,000}{18\,000} = 1.06 \text{ kg heating steam/kg vapour.}$$

In contrast, the steam consumption calculated for incoming solution at 20°C is

$$V_s = \frac{18\,000(2675 - 418.7) + 20\,000(418.7 - 83.7)}{2730 - 4.187 \cdot 138} = 22\,000 \text{ kg}$$

and the specific steam consumption is

$$d_s = \frac{22\,000}{18\,000} = 1.22 \text{ kg heating steam/kg vapour.}$$

The example shows that heating steam can be saved by preheating the solution to be vaporized. In a single-stage vaporization plant energy savings can be made only by preheating the fresh solution by the recycled liquid being evaporated. Figure 4.60 [4.142] shows a scheme of such a plant. Some of the liquid being vaporized is led into a condenser, where it preheats the fresh solution. Such plants are limited to small capacities and to those

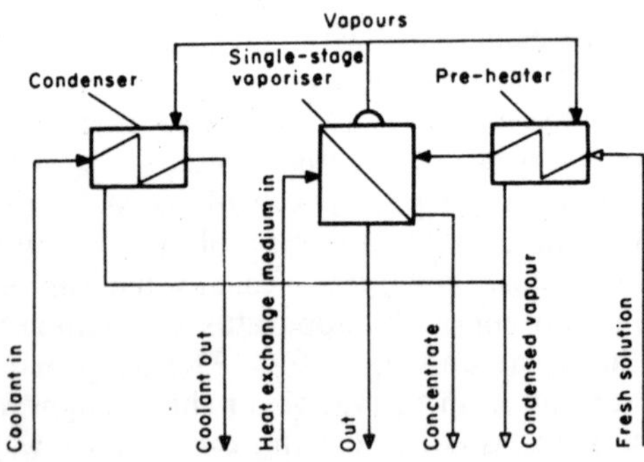

Figure 4.60 Scheme of a single-stage vaporization plant with pre-heating of the fresh solution.

problems where the product must not be exposed to high temperatures for long periods, as is the case in multistage plants. An energy saving of about 13% is achieved by using the condensation of the vaporized liquid. Multistage vaporization and compression of the vaporized liquid are two more ways of making further energy savings.

4.4.2.2. Multistage vaporization

The example in Section 4.4.2.1 has shown that approximately 1 kg heating-steam is required for the single-stage vaporization of 1 kg of solution. In multistage vaporizers the vaporizing liquid from one vaporizer is used as heating medium for a second vaporizer, the vaporizing liquid from which in turn heats a third vaporizer, etc. This produces multistage vaporization plants. The steam or vapour consumption figures of the successive stages are $1:1$, $1:\frac{1}{2}$, $1:\frac{1}{3}$ etc. An ever-decreasing saving of steam or vapour thus occurs with an increasing number of stages. Figure 4.61 shows this in the form of a graph [4.145].

The decrease in the amount of energy required per kg of solution is offset by an increase in the required heating area and hence by an increase in plant costs (Fig. 4.61). As the plant costs increase approximately in proportion to the number of stages, plants with more than six stages are even now still rare. Figure 4.62 [4.145] shows a scheme of a multistage plant. Multistage plants can operate in continuous flow, in countercurrent or in parallel flow, as is shown in Fig. 4.63.

In practice, continuous flow supply, in which the liquid is pumped first into the stage with the highest pressure, is the type most commonly used. The solution then flows in the direction of the flow of vapour from the vaporizing liquid from stage to stage without being subjected to pressure, and leaves the plant as a thickened extract. In countercurrent supply the solution enters the

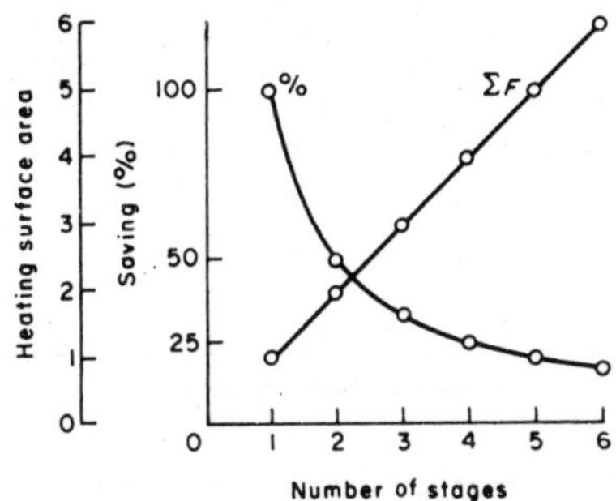

Figure 4.61   Decrease in theoretical specific steam or vapour consumption (%), and increase in approximate total heating area, with the number of stages.

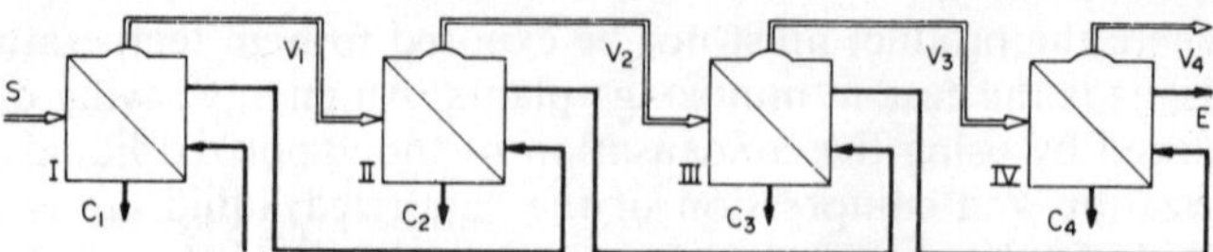

Figure 4.62 Scheme for a four-stage evaporation plant. The volume of vapour formed in each stage is approximately equal to the volume of the heating system. Thus the (theoretical) consumption of steam is of the order of a quarter of the total amount of water vaporized. The patterns of heat flow are similar from stage to stage. V, vapour; S, steam; C, condensate; E, extract.

last stage with the lowest pressure and is pumped from stage to stage against the flow of the liquid being evaporated. In parallel supply the solution is divided into individual fractional streams, which are conveyed parallel to the individual stages. Continuous flow supply requires only one pump and moreover has the advantage that the most highly concentrated solution is exposed to the smallest amount of heat.

Countercurrent supply has the advantage that the highly concentrated solution vaporizes in the stage with the highest temperature, where its viscosity is lower and thus the heat-transfer is better. One disadvantage here is that a separate pump is required for each stage.

Parallel flow supply is used where the solubility limit is passed during boiling down, as stagewise removal of the crystallized substance is not possible.

4.4.2.3. Vapour compression

In addition to multistage vaporization, vapour compression has increased greatly in importance since the beginning of the energy conservation cam-

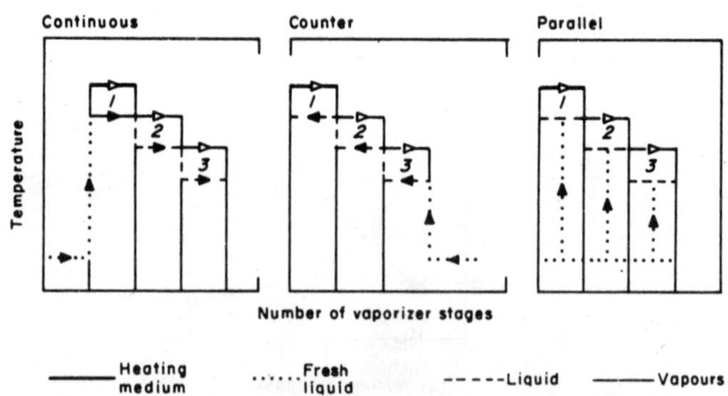

Figure 4.63 Qualitative temperature scheme of a three-stage vaporizer plant for continuous, countercurrent and parallel flow of liquid and steam.

paign. Its mode of action is characterized in that some or all of the vapour from the chamber of a vaporizer is drawn off and compressed under high pressure. The compressed vapour is then led back into the same vaporizer as the heating vapour. It can also be used in single-stage vaporizers and in favourable cases it can make it unnecessary to use multistage evaporators. The compression is achieved either thermically by steam-jet compressors or mechanically by turbo-compressors.

Steam-jet compressors. The steam-jet compressor has acquired great practical importance. Figure 4.64 [4.142] shows the scheme of this apparatus.

The compressor consists of a head with a jet driven by fresh steam and of a mixer-nozzle with diffuser. The fresh steam expands in the jet and is mixed in the mixer-nozzle with the vapour which has been drawn off from the vaporizing liquid. The mixture passes at high velocity into the diffuser, where it is slowed down by increased pressure.

Steam jet compressors work better when the pressure-drop in the jet is large and the increase in pressure exerted in the diffuser is small. Under favourable conditions 1 kg of vapour can be drawn off per kg of forced steam. This results in 2 kg steam-vapour mixture, with which 2 kg water can be vaporized. The principal advantages of steam jet compressors are that they contain no moving parts and can be made of a wide variety of materials, which makes them useful for tackling a wide range of problems.

Mechanical turbocompressors. Turbocompressors are used mainly for compressing large quantities of vapour or steam. They are driven by electric motors or steam turbines. Whereas often only part of the vapour can be compressed with steam jet compressors, all of the vapour is compressed with turbocompressors. Their disadvantage lies in their high initial cost.

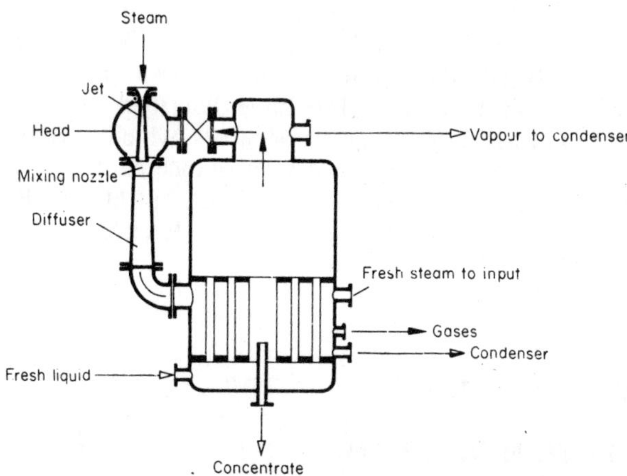

Figure 4.64 Single-stage vaporizer plant with steam jet vapour compressor.

4.4.3. USE OF VAPORIZERS

Vaporizers are widely used in the manufacture of drug extracts. The extraction is normally followed by a concentration stage. This concentration can have the following objectives:

> Concentration to a specified dry substance content.
> Preparation of spissum extracts.
> Preparation of concentrates as intermediate stages for medicines.
> Concentration as a preliminary stage for the preparation of dry extracts.

A concentration which preserves the product is especially important for many pharmaceutical products. *DAB 8* demands a bath temperature of 70°C for this operation, where the temperature of the product must not exceed 50°C. Short exposure to as low a temperature as possible is generally required if the drug is to be conserved during preparation of the extract.

In addition to the sensitivity of many active substances to heat, foaming and deposition of resins in drug extracts in vaporizer plants cause particular difficulties. The often differing batch sizes of various products and the great diversity of products represent a further problem.

Various vaporizers have proved particularly useful here for the pharmaceutical industry:

> Tube vaporizers with automatic recycling.
> Tube vaporizers with forced recycling.
> Falling film vaporizers.
> Thin layer vaporizers with rotary installations.
> Centrifugal rotary vaporizers.

The types of vaporizer described here are discussed in Chapter 5. Publications dealing specially with the problem of the concentration of pharmaceutical extracts are rare. Visher *et al.* [4.146] have described the boiling down of plantain extract in centrifugal rotary evaporators, which have become more used in recent times. Extracts with a dry substance content of 1–4% were used as starting materials here and produced concentrates of 40–60%. The yield of concentrate was investigated in relation to temperature and throughput.

4.5. Drying of extracts

4.5.1. DEFINITION OF DRYING

Drying generally means the removal of a solvent from the material being dried. Here solvents can be water or any other volatile liquid. The dry material is produced either as powder or as a dry, friable mass, depending on the drying process used. There are various types of drying.

Evaporation drying. The material is dried at temperatures below the boiling point of the solvent or solvent mixture to be removed. The liquid is removed purely by evaporation, i.e. other gases as well as the solvent molecules are present in the vapour space. The process can be accelerated by application of a partial vacuum, where the pressure reduction must be controlled so that no boiling takes place.

Vaporization drying. The material is dried at or above the boiling-temperature of the solvent. The liquid comes off by vaporization, i.e. only molecules of the solvent are present in the vapour chamber. The temperature can also be lowered here by reduction of the pressure.

Sublimation drying. Here the solvent comes off not from the liquid but from the solid phase by sublimation. Figure 4.65 [from Ref. 4.147] illustrates the processes with the example of water. Below the triple point, water molecules come directly out of ice by sublimation. At the triple point ice, water and water vapour coexist. Above the triple point the liquid evaporates with rising temperature up to the boiling point. Above the boiling point pure vaporization takes place.

Another classification of the types of drying is by type of heat supply.

Convection drying. A hot gas (usually air, occasionally superheated steam) transfers the heat to the material being dried. It thus absorbs the moisture and takes this away. The air can flow either over stationary material (e.g. in a vortical laminar/layer dryer). Belt dryers, for example, assume an intermediate position; in these, air flows over a layer of material in which the particles remain stationary in relation to each other but the whole of which is at the same time moving.

Contact drying. Here the heat is supplied to the material via heated plates. Examples of contact drying are roller drying and freeze-drying, in the latter only when the heat is supplied via heated plates.

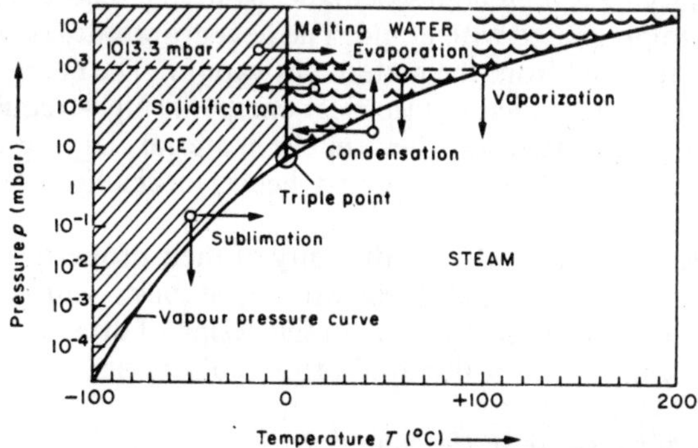

Figure 4.65   Section from the phase diagram of water (logarithmic pressure scale).

Radiation drying. The energy necessary for the drying is supplied to the material as radiation. Here the energy is transmitted in the form of electromagnetic waves through a vacuum or in a chamber filled with radiation-permeable gases. Generally only infrared drying at wavelengths of 0.8–400 µm is included in the literature under the heading of 'radiation drying'. The high surface temperatures produced are a disadvantage for the drying of plant extracts. It is therefore mainly used for water content determination. In addition to IR radiation, microwaves with wavelengths of 0.04–25 cm must also be included for radiation drying. In addition to the drying of foils and other strip-shaped materials, microwaves have also recently come into use for drying extracts. A frequency range of 2450 MHz ± 50 MHz is used for industrial purposes.

Dielectric drying. The moist material is placed as a dielectric between two capacitor plates in a rapidly changing electric field. As the dielectric constant of solids is usually much smaller than that of water, the water dipoles are set in oscillation by the high-frequency alternating field. Heating takes place evenly through the whole product. The higher the water content, the stronger is the heating. The frequency range used extends from 2 to 100 MHz. The high-frequency drying of tablets has been described by Schepky in the pharmaceutical literature [4.148]. The Joule heat produced in an ohmic resistance by an alternating current, also called current heat, also comes under this heading, though it is generally not used for drying.

Another type of classification of various types of drying is by name of the apparatus used: spray drying, freeze drying, lattice (wickerwork) drying, vortical layer drying, vacuum drying, etc. As these designations can all be traced back to basic types of drying, they will not be used as classification criteria.

4.5.2. PRINCIPLES OF DRYING

The principles of drying have been described in detail by a number of authors [4.149–4.151]. Drying is presented schematically in Fig. 4.66. Heat is supplied to the material impregnated with liquid. The heat evaporates or vaporizes the liquid, which is released from the material as vapour. The energy can also be supplied in other forms, such as electrical energy in high-frequency drying the energy being converted into heat in the material. Drying is always an intensely energy-consuming process and hence mechanical drying is preferred where possible.

The multiplicity of materials, the diversity of their properties (i.e. consistency, hygroscopicity, etc.) and their varying stability during the drying process have resulted in a great variety in the design of dryers. The following parameters are important for the total process of drying:

The types of moisture in the material.
The motion of moisture in the material.

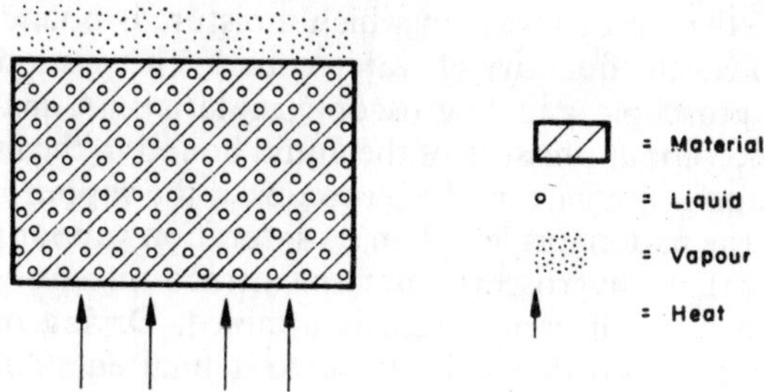

Figure 4.66   Schematic representation of drying.

The properties of the moist air.
The h,X graph of the moist air.
The course and outcome of the drying.

The introduction of the term 'moisture' implies that the water here is removed as liquid. This is also the medium to be removed by drying in the majority of cases. The laws of drying are applicable to other liquids, though the necessary calculations must then be made with the physical constants of the particular solvent. The Mollier h,X graph applies only to moist air and hence to the removal of water as solvent.

4.5.2.1.   Types of moisture in the material

Schicht [4.151] defines four ways in which moisture is bound in the material:

(1)   Adsorbed liquid. This forms a coherent film of liquid on the external surface of the material. Its vapour pressure equals the saturation pressure of the liquid.
(2)   Capillary liquid. This is liquid retained in the pores of porous substances which is driven to the surface by capillary forces during drying. Whereas in many substances the vapour pressure of the capillary liquid equals its saturation pressure (non-hygroscopic behaviour), it is lower than this in other substances below a critical moisture content. The material is then hygroscopic.
(3)   Swelling liquid. This makes the material swell up and hence undergo an increase in volume. Whereas absorbed and capillary liquid wet only the external and internal surfaces of the material, swelling liquid is an actual constituent of the material phase, which it completely permeates.
(4)   Chemically bound liquid (water of crystallization). Its removal is possible only above the decomposition temperature of the material and is generally no longer described as drying.

In addition to the various ways in which moisture is bound, the material itself also influences the quantity of water bound. There is a difference here between non-hygroscopic and hygroscopic materials. A material is non-hygroscopic if the vapour pressure of the liquid bound to the material equals the saturation vapour pressure, and hygroscopic if the vapour pressure of the liquid bound to the material is less than its saturation vapour pressure. This means in practice that a hygroscopic material attracts moisture out of a moist airstream until a state of equilibrium is attained. Drying of hygroscopic materials is therefore possible only down to a final equilibrium moisture content which is dependent on the vapour pressure of the liquid bound to the material.

4.5.2.2. Movement of moisture in the material

The motion of moisture in the material during drying can take place in two ways:

> Liquid movement, caused by capillary and surface forces.
> Vapour movement, produced by the partial pressure gradient in vapour-filled pores of the material.

Only empirical mathematical formulae exist for the quantification of these two types of movement of moisture [see Ref. 4.151], which is attributable to the unknown size distribution of the capillaries in the material being dried.

The behaviour of a wetting liquid in a capillary can be expressed by the formula

$$H = \frac{2\sigma}{\rho_L g r_0}$$

where H = rise of liquid in a capillary (m), σ = surface tension (N/m), ρ_L = density of liquid (kg/m^3), g = acceleration due to gravity (9.81 m/s^2), and r_0 = pore radius (m).

It can be deduced from this formula that the rise H in the capillary for a wetting liquid will be the greater, the smaller the radius r_0. This means in practice that during drying the small capillaries exert a large 'evacuating suction' force which constantly drives fresh liquid to the surface of the material.

The capillary condensation, as well as the above-mentioned parameters, is important for the value of the rise H of the liquid in a capillary. This is observed as soon as the partial pressure of the water vapour in the carrier gas approaches its saturation vapour pressure. A film of liquid, which becomes thicker with increasing partial pressure of water vapour, thus appears in the capillaries at points on the capillary wall where certain areas of molecules strongly adsorb layers of water vapour. The next stage is the formation of a concave liquid meniscus with reduced vapour pressure at the bottom of the capillary. Further vapour consequently condenses there, whereupon the level of liquid in the capillary rises. This reduces the curvature of the meniscus and hence the tendency to condensation. Equilibrium is established when the vapour pressure above the meniscus is equal to the partial pressure of the surrounding medium. The conditions are described quantitatively by Kneule [4.149].

Four different types of motion of the vapour in the material can be distinguished [4.147]:

Molecular motion is defined by Knudsen. When a material is dried under high vacuum, as is the case, say, in freeze-drying, the water vapour molecules in the pores of the material collide with each other much less frequently than with the walls of the pores. They rebound elastically from the pore-walls and follow paths determined solely by the angle of impact (Fig. 4.67).

Laminar vapour flow. At high pressures the vapour molecules collide with one another considerably more frequently than with the pore-walls and the laws of classical flow theory apply. The small diameter of the pores frequently induces laminar flow, in which the individual particles flow in an orderly manner in relation to each other.

Turbulent vapour flow. This is characterized by the predominance of forces of inertia over the forces of viscosity. The vapour moves in the form of eddies of varying size. Turbulent flow is observed in narrow pores only in quite extreme cases. However, turbulent-type flow of vapour is a frequent occurrence.

Turbulent type vapour flow. The many constrictions, expansions and changes in direction in the pores of a material cause the forces of inertia to come into effect when laminar, i.e. viscosity controlled flow, would otherwise have been prevalent.

4.5.2.3. Properties of moist air

In most cases air serves as a carrier medium for the moisture being removed

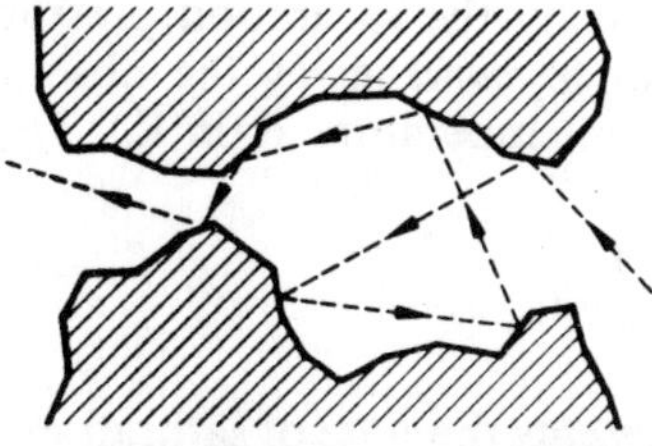

Figure 4.67   Path of a water vapour molecule in a pore of material in Knudsen flow.

in drying. The heat required for the drying is also often supplied via air. For this reason the properties of moist air will be examined. The following factors are important for drying: absolute moisture content, relative humidity, density and enthalpy.

4.5.2.3.1. Absolute moisture content

According to Grassmann [4.151] it is advantageous to choose dry air as reference value rather than moist. The mass of dry air remains constant when it flows over the moist material. The *absolute moisture content X* of the air is therefore given as loading in the form

$$X = M_{\mathrm{wv}}/M_{\mathrm{A}}$$

where M_{wv} = mass of water vapour (kg) contained in the moist air, and M_{A} = mass of dry air (kg).
   With

$$\frac{P_{\mathrm{wv}}}{P_{\mathrm{A}}} = \frac{P_{\mathrm{wv}}}{P - P_{\mathrm{wv}}} = \frac{M_{\mathrm{A}}}{M_{\mathrm{wv}}} X$$

where P = total pressure (bar), P_{wv}, P_{A} = partial pressure of water vapour and of air (bar), and M_{wv}, M_{A} = molar mass of water vapour and of air (kg/kmol).

we obtain $X = \dfrac{M_{\mathrm{wv}}}{M_{\mathrm{A}}} \cdot \dfrac{P_{\mathrm{wv}}}{P - P_{\mathrm{wv}}}$

M_{wv} = 18 kg/kmol and M_{A} = 29 kg/kmol apply specially to the system of water vapour/air, hence

$$X = \frac{18}{29} \cdot \frac{P_{\mathrm{wv}}}{P - P_{\mathrm{wv}}} = 0.622 \frac{P_{\mathrm{wv}}}{P - P_{\mathrm{wv}}}$$

The maximum quantity of water vapour which air can absorb at a certain temperature is determined by the saturation pressure P_{wv} existing at this temperature. The *saturation curve* gives the absolute vapour content of saturated air in relation to temperature T (in K). The equation

$$X_{\mathrm{S}}(T) = 0.622 \frac{P_{\mathrm{wv}}(T)}{P - P_{\mathrm{wv}}(T)},$$

in which the index S represents the state of saturation, therefore applies here.

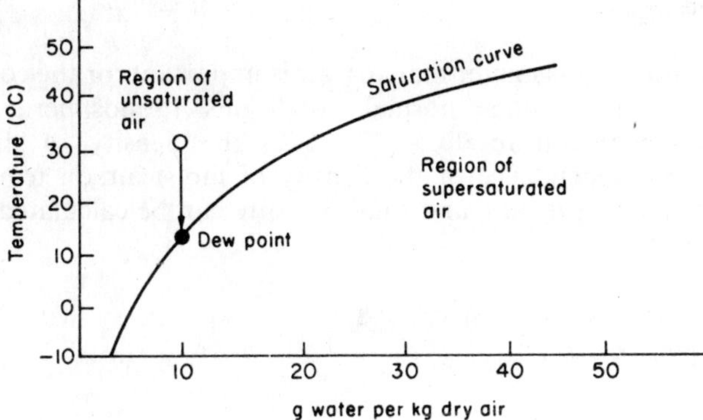

Figure 4.68   Capacity of air to absorb water vapour.

The capacity of air to absorb water vapour increases with increasing temperature. The conditions are illustrated in Fig. 4.68.

When unsaturated air is cooled, the dew-point is reached at a certain temperature and water begins to condense in the form of droplets. A definite dew-point exists for each temperature and each absolute moisture content. All dew-points together form the saturation curve.

4.5.2.3.2.   Relative humidity

The relative humidity φ is defined by the ratio of the partial pressure of the water vapour to the saturation pressure at the temperature T.

$$\varphi = \frac{p_{wv}}{P_{wv}} = \frac{X}{X_s} \cdot \frac{p - p_{wv}}{P - P_{wv}}$$

where p_{wv} = partial pressure of water vapour (bar), P_{wv} = saturation pressure (bar), X = absolute moisture content of air (kg/kg), X_s = saturation moisture content of air (kg/kg), and P = total pressure (bar).

On the basis of this definition the values for the relative humidity must run from zero to one. The relative humidity is frequently expressed as a percentage:

$$\varphi = \frac{p_{wv}}{P_{wv}} \times 100 \ (\%).$$

4.5.2.3.3. Density

The fact that moist air is lighter than dry air is important for the conveyance of air in a dryer. Under normal conditions (atmospheric pressure 1013.25 mbars, temperature $0°C = 273.15\,K$) the density of dry air is $1.293\,kg/m^3$. The dependence of the density of moist air on temperature, partial water vapour pressure and total pressure can be calculated with the following formula:

$$\rho = 1.293 \cdot \frac{273.15}{273.15 + \delta} \left(1 - \frac{0.378 \cdot P_{wv}}{P} \right)$$

where ρ = density of air (kg/m^3), δ = temperature (°C), P_{wv} = partial pressure of water vapour (bar), and P = total pressure (bar)

4.5.2.3.4. Enthalpy

Enthalpy is the product of specific heat and temperature:

$$h = c_p T$$

where h = enthalpy (J), c_p = specific heat ($J\,kg^{-1}\,K^{-1}$), T = temperature (K)

As the temperature here is the absolute temperature, the enthalpy for any temperature above 0 K is greater than zero. However, for technical purposes it was agreed that the enthalpy for dry air at 0°C shall be zero:

$h = 0$ for dry air at 0°C.

The enthalpy of moist air is composed of the proportions of dry air and water vapour:

$$h = h_A + H\,h_{wv}.$$

When $h_A = C_A T$ and $h_{wv} = X\Delta H_v + X C_{wv} T$ the enthalpy h of moist air is

$$h = C_A T + X(\Delta H_v + C_D T),$$

where C_A = specific heat of air, C_{wv} = specific heat of water vapour, X = absolute moisture content, ΔH_v = heat of vaporization of water, and T = temperature.

In this way of considering enthalpy the variation of specific heat with temperature is negligible.

4.5.2.3.5. The Mollier h,X graph of moist air

In 1923 Mollier proposed a right-angled graph with the enthalpy h as ordinate and the absolute water content X as abscissa. In this type of graph the range of unsaturated air of interest in drying is restricted to a small section (Fig. 4.69). For greater clarity the graph is now used as an oblique-angled coordinate system (Fig. 4.70), with the region of unsaturated air extended, and hence the graph is more accurate. In the oblique-angled graph the actual ordinate is formed by the temperature T, where the line for 0°C coincides with the horizontal abscissa. The essential components of the h,X graph are shown in Fig. 4.71.

Either the values for the enthalpy h or the temperature values or both are plotted on the ordinate. The horizontal abscissa, which coincides with the temperature line for 0°C, contains the values for the absolute moisture content. The saturation line is made prominent as a bold curve. The lines of

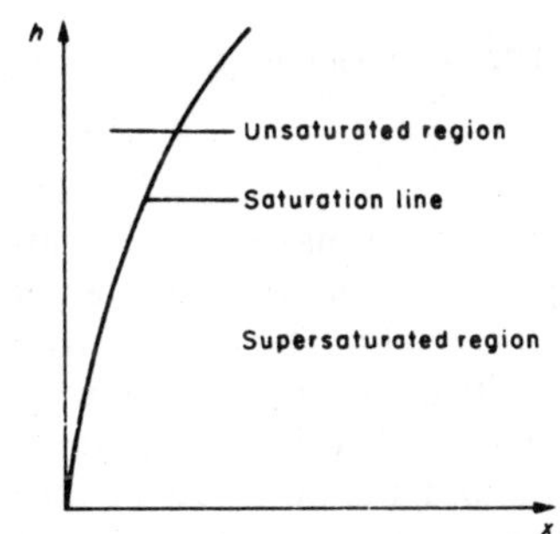

Figure 4.69   Right-angled Mollier h,X graph (schematic).

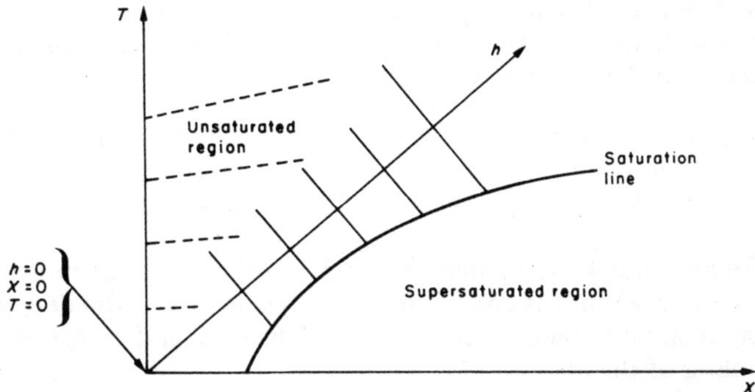

Figure 4.70   Principle of Mollier h,X graph.

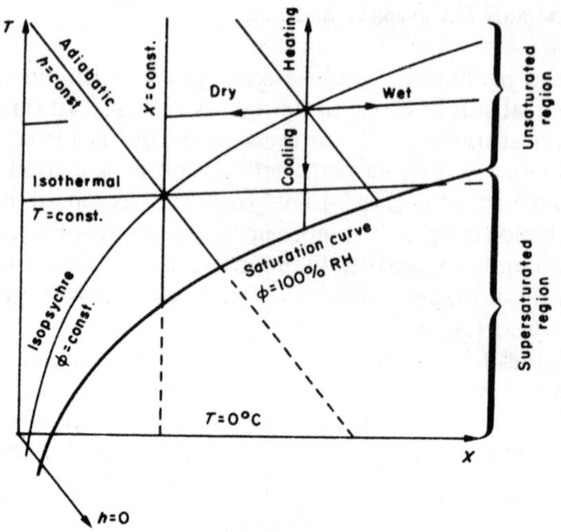

Figure 4.71   Components of Mollier h,X graph.

equal enthalpy (also known as isenthalps or adiabatic curves) run from the upper left to the lower right. The lines of equal temperature (isotherms) run from the ordinate from left to right. The lines of equal relative humidity are shown as a group of curves (isopsychres) above the saturation curve).

Depending on the region used, h,X graphs are used for different regions of absolute moisture content and temperature in order to obtain optimum accuracy in the region concerned. Graphs for temperatures up to about 150°C and absolute moisture contents up to 150 g/kg of air are useful in drying technology. Figure 4.72 [4.147] shows such a graph.

The graph applies to an atmospheric pressure of 980 mbars, which corresponds to the mean value for a town 200 m above sea-level. The processes important for drying, namely warming of moist air, cooling of moist air and mixing of two air-streams, can be clearly and quantitatively represented with the h,X graph.

Warming and cooling of moist air. Figure 4.73 shows the processes involved in the warming and cooling of moist air [4.152].
The alterations occurring here can be summarized as follows:

(1)   The absolute water content of air does not alter when moist air is heated or when it is cooled down to the vicinity of the dew-point.
(2)   The relative humidity decreases upon heating and increases upon cooling of the air.
(3)   The enthalpy content of the air is increased by the energy supplied during heating.

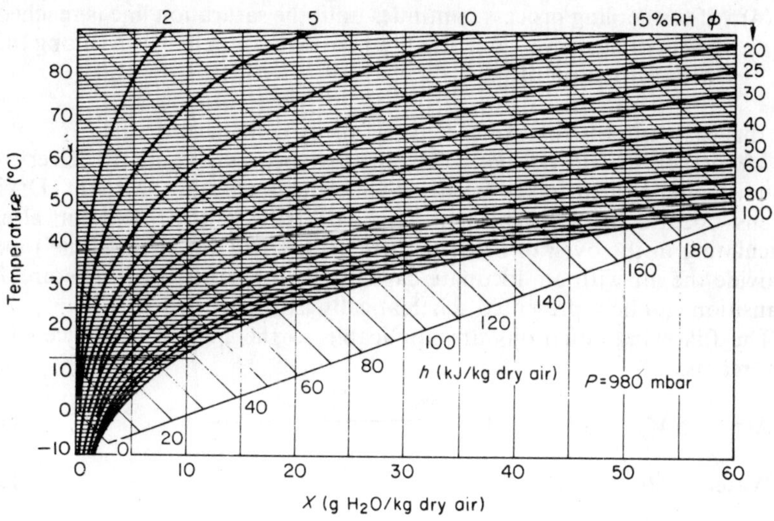

Figure 4.72 h,X graph for moist air for total pressure $P = 980$ mbar.

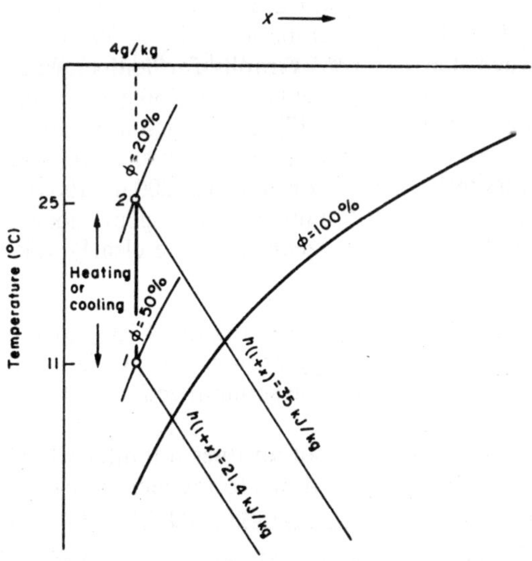

Figure 4.73   Warming and cooling of moist air.

(4)   If the cooling process continues until the saturation line is reached, the cooling curve, after reaching the saturation line, runs along this with formation of mist.

In this case water is removed from the air by condensation.

Mixing of two airstreams. This process occurs frequently. Every air-conditioning system mixes fresh air with air already in the room. Drying-ovens operate on a mixture of fresh air from outside and air already circulating in the oven in order to save energy and yet at the same time to provide the air with an adequate capacity for absorption of moisture. The transition can be represented schematically as shown in Fig. 4.74.

The following equations are applicable to the quantity balance of the airstreams:

Air: $\dot{M}_{A,1} + \dot{M}_{A,2} = \dot{M}_{A,m}$ (1)

Water: $\dot{M}_{A,1}X_1 + \dot{M}_{A,2}X_2 = \dot{M}_{A,m}X_m$ (2)

The absolute moisture content of the mixed air is obtained by substituting for $\dot{M}_{A,m}$ in (2) and solving the equation for X_m:

$$X_m = \frac{\dot{M}_{A,1}X_1 + \dot{M}_{A,2}X_2}{\dot{M}_{A,1} + \dot{M}_{A,2}}$$ (3)

The equation

$$\dot{M}_{A,1}h_1 + \dot{M}_{A,2}h_2 = \dot{M}_{A,m}h_m$$ (4)

is obtained in the same way for the enthalpy balance and with (1) forms

$$\dot{M}_{A,1}h_1 + \dot{M}_{A,2}h_2 = (\dot{M}_{A,1} + \dot{M}_{A,2})h_m$$ (5)

and, after conversion,

$$h_m = \frac{\dot{M}_{A,1}h_1 + \dot{M}_{A,2}h_2}{(\dot{M}_{A,1} + \dot{M}_{A,2})}$$ (6)

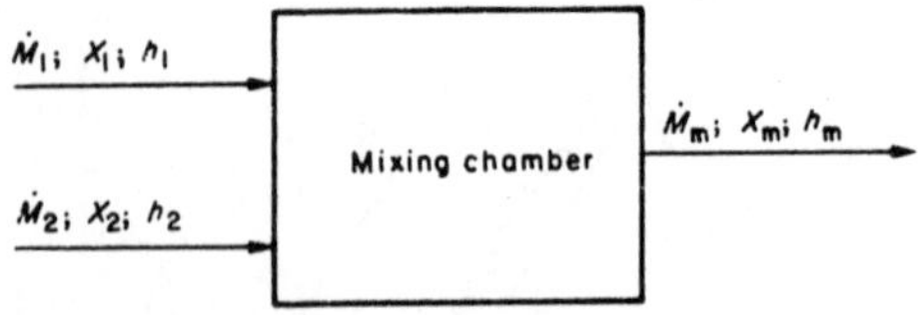

Figure 4.74   Mixing of two airstreams. $\dot{M}$ = Air flow (m³/h); X = Absolute humidity (kg per kg air); h = Enthalpy of the air (J); 1, 2, air in ; m, mixed air out.

It follows from (3) and (6) that both the absolute moisture content and the enthalpy of the mixed air are produced solely from principles of energy conservation, i.e. the $\dot{M}_{hm};X_m;h_m$ mixture point lies on the straight connecting lines of the initial states in the h,X graph (Fig. 4.75). The position of the mixture point is determined by the quantities of the airstreams. If both quantities of the airstreams containing different quantities of moisture are the same, the mixture-point lies half-way between the two starting points in the h,X graph.

The following cases may be distinguished (Fig. 4.75):

(1)   When warm dry air (A) is mixed with moist air (B) of about the same temperature, the relative humidity of the resultant air mixture lies between the initial humidities.

(2)   The relative humidity of a mixture of cold moist air and warm moist air can in certain proportions be higher than either of the initial humidities (example CD).

(3)   In the extreme case the humidity can go beyond the saturation line here and mist droplets can be formed, whereupon the system comes out of equilibrium (example EF).

4.5.2.4.   Course and outcome of drying

4.5.2.4.1.   Cooling limit temperature

When air flows over a damp material it absorbs moisture and is cooled down until it reaches its saturation point. If the surface of the damp material is completely wetted with moisture, i.e. the entire surface is covered with adhering water, a constant temperature, the so-called cooling limit temperature, is established on the surface. This depends solely on the initial state of

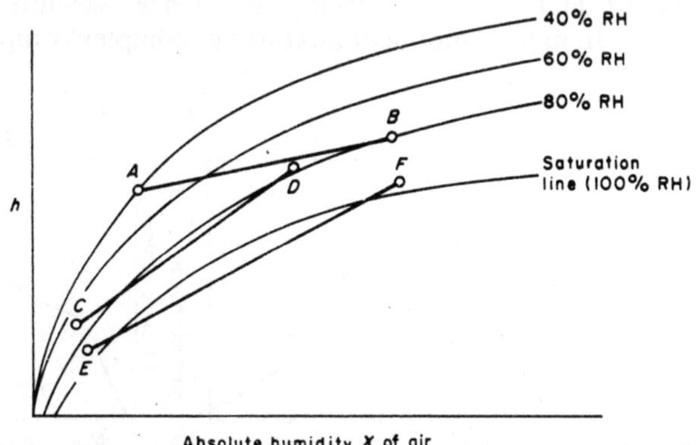

Figure 4.75 h,X graph of two mixing airstreams.

the air but not on the material. These statements are therefore true only in the first stage of drying (see below).

4.5.2.4.2. The stages of drying

Kneule [4.149] distinguishes between drying under steady conditions and under changing conditions (technical drying process). The following statements relate exclusively to drying under steady conditions. We refer to the appropriate specialist literature [4.149 and 4.150] for changing drying conditions.

Steady drying conditions exist when temperature, humidity and rate of flow of the drying-air are kept constant during the drying process. Here the rate of the airflow must be adjusted so that its state is practically unaffected by the heat given up to the material and by the moisture absorbed from the material. Under these conditions the course of drying with time can be plotted and classified into three different types [4.149] (Fig. 4.76).

Figure 4.76a already shows that the moisture content of the material being dried decreases constantly over a certain time. Plotting of the rate of drying against time or against the moisture content of the material (Fig. 4.76(b and c)) clearly reveals the existence of a phase of constant drying-rate which is called the first stage of drying. During this first stage of drying only the water adhering to the completely wetted surface evaporates from the material which is being dried. As soon as the surface of the material is no longer completely wet the drying rate decreases, as water must be transported to the surface by capillary forces from inside the material for further evaporation, which takes longer than pure evaporation at the surface.

The transition from the first to this second stage of drying is denoted by the so-called 'kink point', the position of which is determined at a given temperature and initial moisture content by the thickness of the material, as shown in Fig. 4.77 [4.151].

It is possible for the first stage of drying, and hence a definite kink point, not to be observed in multicomponent mixtures of complex composition and

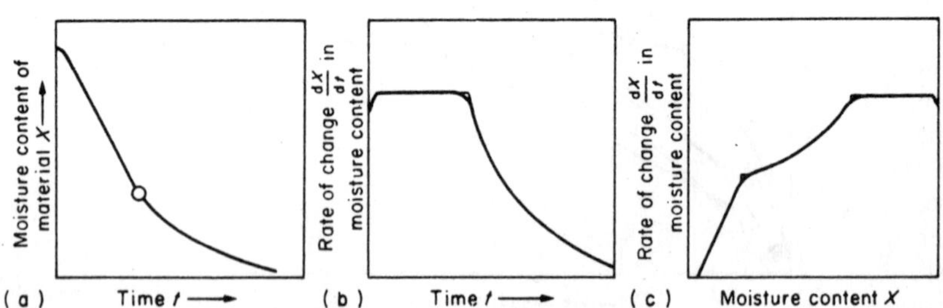

Figure 4.76   The drying process.

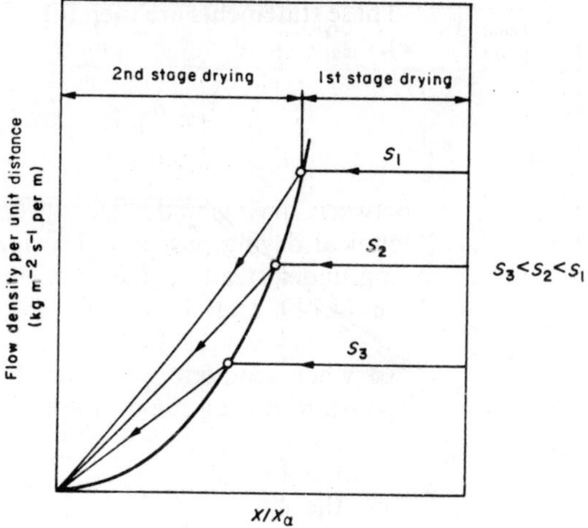

Figure 4.77 Displacement of the kink point by varying thicknesses of the material being dried. X, absolute moisture content (kg per kg air); X_a, initial absolute moisture content (kg per kg air); s, thickness of material (m).

in products in which surface is not completely wet with adhering water even at the commencement of drying. The product is then in the second stage of drying right from the commencement of the drying process. As already mentioned, the drying rate decreases while the temperature of the surface of the material increases in this second stage of drying. Water evaporates from the capillaries, whereupon narrow capillaries withdraw water from wide capillaries by capillary suction. The drying 'front' thus retreats ever further back into the material.

The degree of drying attainable depends on the temperature and humidity of the air used for drying and on the hygroscopicity of the material. In non-hygroscopic materials drying is completed after the second drying stage, whereas hygroscopic materials require a third drying stage. The transition to the third drying stage is denoted by a second kink point. This does not occur in non-hygroscopic materials. The rate of drying of hygroscopic materials decreases rapidly towards zero after the second kink point has been reached, as complete drying is not possible in these materials. The conditions are summarized once again in Fig. 4.78.

The transitions between non-hygroscopic and hygroscopic materials are fluid and hence we may logically speak of fairly hygroscopic materials. The degree of drying will be governed less by the description 'hygroscopic' and rather by the attainable water content. The question of how a residual moisture content, once it has been attained, can be maintained, i.e. how absorption of water by a material during storage and subsequent use can be prevented, is often much more important.

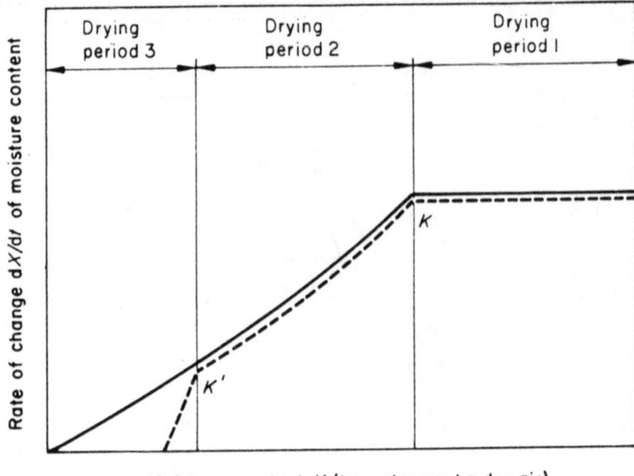

Figure 4.78 Course of drying of hygroscopic and non-hygroscopic materials. K is the kink point and K′ is the kink point for hygroscopic materials. Dashed line, hygroscopic; solid line, non-hygroscopic.

4.5.3.　APPLICATION OF VARIOUS TYPES OF DRYING

The choice of a suitable method for drying certain products is determined by many factors such as properties of the original product, water content and quantity, stability of the constituents of the product, etc. No general guidelines exist for making this choice. It is, for example, quite possible for the industrial conditions for a drying process to alter in the course of time after it has been commenced, and hence the process must be reconsidered. This last point in particular has become very significant as a result of the energy crisis. Whereas, for example, the energy costs in the spray-drying of pharmaceutical products were formerly barely significant, they are now a critical factor in the calculation of the total production costs. Such a development leads automatically to the application of other methods in which this particular substantial disadvantage is perhaps less significant.

The following factors are therefore involved in the choice of a drying method:

The nature and consistency of the starting-product which is being dried.
The quantity of moisture to be removed.
The quantity of product to be processed per time unit.
The hygroscopicity of the end-product.
The stability of the active substances contained in the product.
The desired physical consistency and constitution of the end-product.

The outlay costs
The running production costs.

The application of the more important methods for phytopharmaceuticals and the properties of the products resulting from these will be discussed next.

4.5.3.1. Drying as per the pharmacopoeias

General directions for drying extracts are given only in *DAB 8*. According to this, 1 part of the drug material is percolated until 3 to 4 parts of percolate have run off, or the prescribed quantity of solvent has been consumed. Unless otherwise prescribed, the percolate is dried under reduced pressure at a water bath temperature not exceeding 70°C. The temperature of the extract being dried must remain below 50°C here. Further drying after pulverization is carried out in a desiccator.

The drying of, for example, belladonna extract differs from these general directions in that the water bath temperature must be at most 40°C.

Neither *USP XX* nor *BP 80* gives general directions for drying of extracts, but state the drying conditions for individual products in the pertinent monographs. Table 4.34 summarizes these.

4.5.3.2. Spray drying

Spray drying is now widely used in the preparation of phytopharmaceuticals, namely in the preparation of spray-dried tea and infusion preparations, in the preparation of spray-dried individual extracts as end or intermediate products, and in the 'microencapsulation' of ethereal oils.

Table 4.34   Drying procedure for individual extracts as given in the pharmacopoeias

Name	Drying conditions	Diluent for adjusting content
Powdered belladonna extract (*USP XX*)	60°C Under reduced pressure	Starch
Belladonna dry extract (*BP 80*)	60°C under reduced pressure, further drying at 80°C	Powdered drug
Cascara sagrada extract (*USP XX*)	Drying *in vacuo*. Further drying at 100°C	Starch
Cascara dry extract (BP 80)	Boiling down to dryness	—
Hyoscyamus dry extract (BP 80)	60°C under reduced pressure, further drying at 80°C	Powdered drug

Table 4.35　Examples of spray-dried drug extracts with literature citations

Author	Spray-dried extract	Remarks
Bullock and Lightbrown ([4.154]	*Senna* leaves, rathany root, gentian root, *cinchona* bark	
Prokofiev and Solonouz [4.155]	*Adonis* (pheasant's eye) alder buckthorn bark valerian root	
Realdon [4.156]	Belladonna leaves Nux vomica seeds Liquorice	
Ganchukov and Davituliani [4.157]	Solution	Investigation of dependence of particle size on initial water content of the spray
Baraschkov [4.158]	*Solanum robulatum*	Recovery of solasodin

The preparation of 'microencapsulated' ethereal oils, which should more correctly be called spray-embedding, yields precursor or intermediate products for medicinal preparations.

Rinkel [4.153] gives a comprehensive survey on the use of spray drying of phytopharmaceuticals and Table 4.35 cites examples of spray-dried extracts described in the literature in addition to which there are numerous as yet unpublished experiences of spray-drying of drugs. The number of drug extracts on the market in drinkable tea or infusion preparations alone is considerable.

The various parameters which influence the result may be discussed by means of a scheme (Fig. 4.79). According to this scheme the composition of the spray solution affects such properties of the medicinal preparation as viscosity, solid content and surface tension. The temperature at which the spray mixture passes through the nozzle affects these parameters, with the exception of the solid content.

A further group of factors affecting the result of spray drying is the various atomization or spraying systems.

Finally, the conditions, i.e. temperature of the incoming and outgoing air and the duration of stay of the materials, prevailing in the spray drying tower, are of great significance for the result of the drying process. The individual parameters are discussed below.

As well as the extractive substances, the formula of the spray solution contains additives, which are especially important when the spray product represents an end product. In this case, content, colour, bulk density, etc. are adjusted with additives. Flavourings and colourings are necessary for improving flavour and for standardizing the appearance of different batches of product. Colour variations may be expected in different batches of manufactured product as a result of the natural origin of drug extracts and hence the use of colourings, in the case of infusions usually malt beer

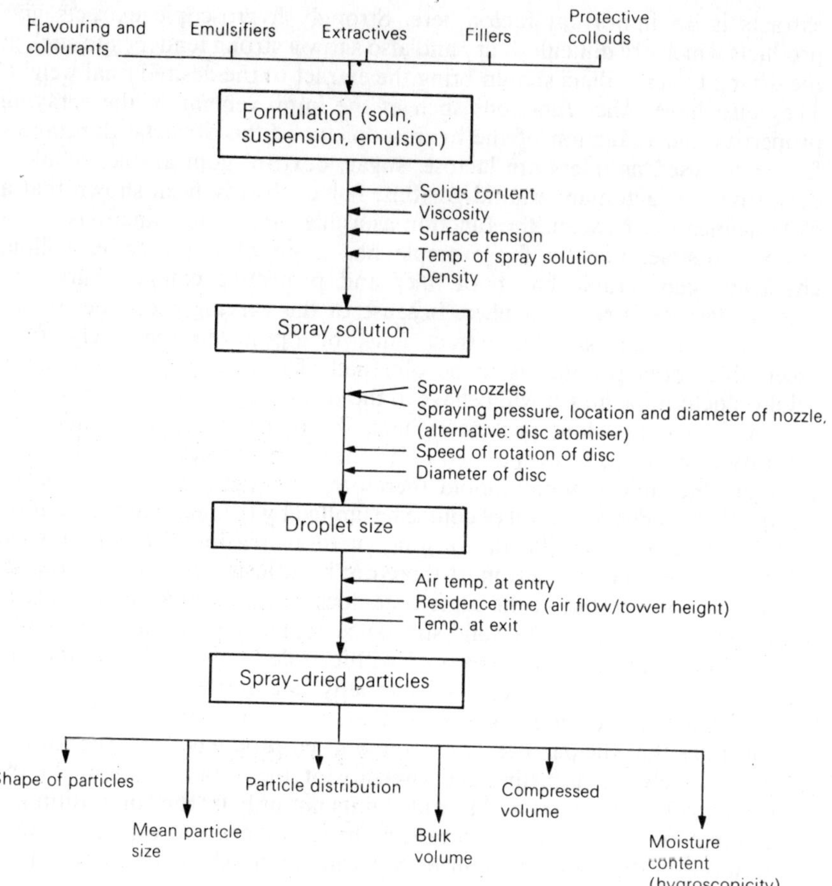

Figure 4.79 Factors affecting the result of spray drying.

colouring, is indispensable for the uniform appearance of a product. Emulsifiers are sometimes necessary for solubilization when lipophilic flavourings or ethereal oils are used. Junginger's demonstration that the proportion of amorphous material in the resultant embeddings can be increased by addition of suitable emulsifiers in the spray drying of acetylsalicylic acid [4.159] certainly cannot be applied directly to phytopharmaceutical preparations. However, he did show that the addition or the natural presence of surfactants can have a considerable influence on the quality of the end-product.

The extractive substances form the principal constituent of the extract. Their proportion comes second only to that of the fillers. They also decisively influence the spraying properties of a solution. The hygroscopicity of the extracts is an important factor here. Strongly hygroscopic extracts give products which are difficult to dry and also show a strong tendency to stick in the drying tower. Fillers should bring the extract to the desired final weight. They also have other functions such as the improvement of the spraying properties and reduction of the hygroscopicity of the extracts. Substances frequently used as fillers are lactose, sugar, dextrin, gum arabic, cellulose derivatives, galactomannan and gelatins. It has already been shown that a clear delineation between the function as a filler and other functions is not always possible. Gelatin for example has more of a protective colloid character, gum arabic has both filler and protective colloid characters, whereas lactose is purely a filler. Because of the varying properties of the extractive substances, each extract must be optimized separately if an acceptable spray product is to be obtained. The properties of the spray solution determine how it will behave in the spraying process. The total solid content is an important parameter here; the higher it is, the smaller the quantity of water to be evaporated off in the drying process. For reasons of economy the solid content should therefore be made as high as possible, though these endeavours will of course be limited by the fact that the solution becomes increasingly difficult to spray with increasing content of solid substance. The solid content must therefore be adjusted so that the product can still be sprayed but at the same time does not contain an unnecessarily large quantity of water. The solid substance contents are usually 15–40%.

The viscosity, which may be varied within wide limits by the addition of viscosity-increasing substances, substantially affects the droplet size and hence the granule or particle size of the spray dried product. Nürnberg [4.160] states that the particle size of spray dried products can be controlled quite accurately by adjusting the viscosity of the spray solution. As the viscosity can be influenced within wide limits not only by the composition of the product but also by the temperature, the temperature at which the spray solution passes through the atomizer, a parameter to which little attention is paid in practice, becomes very important.

The surface tension of the solution likewise affects the droplet size. The conditions prevailing in a spray jet have so far made it impossible to calculate this effect. The drying-rate is influenced by the droplet-size and density of the spray solution. According to Kneule [4.149], the speed at which a droplet falls in a stream of gas may be calculated by Stokes's law as follows:

$$v = \frac{\mathrm{d}l}{\mathrm{d}t} = \frac{d^2(\rho_P - \rho_A)g}{18\eta_A} + \bar{v}_A$$

where v = rate of fall of droplet (m/s), l = path length (m), d = diameter of droplet (m), ρ_P = density of droplet of spray solution (kg/m^3), ρ_A = density of air (as carrier gas) (kg/m^3), g = acceleration due to gravity (kg m s^{-2}), η_A = viscosity of air (Pa s), and $\bar{v}_A$ = mean velocity of airflow (between 0.05–0.6 m/s in industrial spraying towers).

Depending on whether the dryer operates by direct (continuous) flow or by countercurrent, an expression for the velocity of the air must be added to or subtracted from the fall velocity of the droplet in stationary medium. The equation affects the fall velocity and hence the time required for drying. The greater the density, the greater will be the fall velocity and the shorter will be the time available for drying. A dryer must be designed so that the fall path is sufficiently long for drying of the largest droplets.

The process of atomization or spraying from nozzles is governed by the parameters:

Spraying pressure.
Type of nozzle or jet.
Diameter of nozzle or jet, determined in rotating discs by speed of rotation.
Diameter and design of the disc.

The characteristics of the various systems are described in Section 5.7.1.2. Here we shall discuss only the differences in the resultant droplet size distribution and the differences in the spray products resulting from this. Table 4.36 summarizes these differences.

Table 4.36 Differences in spray-dried product according to the type of atomizer

Atomizer	Operating conditions	Particle sizes attainable	Remarks
One component nozzle	Spray pressure: 7–500 bar [4.172] Generally 20–60 bar [4.161]. Output: up to 5 tons/h	8–800 μm	Relatively coarse product consisting of hollow spheres, small particle size range
Two component nozzle	1.5 to 6 bar Generally 5–7 bar [4.161]. Output: up to 100 kg/h	30–250 μm	Finer product, substantially broader particle size range, higher proportion of fine material than with one-component nozzle, also irregular particles and fragments of hollow spheres in addition to intact hollow spheres
Rotating disc	3500–5000 rev/min [4.172] Generally 5000–25000 rev/min [4.161]. Output: 30 kg–7 tons/h [4.172] Up to 200 tons/h [4.161]	25–950 μm	Free-running product of hollow spheres, very small particle size range, little dust

The differences in the spray dried products resulting from the use of various spraying systems have been confirmed by several authors. Graf and Nürnberg [4.172] report a 'very much broader particle-size distribution' with a 'higher proportion of powder' for a nozzle atomizing two substances than for a nozzle atomizing only one substance. Ritschel [4.173] obtained a finer product with the rotating disc than with atomizer jets, which is somewhat contradictory to the findings of other authors.

In a further paper Ritschel [4.174] reports that rotating discs yield a consistent product only at low delivery speeds, as the scattering of the droplet sizes increases with increasing delivery speed. Rinkel [4.153] shows the differences in the spray dried products from one-component and two-component (binary) nozzles very clearly by means of an illustration. The relation between particle size and distribution of the spray dried product is shown in Table 4.37.

According to Kröll [4.161], the drop size distribution obtained from rotary sprayers can be described with the normal logarithmic distribution. The square root distribution can be used for the drop size range from one-component nozzles, whereas special distributions are used for two-component nozzles. Troesch [4.162] has given a drop distribution curve (Fig. 4.80). In this curve the most frequently occurring drop size has a diameter of 1/10 of that of the largest drops which occur. A mean diameter is also defined, above and below which lie 50% (numerically) of all the drops. This diameter (after Sauter) gives the drop size at which the ratio of surface area to volume is the same as for the sum total of the drops.

The course and result of the drying of droplets in the tower is influenced by

(1)   The temperature of the incoming air.
(2)   The length of time for which the droplets remain in the tower.
(3)   The temperature of the outgoing air.

The temperature of the incoming air must be such that the supplied energy is sufficient for drying. The higher the temperature of the incoming air, the shorter will be the drying time under otherwise constant conditions. In addition to the higher energy costs associated with this, the stability of the

Table 4.37   Particle size and range of particle size shown diagrammatically according to the spray drying atomizer used

Atomizer	Particle size	Range of particle size
Rotating disc	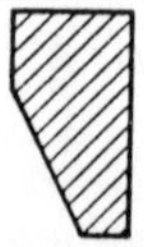	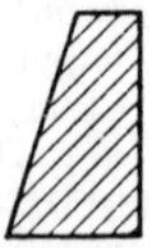
One-component nozzle		
Two-component nozzle		

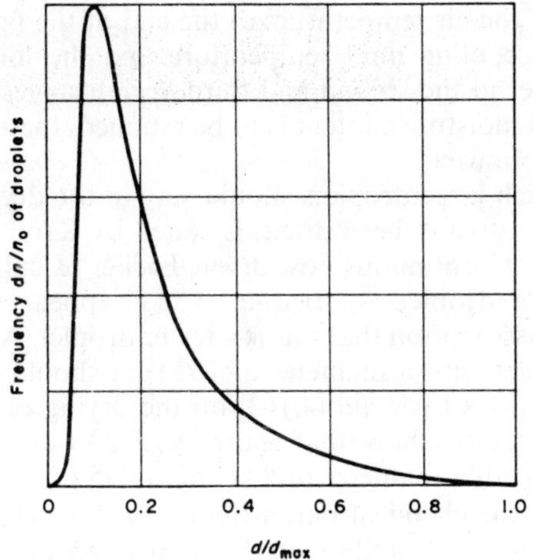

Figure 4.80 Drop count distribution curve for a specific case (from Troesch, 4.162).

product limits any increase in the temperature of the incoming air. Temperatures of 180–230°C are generally used.

The time for which the droplets remain in the tower depends on the air throughput (quantity of air per time unit) and on the design of the tower. An increase in the air throughput results in shorter times for which the droplets remain in the tower and the attendant risk of incomplete drying.

Two zones, the 'nozzle zone' and the 'fall zone', are to be distinguished with regard to the velocity of the droplets. The high velocities in the nozzle zone result in only partial drying of the finest droplets, whereas the actual drying process takes place in the fall zone. The fall path and, hence, the time for which a drop remains in the tower, depend on the diameter of the drop and also on the density of the spray solution. Kneule [4.149] gives a detailed discussion of these influences.

The temperature of the air issuing from the drying tower should be as low as possible so that the product is well conserved. It depends mainly on the temperature of the incoming air, on the quantity of water evaporated and, as mentioned above, on the throughput of the air. The temperature of the product at the exit of the spraying tower is generally 15–20 K lower than the temperature of the outgoing air. The temperature of the product can also be rapidly lowered by passing the product through a cooler attached to the tower.

As mentioned in Section 4.5.2.4.2, the duration and severity of the exposure of the product to high temperature depend on the course of the drying process. In the first drying stage the temperature of the product rises only to the cooling limit temperature. After the first kink point further drying is associated with a rise in the temperature of the material. This rise is the greater, the further the air temperature at the end of the first drying stage is removed from the cooling limit temperature and the longer the product remains in the tower in the second and third drying stages.

As a certain final moisture content is to be attained, these conditions must also be carefully optimized.

The time for which large droplets should stay in the drying tower (as the small droplets have already been dried) is stated by Kröll [4.161] to be 10–15 s for a large direct continuous flow dryer. Earlier calculations, according to which a 100 μm droplet is dried in 0.32 s, appear to be somewhat unrealistic, on the assumption that this is a water droplet. According to these calculations a droplet with a diameter of 1000 μm should be dried in 3.2 s. However, investigations by Wahl [4.163] on the drying of aqueous sodium dodecylsulphate solution show that more than 20 s are required for the drying of droplets with a diameter of 800 μm at 235°C.

Physical and physicochemical parameters as well as chemical properties such as the content and stability of the active substances are especially important for further processing of a spray dried product. As mentioned above, the shape of the particles can vary from spherical to rather irregular forms and also fragmented hollow spheres. A product consisting of spherical particles has a pleasing appearance, though it is often not very suitable for mixing with other substances of different particle shape and density. The mean particle size and the particle size range are criteria which decisively affect the free-running properties and miscibility of the product.

As the doses of most spray dried products are measured by volume (e.g. by teaspoon), the bulk factor is surprisingly important. Maintenance of a certain bulk factor is achieved by ensuring that both the composition of the spray solution and the spraying conditions are correct. The bulk factor can also be altered by gassing the spray solution with carbon dioxide immediately before spraying. The tamping volume is important as it provides information on the hardness or mechanical strength of the individual spray dried particles. A large difference between bulk and tamping volume indicates an unstable particle structure.

The water content of the end product of spray dried extracts must not be more than 3–4%, as otherwise agglutination occurs. This value is easily attained by spray drying (the values normally being about 1%), though it is difficult to maintain, as the hygroscopicity of the drug extracts leads to the reabsorption of water: water is present in varying degrees, depending on the product, though in principle it is present in all products.

As a general rule it can be said that spray drying is an economical process, another advantage of which is that the product does not need to be ground any further. The extent to which temperatures of 50–70°C can be tolerated in the product depends on the individual cases. These temperatures, which are higher than those in vacuum drying, lessen the advantage of the considerably shorter exposure to temperature stress obtained with these higher temperatures.

In addition to spray drying of extracts, two further special forms of spray drying, which are frequently carried out with the aid of raw plant substances, namely spray embedding of ethereal oils and spray embedding of difficultly soluble active substances, must also be mentioned.

Spray embedding of ethereal oils was formerly [4.153, 4.170] regarded as a microencapsulation process. However, whereas in microencapsulation a particle, solid or liquid, is completely enclosed in a uniform shell, in spray embedding the ethereal oil exists as microdroplets distributed evenly among the particles of the spray dried product. This means that ethereal oil is also present on or near the surface of the particles. This constituent can be determined quantitatively by determination of the so called washability, where the product is briefly treated with a solvent (e.g. hexane) which, however, must not alter the coating material. These products therefore do not correspond to microcapsules. Nürnberg [4.160] gives the following definition:

'Embeddings are products in which particles of active substance(s) in very finely divided form are distributed homogeneously in an inert— usually highly polymerized—adjuvant substance.'

The size of the oil droplets existing as an internal phase in the emulsion which is to be sprayed should be about 1–2 μm in order to achieve as complete an embedding as possible. The maximum obtainable oil content of such embeddings is 15–20%. The object of the spray embedding of difficultly soluble active substances is to improve their biological availability. This topic was studied by Nürnberg. Aqueous solutions and dispersions and organic solutions were sprayed, and the active substances tested ranged from digoxin and digitoxin through dexamethasone, prednisolone and prednisolone acetate, chlormadinone acetate, acetylsalicylic acid and tetracycline [4.160] to griseofulvin [4.171]. It was found that the active substances exist in an amorphous form in the spray embeddings, which upon dissolution produces supersaturated solutions. The blood levels obtainable with these preparations are frequently higher than those of the corresponding preparations of crystalline substance.

4.5.3.3. Vacuum belt drying

Whereas temperature stressing of the material being dried at 50–70°C for a maximum of 10–15 s must be expected in spray drying, vacuum belt drying operates at lower temperatures with longer exposure times. The possibilities for variation of temperature, exposure time and of the reduced pressure applied are so numerous that we shall restrict ourselves here to general guidelines on approximate values.

A starting product can also be dried in different ways to a defined end-product with a defined water content in vacuum belt dryers. A short drying tunnel suffices with higher temperatures whereas a longer tunnel is required for low temperatures. The fundamental decisions on the length of the drying-tunnel must be made with regard to the operating temperature before the equipment is installed or any vacuum drying is carried out.

Vacuum belt dryers are used for pumpable concentrates, extracts and thick suspensions. The concentration of solids in the starting material is usually higher here than in spray drying, i.e. less water generally needs to be evaporated. The output capacity of the plant now on the market ranges from $1.2\,\mathrm{kg\,m^{-2}\,h^{-1}}$ minimum through a mean output of $2.4–4.0\,\mathrm{kg\,m^{-2}\,h^{-1}}$ to a maximum output capacity of $10\,\mathrm{kg\,m^{-2}\,h^{-1}}$. The output capacity is determined by the area of the belt $(\mathrm{m^2})$ which can move through the dryer per hour.

Whereas spray drying is a convection drying process, vacuum belt drying is purely a contact drying process, in which the necessary heat is transmitted through the heating elements immediately underneath the conveyor belt.

The drying material continuously applied over the whole breadth of the heated belt from swivelling tubes froths up in the partial vacuum and releases its moisture as water vapour from the bubbles which are formed. The dry material left behind is light, friable, porous and voluminous. Its properties resemble those of lyophilized (freeze dried) substances.

Vacuum belt drying is widely used in the food industry for the preparation of vegetable extracts, soups and soup ingredients, fruit juice concentrates and cocoa, tea and coffee extracts. Basic materials for the production of beer, such as malt and hop extracts, are also prepared by this process. In the pharmaceutical industry valerian, alder buckthorn bark, aloe and horse chestnut extracts have also been dried by this method.

Table 4.38 shows a possible typical course of vacuum belt drying and the differing treatment of the product in the three heating zones.

The products are noted for good solubility, such as is for example demanded in the foodstuffs industry for instant products.

The disadvantage of vacuum belt drying undoubtedly lies in the fact that the resultant products must be ground or pulverized. A considerable degree of comminution is of course already possible in the dryer, though in most cases this is not sufficient. For the comminution of dried extracts see Section 4.6.

4.5.3.4. Roller drying and oven drying

The continuous roller drying process gives solid, hard products which must in every case undergo further comminution. Temperature stressing during the drying process is generally high and damage to sensitive products can therefore be expected.

Drying in drying ovens without vacuum has the same disadvantages and hence both processes are now of little importance. The conditions are substantially improved when vacuum is used. In roller drying the softening point of extracts, which sometimes lies at low temperatures, can cause difficulties when the product has to be scraped off the rollers.

Table 4.38 Operation of the VBT 60 vacuum belt drying plant

1. *Operating conditions*

Operating vacuum (mbar)	15
Concentration of material (% dry substance)	86
Quantity of material (kg/h)	493
Quantity of dried cake (kg/h)	435
Coefficient of flow (L/h)	2.15
Time for which material stays in dryer (min)	28
Specific dried cake quantity (kg m^{-2} h^{-1})	7
Thickness of dried cake (mm)	12–13
Volume of dried cake (cm^3/m^2)	12200
Bulk density (g/L)	300

2. *Drying conditions*[a]

	I	II	III
Heating zones (m^2)	18	18	12
Heating-water temperature (°C)	145	135	120
Product temperatures (°C)			
over vacuum (solid/gas interface)	25	25	25
over belt (solid/solid interface)	65	70	75
Heating area loading (kJ m^{-2} h^{-1})	6700	4200	1050
Quantity of water vaporized (kg/h or kg m^{-2} h^{-1})	31.0	24.8	6.2
	1.72	1.38	0.52
Concentration range (% dry substance)	86–91.5	91.5–96.3	96.3–97.6
Thermal conductivity of product (W m^{-1} K^{-1})	0.5583	0.3102	0.1047
Effective conductivity index or heat transfer coefficient (W m^{-2} K^{-1})	15.5	10.6	4.6

[a] *I, II* and *III* are various heating zones of the dryer

4.5.3.5. Other methods of drying

These include microwave drying and freeze drying. Microwave drying, which is carried out in the frequency range of 2400–2500 MHz permitted for this, has so far been used especially in the foodstuffs industry. Its advantages lie in the rapid removal of water and in the reliable drying out of pockets of moisture. The structures of the products are similar to those obtained with vacuum belt drying, as microwave drying also operates under reduced pressure. Local overheating can occur at points of high moisture concentration. It must also be remembered that sensitive products can be damaged by this high energy radiation. Its applicability must therefore be accurately tested for each individual case.

Freeze drying has occupied an important place in the food industry for a long time. The first freeze drying processes were carried out as long ago as 1813. Its industrial application began somewhere around 1935 and we should be reminded that the isolation and widespread use of penicillin became possible only through freeze drying. It is the method of choice where substances in solution are unstable or susceptible to oxidation. As the resultant products can easily be redissolved, and are therefore lyophilic, the term lyophilization was coined. Freeze drying is used in medicine for the conservation of blood, blood fractions and transplants. Technically the process is divided into freezing, primary drying and redrying stages.

The conditions under which the material is frozen already affect the consistency of the product. Rapid freezing at low temperatures produces numerous crystallization seed points and many small ice crystals are formed. A lyophilizate of this type has a high rate of dissolution. Slow freezing produces a few large crystals, which of course sublimate more rapidly, as the water vapour which is produced can escape quickly through the relatively large pores. The fine structure of the product will however turn out differently here than with rapid freezing.

There also exists the possibility of slow cooling in the vicinity of the freezing point and hence of the formation of a supercooled solution. Fine crystals are likewise produced by rapid cooling of a supercooled solution. Rapid cooling to low temperatures is however more reliable. Temperatures of about $-40°C$ are sufficient in practice. When a solution is cooled the solvent crystallizes first (e.g. to ice), which concentrates the solution (freeze concentration) until the eutectic mixture precipitates when the eutectic point is reached.

If the product is to be frozen completely it must always be cooled to below the eutectic temperature. This temperature can be determined experimentally by measuring the electrical resistance of the frozen product as it is being warmed up. As soon as the ice begins to liquefy the electrical resistance decreases sharply.

It should however be noted that conditions in complex extracts and also in foodstuffs are sometimes indefinite and unpredictable and so it is better to work with the empirical freezing temperature of $\sim -40°C$. Flink [4.165] showed with the example of coffee that the freezing conditions affected the colour of the freeze dried coffee, which clearly proves the importance of the correct choice of freezing process.

Heat must be supplied during primary drying, as otherwise the material being dried would be further cooled by sublimation of the ice, which consumes heat, and drying would gradually cease. The heat supply must be controlled so that the temperature of the material remains far below the eutectic point. The highest drying rate is achieved in this region.

Sublimation is accelerated by the application of vacuum; the lower the pressure, the better it proceeds. However, as heat transfer deteriorates in a space devoid of air, it is best to operate in practice with pressures of 0.01–1 mbar. Primary drying is completed as soon as all ice has sublimated, which is recognized by a temperature rise in the product.

Secondary drying removes water bound by adsorption. It is carried out at high temperature and low vacuum, where care must be taken to ensure that the product does not soften or even melt and thus undergo structural alteration.

Rahm [4.166] has given an example of the entire course of the temperature in a freeze drying (Fig. 4.81).

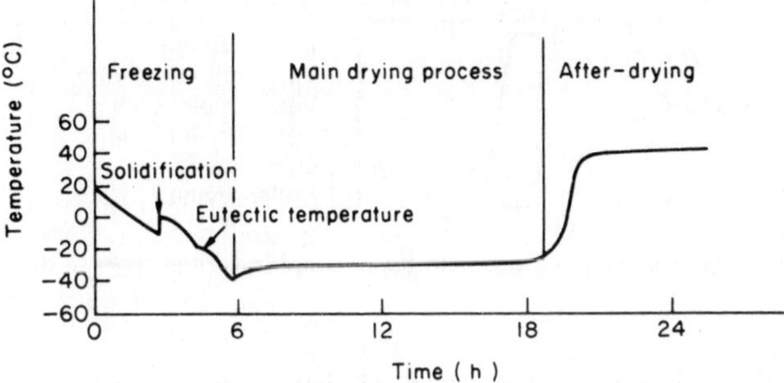

Figure 4.81   Temperature curve in freeze drying.

It is surprising that, apart from the above mentioned areas of use in the pharmaceutical industry, freeze drying has not become widely used; this is all the more surprising as expensive methods are nearly always used in the production of quality foodstuffs. However, other factors as well as high costs may also be responsible for this. Some loss of ethereal oils is inevitable with the very low pressures. These must therefore be collected separately in a cooling trap and then added to the product again. Relatively few authors have reported on freeze drying of plant extracts. Graf and Bornkessel [4.157] prepared dried valerian extract both by spray and by freeze drying. Whereas spray drying caused the valerate content to fall, stable dried extracts were obtained by freeze drying.

Chebak *et al.* [4.168] obtained yields of active substance in the freeze drying of an extract of *Ammi visnaga* which were substantially better than those obtained by normal vacuum drying. Chernov *et al.* [4.169] describe the freeze drying of certain plant extracts, among others Indian cress (*Nasturtium*) seeds, plantain and celandine. Chernov and Shebanova [4.170] investigated the dependence of the drying curve on the initial water content in extracts of the fruits of *Ammi visnaga* and *Solanum aviculare*. There is as yet no known industrial application of freeze drying to plant extracts.

4.5.3.6. Comparison of various types of drying

The various types of drying can be compared with one another from technological, industrial and economic viewpoints and with regard to the quality of the products obtained. In many cases the desired quality demands a certain technique and thus at the same time determines the costs.

Drying temperature and time are of particular interest. Figure 4.82 shows the relationship of these two parameters to the type of drying. As a great simplification it can be said that the drying time increases as the temperature

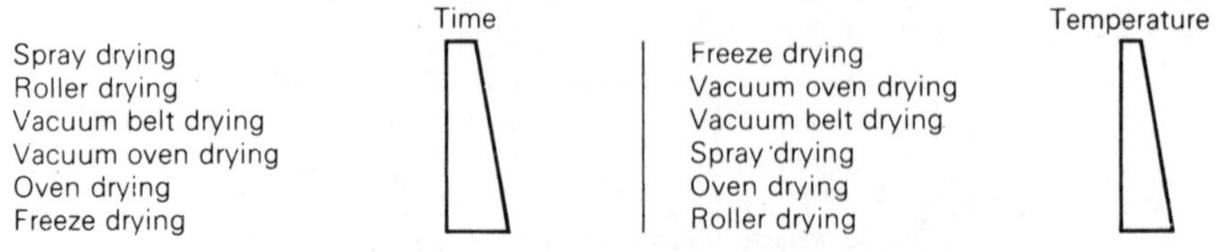

Figure 4.82 Comparison of drying time and temperature in various types of drying process.

decreases. There is no ideal method of drying in a short time at a low temperature.

An economic comparison of various methods must be made separately for each problem if exact results are required. It can be generally stated that freeze and spray drying are affected by increasing energy costs more than the other drying processes. Freeze drying demands an energy supply for the maintenance of the vacuum and the removal of heat over long periods. This may be the main reason why freeze drying has so far not been widely adopted in the pharmaceutical industry. The structure of freeze dried products may also be a disadvantage (e.g. in tabletting). In spray drying, very large quantities of air must be heated and this puts a great strain on the products. This is perhaps the reason for the great interest in vacuum belt drying, which is in a median position as regards time and energy consumption. The possibility of being able to work with very high initial concentrations may also be an economic advantage.

Any comparison of the resultant products will always be governed by the technology used, where both physical and chemical parameters must be considered.

Among the chemical parameters, stability plays an important part for pharmaceutical products. Graf [4.167] showed with the example of valepotriates that the type of drying, together with the adjuvants used, can have a decisive influence on stability. In his investigations a freeze dried extract, with microfine cellulose as adjuvant, had the best stability, whereas lyophilized extracts with Aerosil or Tylose MH were unsatisfactory.

Parameters such as appearance, structure, bulk density and tamping (compaction) density can also be decisive for the choice of the drying method for extracts containing stable active substances. It must however be emphasized that spray drying is the only type of drying which produces the end-product directly. In all other cases the resultant products must undergo grinding. Roller-drying gives the product with the highest bulk density. Dense products can also be obtained with oven drying, though this depends on the composition and initial concentration of the liquid or pastey extract. On the other hand vacuum processes generally give light, porous products.

4.5.4. COMMINUTION OF DRIED EXTRACTS

Dried extracts have greatly differing structures. Roller dried extracts usually consist of hard but friable lumps of compact structure and are of a similar appearance to extracts from drying chambers. Freeze dried extracts have a loose, light structure and can be comminuted easily. They are therefore similar to extracts prepared in vacuum belt dryers. Only spray dried extracts are directly produced as ready-to-use powders.

Comminution of the dried extracts is necessary for most areas of use. The hygroscopicity of extracts makes their grinding very problematical. Pin mills are the most suitable machines for this. However, when comminution is carried out without any further precautions the spaces between the grinding pins very quickly become clogged with material, which results in frequent interruption of the grinding process. This can be kept to a minimum either by using cold milling or by predrying the air passing into the mill or by working with additives.

Predrying of the air has the smallest effect here and is therefore seldom used in practice. Cold milling proves substantially better, as it not only permits conservation of the product if the mill has to stand idle for a long time but also generally provides a larger throughput. The disadvantage of cold milling lies in the higher costs brought about by the use of expensive milling plant and the consumption of nitrogen or carbon dioxide.

Where such plant is not available, the process can often be assisted by the use of suitable additives, which are admixed with the extract before it is ground. Colloidal silicic acid is described as such an additive in the literature. It not only improves the free-running properties of the extract but also coats fresh surfaces with a thin film of silicic acid, though it does not reduce the hygroscopicity of the extracts. The use of magnesium stearate as a grinding aid brings an improvement in this respect. The hydrophobic (water repellent) effect of magnesium stearate reduces hygroscopicity. Surface films which improve the free-running properties are also formed.

Microtalc, micronized saturated triglycerides (e.g. of the Dynasan series) and hardened castor oil (Cutina HR) are further possible milling aids with similar properties.

4.6.   Recovery of ethereal oils

Ethereal oils (essential oils) are products of secondary metabolism in plants, i.e. they are complex mixtures of fairly volatile plant constituents. At the time of harvesting they usually exist in a free form in the plant. Only a few exist bound to glycosides, from which they can be released by hydrolytic cleavage. As can be inferred from their name, they are oils, though the only thing they have in common with fatty oils is the fact that they are lipophilic. The chemical structure of their components differs greatly according to their type,

Table 4.39 Classification of oils in the perfume industry

Starting material	Method of recovery	End product
Fresh flowers	Extraction (organic solvent)	'Absolues'
	Extraction (warm; fats or oils) enfleurage	'Concentrés de pommades'
Dried (more rarely fresh) plant parts	Extraction (organic solvent)	'Résinoïdes' (oleoresinae)
Fresh citrus fruit peel	Expression	
Fresh or dried plants or plant parts	Steam distillation	'Huiles essentielles' (ethereal or essential oils)

nature and origin. Ethereal oils are produced in the plant's secretory cells, which are also known as glands. There are two different types of gland— exogenous glands, which arise from the epidermis, and endogenous glands, which are situated as isolated cells or tubular secretory ducts inside the plant.

Hegnauer [4.175] has reported in detail on the systematic classification of ethereal oils, including a classification by starting-material and method of recovery for the common ethereal oils used in the perfume industry (Table 4.39). This summary shows the more important methods of recovery of ethereal oils, namely expression, extraction, enfleurage and steam distillation.

Ethereal oils are extracted from fresh or dried drug plants by the methods described in Section 4.2 and hence only the enfleurage process and steam distillation, which have been described in detail by Treibs [4.176], are to be added here.

4.6.1. THE ENFLEURAGE PROCESS

This is based on the ability of fats and fatty oils to absorb, dissolve and bind ethereal oils, principally from flowers. There are two different types of enfleurage—that without the use of heat (enfleurage à froid or cold flower extraction) and hot enfleurage (enfleurage à chaud or hot flower extraction).

4.6.1.1. Enfleurage à froid

Purified pork fat and beef tallow are particularly used for the extraction of ethereal oils from plant parts. As pork fat alone is too soft and pure beef tallow is too hard, mixtures of the two fats are used. The mixture ratio chosen here depends on the temperature of the surroundings. One part of beef tallow to two parts of pork lard are used at normal outside temperatures and a 1:1 mixture ratio is used at high temperatures, such as are prevalent during the summer months in the south of France, where the process is mainly used.

The quality of the recovered extract depends especially on the purity of the fats used; the way in which these are prepared is therefore particularly important. Despite numerous attempts, the mixture of beef tallow and pork lard has so far not been replaced by any other fat or by other vehicles.

The process is carried out by smearing both sides of glass plates with the fat mixture and increasing the surface area of the fat by making furrows in it with a spatula. The glass plates are placed in wooden racks and the plant parts which are to be extracted are laid on the fat. By placing a number of these wooden frames ($\sim$ 30–40) one on top of the other the fat layers on the under and upper sides of the glass are made to absorb ethereal oil. After an exposure time of 24–72 h the plant parts are removed, the glass plates are turned over and fresh plant parts are laid on them again. This reversal of the glass plates ensures that all the fat is evenly impregnated with ethereal oil. The process is repeated up to 30 times. The fat extract thus obtained, which is known as pomade, is the more valuable, the more often the treatment has been repeated.

Flowers are used in by far the majority of cases for this extraction. The ratio of weight of flowers to weight of fat mixture used is called the 'flower value'. The number of treatments is given as the number. Hence pomade No. 30 signifies that this pomade was obtained by 30 treatments with flowers.

4.6.1.2. Enfleurage à chaud

Oils or molten fats obtained from plants or animals are used for this process. In addition to beef tallow and pork fat used in the process without heat, these can, for example, be olive oil, paraffin oil or even molten paraffin wax. The plant parts are extracted at 50–70°C by being placed into the fat either loose or in linen sachets. As in enfleurage à froid, fresh plant material is repeatedly extracted with the same fat, the total weight of the plant parts amounting to 5–10 times that of the extraction agent.

The use of this method is limited by the volatility of the constituents of the ethereal oils and by their increasing instability with rising temperature.

4.6.1.3. Processing of pomades

Pomades obtained by the methods described in Sections 4.6.1.1 and 4.6.1.2 are extracted with ethanol to concentrate the ethereal oils. Although methanol, propanol and butanol are cheaper, ethanol is still preferred, particularly for sensitive flower constituents. The extraction is carried out in mixers and the separation by cooling, centrifugation or filtration. These can be followed by further purification stages such as distillation, extraction with petroleum ether, etc.

4.6.2.　STEAM DISTILLATION

The dominant method for recovering ethereal oils is steam distillation. The steam is used here as a 'tractor' for the ethereal oils. The drug plant material is put together with water into a retort, into which steam is also passed. The retort is also heated to prevent the steam from condensing prematurely. In the ideal case the vapour issuing from the liquid is saturated both with steam and also with the vapour of the ethereal oil. The composition of the liquid condensing in the condenser thus corresponds to the composition of the vapour. The composition of the vapour will not alter during the distillation so long as all the substances are present in the retort.

Steam distillation offers the following advantages for the recovery of ethereal oils:

(1)　The boiling point of the steam–ethereal oil mixture is almost 100°C at atmospheric pressure and guarantees adequate volatility of the components to be distilled without the risk of overheating, as the saturation temperature is not exceeded during the distillation process.

(2)　The condensation of both oil and water in the condenser is at the same time an effective method of separation by virtue of the poor solubility of the oil in water. The losses caused by the low solubility of the oil in water can be minimized by reusing the distilled water to produce steam.

(3)　Steam protects the ethereal oil from oxidation.

(4)　The plant parts are swollen up by the steam, thus facilitating the release of the oils through the dilated pores.

(5)　The process is cheap, technically simple and optimal as regards operational safety, as the use of water provides protection against explosion and fire.

The only disadvantages are the risk of decomposition by heat and hydrolysis, though the desirable hydrolytic cleavage of glycosidically bound ethereal oil components can also be advantageous here. The conversion of matricin into chamazulene is mentioned as an example of a conversion of a substance into a steam-volatile component under the conditions of steam-distillation:

Matricin

Chamazulen

4.6.2.1. Principles of steam distillation

The steam distillation of a defined, steam volatile component can be described quantitatively on the basis of Gibbs's phase rule by Dalton's law. This ideal case of steam distillation requires the components involved to be brought into equilibrium. If equilibrium cannot be established it is termed inhibited steam distillation; this is practically always the case in the recovery of ethereal oils from plants or plant parts.

The principles are first discussed with the example of ideal steam distillation.

The distillation of a pure liquid is obviously governed by pressure and temperature. At a given pressure it boils at a certain temperature and vice versa. Hence water boils at 100°C at atmospheric pressure. The system thus has only one degree of freedom. The phase (Gibbs) law

$$f = c + 2 - \varphi \tag{1}$$

applies, where f = number of degrees of freedom, c = number of components, and φ = number of phases.

According to Treibs [4.176], a phase is the sum of all particles in a space which are at the same pressure, temperature and concentration as all the material components and are delineated from other phases by sudden alterations.

The phases present in the ideal steam distillation of an ethereal oil are the oil phase, the steam/vapour phase and the water phase. The components are oil and water. The number of degrees of freedom thus produced according to Gibbs's phase rule is $f = 2 + 2 - 3 = 1$. Only one degree of freedom is therefore available. The boiling temperature of the mixture is thus fixed if the steam distillation is carried out at atmospheric pressure.

However, the composition of the phases is also fixed. In practice this means that the composition of the vapour phase does not alter during the distillation process so long as all three phases are present.

The composition of the vapour phase and hence of the distillate obeys Dalton's law:

$$p = \sum_{i=1}^{i=n} p_i \tag{2a}$$

$$pV = nRT \tag{2b}$$

The sum of the components involved is

$$pV = (n_{H_2O} + n_{oil})RT \tag{3}$$

where $n =$ mole fraction, $p =$ total pressure (of vapour distilling off) (bar), $p_i =$ partial pressure of one component of the vapour mixture (bar), $V =$ volume (m^3), $R =$ general gas constant [J K^{-1} mol^{-1}), and $T =$ temperature (K). It is thus still not fixed as to how the product pV is distributed over the two components.

The partial pressures of the individual components can be calculated as fractions of the total pressure p from their mole fractions:

$$p_{H_2O} - \frac{n_{H_2O}}{n_{H_2O} + n_{oil}} p \tag{4}$$

$$p_{oil} = \frac{n_{oil}}{n_{H_2O} + n_{oil}} p \tag{5}$$

$$p_{H_2O} + p_{oil} = p \tag{6}$$

As regards the individual components,

$$p_{H_2O} = \frac{n_{H_2O}}{V} RT \tag{7}$$

$$p_{oil} = \frac{n_{oil}}{V} RT \tag{8}$$

In practice the partial pressures of steam and oil are of much less interest as part of the total pressure, and hence of the composition of the distillate, than the quantity proportions.

Division of (7) by (8) produces

$$\frac{p_{oil}}{p_{H_2O}} = \frac{n_{oil}}{n_{H_2O}} \tag{9}$$

The mole number is

$$n_{H_2O} = \frac{m_{H_2O}}{M_{H_2O}} \tag{10}$$

$$n_{oil} = \frac{m_{oil}}{M_{oil}} \tag{11}$$

where $m =$ mass, $M =$ molecular weight

Incorporation of (10) and (11) into (9) gives

$$\frac{p_{\text{oil}}}{p_{\text{H}_2\text{O}}} = \frac{m_{\text{oil}} M_{\text{H}_2\text{O}}}{m_{\text{H}_2\text{O}} M_{\text{oil}}} \tag{12}$$

which after transformation gives

$$\frac{m_{\text{oil}}}{m_{\text{H}_2\text{O}}} = \frac{p_{\text{oil}} M_{\text{oil}}}{p_{\text{H}_2\text{O}} M_{\text{H}_2\text{O}}} \tag{13}$$

The molecular weights and partial pressure ratio must therefore be known if the water:oil quantity ratio in the distillate is to be predicted.

The partial pressure ratio can be deduced from graphs containing the logarithm of the saturation vapour pressure of the components as a function of the reciprocal temperature extended by a factor of 1. The type of graph must guarantee a linear course of the logarithm of the vapour pressure plotted against the reciprocal temperature, as shown in the graph for fenchone and water (Fig. 4.83).

For the fenchone + water system

$$\log p_{\text{fenchone}} = 1.60 \text{ and } \log p_{\text{H}_2\text{O}} = 2.88$$

apply *approximately* as

$$\frac{p_{\text{oil}}}{p_{\text{H}_2\text{O}}} = \log p_{\text{oil}} - \log p_{\text{H}_2\text{O}}$$

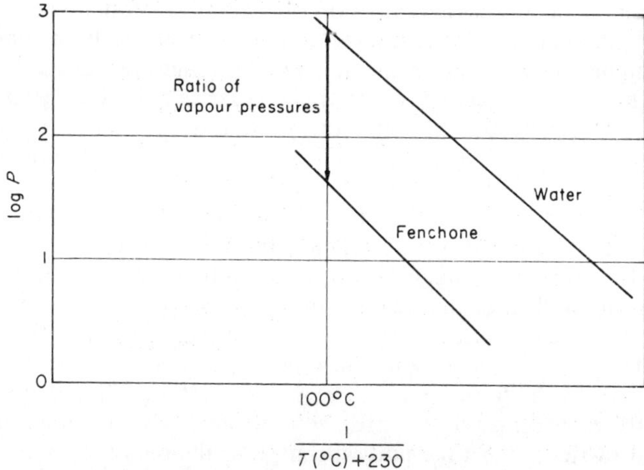

Figure 4.83 Partial pressure plot of fenchone/water.

and hence it follows that

$$\log p_{oil} - \log p_{H_2O} = 1.60 - 2.88 = -1.28$$

and hence

$$\frac{p_{oil}}{p_{H_2O}} = 0.053$$

The molecular weight of water is 18 and that of fenchone is 128. Insertion of these into (13) gives

$$\frac{m_{oil}}{m_{H_2O}} = 0.053 \cdot \frac{128}{18} = 0.377$$

i.e., the distillate is composed of 0.377 g fenchone per 1 g water.

These calculations are applicable only to pure, non-miscible two substance systems. In the case of oil mixtures the boiling point and the composition of the vapour phase alter during the distillation.

4.6.2.2. Application of steam distillation

The conditions of ideal steam distillation as described in Section 4.6.2.1 are not fulfilled in steam distillation. No equilibrium is established, as the oil phase does not normally come into direct contact with the vapour phase because the oil is transported with difficulty through the cell walls by substance exchange processes such as diffusion. Thus there is an excessive consumption of steam for the expulsion of a certain quantity of oil. This can be shown by the fact that the recovered ethereal oil has to be steam distilled again. A higher oil/water ratio in the second distillation indicates that this hindrance had taken place. However, this is the case only so long as the oil is uniform, as otherwise the oil/water ratio alters during the distillation and causes varying steam consumptions. The method has only a limited practical value, as naturally occurring ethereal oils are not uniform and moreover the degree of comminution of the drug plant material also represents an additional influence in the establishment of the equilibrium.

Under ideal conditions the oil/water ratio is independent of the distillation rate. If the rate of flow of the steam is increased during the drug distillation, and hence the time of contact with the drug material shortened, the oil/water ratio will decrease with increasing hindrance to the release of the oil from the drug material. In the limiting case, i.e. at very high distillation rate, the oil content of the steam tends towards zero. Conversely, a reduction of the throughput of steam per time unit will increase the oil/water ratio until the value of the ideal steam distillation is reached in the limiting case of complete saturation of the steam with oil. Long periods of contact of the steam with the drug material are required for this and hence the absolute quantity of oil recovered per time unit is small.

These limiting cases are uneconomical and an optimum, which is different for each drug material, must be sought. It can easily be seen from the morphological differences and the points discussed above that the distillation conditions for camomile flowers will be different from those for pine needles. Depending on the drug plant material, 40–80% of the oil saturation value in steam is generally taken as the working value.

In addition to the coefficient of diffusion, which is determinative for the diffusion of the oil out of the plant parts, the saturation concentration of the oil in steam is especially important for the choice of the distillation conditions. Nitrogen and oxygen containing components of the ethereal oil can be 100 times more soluble than hydrocarbons in hot steeping water and thus rapidly produce a partial vapour pressure which is close to the saturation vapour pressure and, hence, close to the oil/water ratio favourable for distillation. Hydrocarbons, which are only slightly soluble in water, diffuse more slowly out of the plant cells and, hence, go over into the distillate with the steam later than the more polar components.

Diffusion of the oil out of the plant cells is facilitated by previous comminution of the drug plant material, though if it is ground too finely losses can be caused by volatility or oxidation of the ethereal oils. The flow of steam can also be hindered by dense packing of fine particles. The chosen degree of comminution is strongly influenced by the type of drug material and the plant species from which it is obtained.

Leaves, flowers, grasses and herbs with thin stems and small buds are usually steam distilled without first being comminuted. This is possible with seeds, berries and fruits only if the temperature of the steam is sufficient to burst their skins. These medicinal plant parts must otherwise be broken open by crushing or coarse comminution (the 'crushing' for example of aniseed and fennel in the pharmacy for tea infusions is also to be seen from this point of view). Barks, roots, rhizomes and woods must also be comminuted before being steam distilled.

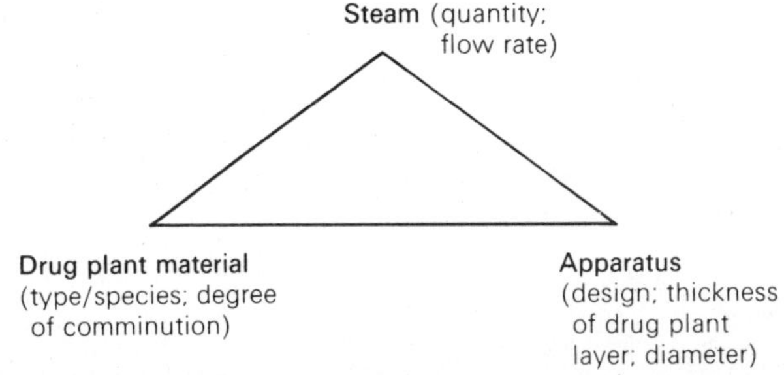

Figure 4.84 Triangular diagram for drug material/steam/apparatus.

The above statements can be summarized in a triangular illustration which reflects the interplay of drug material, steam and apparatus used (Fig. 4.84).

Together with the type and degree of comminution of the drug plant material, the quantity of steam required for complete separation of the ethereal oils plays a major economic rôle.

The parameters of the flow-rate of the steam, together with height, diameter and design of the apparatus used are the most important variables for optimization of steam distillation. In a given apparatus the flow-rate of the steam is the most important variable for adaptation of the distillation parameters to the particular drug plant material being distilled.

Chapter 5

Apparatus and Machinery

5.1. Apparatus for sorting and grading drug plant materials

Apparatus for grading drugs is required for products undergoing no further processing, i.e. the graded product is the end-product, or a part thereof. This is the case with individual drugs sold as infusions or infusion mixtures from whole drug plants.

Machines for grading drugs operate on two principles:

Grading by size according to length or diameter.
Optical sorting, i.e. sorting by light–dark differences or by colour differences.

5.1.1. GRADING BY SIZE

This is mechanical grading. A distinction is made between length and diameter grading. In length grading, material of various sizes is put in a machine in which it is moved along by vibration on cylindrical sieving rods so that it is aligned on its longitudinal axis. When the smaller fragments come to a gap in the rods they fall through, whereas the larger ones continue to move forward. If a number of gaps are arranged in order of increasing size, several length fractions can be obtained. Figure 5.1 shows the principle of a machine with gaps for three fractions. Diameter grading selects by diameter. The material is moved along on vibrating rods or steel rollers revolving about their longitudinal axes in opposite directions to each other from the point of entry through to the end. The material falls down between the rods or rollers when the diameter of the fragment is smaller than the distance between the rollers. The principle is shown in Fig. 5.2.

Optimum results are obtained by a combination of grading by diameter and by length. A machine consisting of a diameter grader and three length graders subsequently connected in series gives a total of nine different fractions, as shown in Table 5.1.

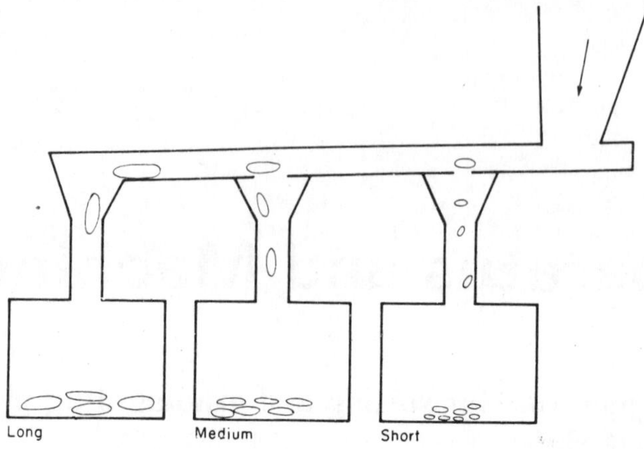

Figure 5.1   Grading by length into 3 fractions.

Sorting by size is used for products graded only by differences in size but not by other characteristics, i.e. colour, surface texture, etc. The sorting of eggs into various market grades is one example of a typical size grading. However, if other characteristics, such as colour, are important, as is the case, say, in the sorting of green from red tomatoes, other methods such as optical or visual grading must be used.

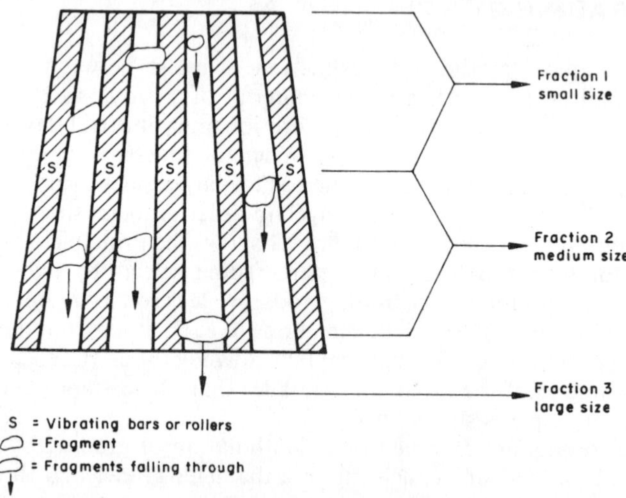

Figure 5.2   Grading by fragment diameter.

Table 5.1 Combined diameter and length grading to yield nine fractions

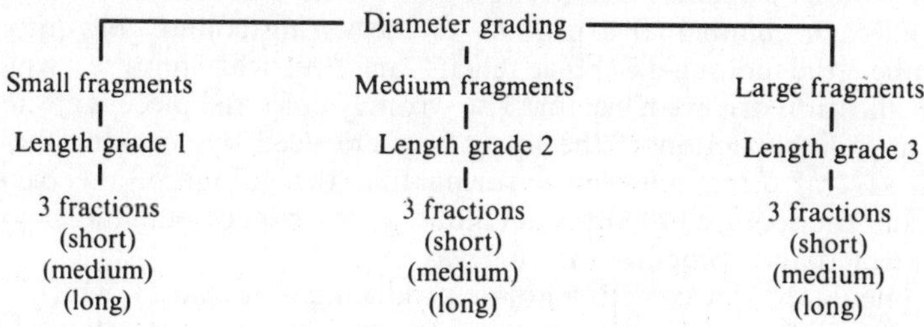

5.1.2. OPTICAL GRADING

In optical grading either light–dark contrast is measured with monochromatic light or a colour is measured with dichromatic light. Monochromatic and dichromatic sorting machines will consequently differ from one another. Figure 5.3 [5.1] illustrates the basic design of a dichromatic machine.

The material is led from the hopper (1) via a vibrating chute (2) on to a V-shaped conveyor belt (3). The speed of the conveyor belt and of the vibration of the chute must be tuned to one another so that the products fall on to the conveyor belt in such a way that they are separate, individual pieces and can fall through the inspection chamber (4) one behind the other in a line. In this way they are sorted optically.

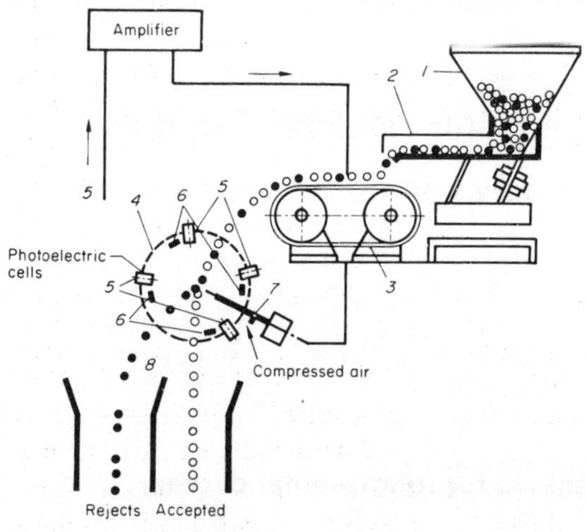

Figure 5.3 Dichromatic grading.

The optics consist of 2–4 inspection channels (5) which test the product from various directions. A background of the theoretical colour is placed behind each channel. If a product of the wrong colour goes into the test chamber the colour-difference elicits an electrical impulse, which after intensification drives an ejector. The wrongly coloured piece is ejected at (8).

Very reliable action of the sorter is guaranteed by checking the product from several directions and ensuring that the colour of the background matches the required theoretical colour. Even maggot punctures in fruit can be detected and recognized in this way.

A pneumatic device with a nozzle producing a flat air jet, which is opened for a brief moment so that the air jet ejects the wrongly coloured piece, is used as the ejector mechanism for larger products such as nuts, almonds, pistachios, raisins, etc.

An electrostatic ejector is used for smaller products such as grains of rice, cereals, ergot, etc. Here the piece which is to be removed is charged electrostatically by a needle carrying a high voltage and then attracted by a negatively charged deflector plate. This change in direction sends the piece of product which is to be removed into the run-off channel for reject material.

In cases where colour differences need not be detected, monochromatic light suffices for the separation. Here the reflection of light from the product is measured with a photocell. The arrangement of the inspection channels is the same as in dichromatic sorting.

Optical sorting apparatus have now attained a precision far greater than that of the human eye. For example, suitable adjustment of the sensitivity makes it possible to detect colour differences hardly perceptible to the eye. The advantage of such sorting machines becomes obvious when it is remembered that the eye tires very rapidly during sorting by hand, whereas sorting machines can work round the clock with unaltering precision. Size grading with optical apparatus can solve practically any grading problem.

5.2. Machinery for comminution of drug plants

5.2.1. MACHINERY FOR CUTTING UP DRUG PLANTS

Cutting up of drug plants is the first stage in the crushing process and normally precedes milling or grinding. Machines for cutting up drug plants must be of robust construction and have a wide variety of cutting devices. In principle a distinction is made between a simple longitudinal cut and the rectangular or cubical cut.

5.2.1.1. Machines for longitudinal cutting

The principle of a machine for making a simple longitudinal cut is shown in Fig. 5.4.

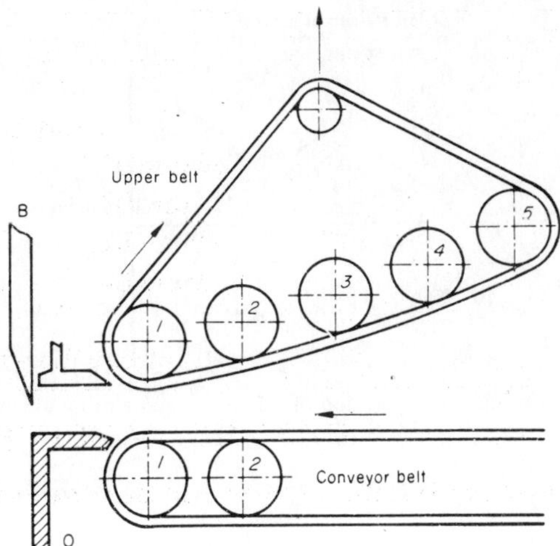

Figure 5.4   Longitudinal cutting machine.

The drug plant is transported on the conveyor belt to the orifice O. The width of the conveyor belt can vary from 250 to 400 mm, depending on the type of machine. The projecting parts of the drug plant are caught by an upper belt running on fluted rollers on its way to the orifice, pressed together by the convergence of the upper belt and the conveyor belt, and the drug plant then passes into the cutting zone in this prepressed form. The cutting blade B makes a vertical cut downwards. On the upward stroke of the blade the head of the blade moves away from the orifice in order not to impede the forward movement of the material for the next cut.

The length of each cut section is determined by (1) the forward speed of the conveyor belt, and (2) the cutting rate of the blade.

The section lengths can be continuously adjusted from 1 to 12 mm by appropriate tuning of both parameters. Special accessories are necessary for section lengths of less than 1 mm.

The drug plants must not slip or slide during conveyance to the orifice. Metal conveyor belts are therefore unsuitable. Belts made of rubber or PVC or so-called brass profile rod belts, i.e. conveyor belts which in cross-section consist of brass profile rods laid against one another, are used.

Defined cuts cannot be obtained with the longitudinal cutting machines described so far, as is shown by Fig. 5.5 with the example of leaves. 'Statistical' sections, in which longitudinal, transverse and oblique cuts occur in a statistically random manner, are produced. Uniform cuts are produced by the rectangular or cubic cutting technique.

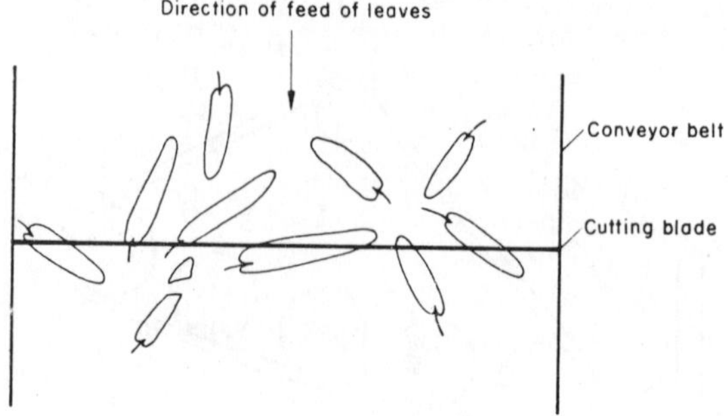

Figure 5.5 Principle of operation of a longitudinal cutting machine.

5.2.1.2. Machines for rectangular and cubical cutting

These are, in principle, the same as machines for longitudinal cutting. So-called slit-blades, which cut the pressed drug plant bundle in a longitudinal direction as it passes through the orifice, are placed only in the region of the orifice or directly behind it, i.e. in front of the main blade. The design is shown in principle in plan view in Fig. 5.6.

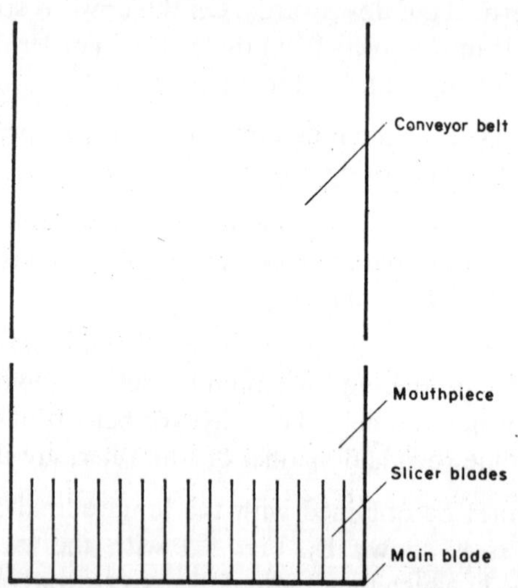

Figure 5.6 Rectangular cutting machine, plan view.

The slit-blades can be either flat steel blades or circular blades. Flat steel blades are used for roots, bark and other hard parts of drug plants, though they can also be used for leaves and soft parts. Circular slit-blades are however more suitable for leaves and soft parts. They consist of circular disc blades fitted at the desired distances on an axle and project from below into the slit by 1/6 of their circumference. The compressed drug plant is cut longitudinally as it passes through the slit orifice. The next main transverse cut produces a rectangular or cubic cut, depending on the adjustment of the speed of the conveyor belt. When the blades in the orifice get blunt they are pushed in by a further 1/6 of their circumference so that the entire circular areas of the blades are used in succession. When all of them have been used the blade axle with the circular blades is replaced by a new one and the blunt circular blades are removed from the machine and resharpened. Even drug plant sections can be made with rectangular and cubic cutting machines.

5.2.2. MACHINES FOR GRINDING DRUG PLANT MATERIALS

An almost limitless number of grinders is available. It is therefore impossible within the limits of this survey to describe all the design types. Classification by the demands made by the material which is to be ground does not appear to be very meaningful, as often no clear distinction can be made between percussion, shearing and cutting requirements. Classification by type and design of the grinder appears more reasonable, although the transitions from one type of grinder to another are also blurred here by the combination of various grinding accessories, such as a fan-beater with a sieve or with a percussion grinding track.

Finally, manufacturers of grinders are increasingly tending to offer apparatus with interchangeable grinding accessories ('universal grinders') which can potentially solve many of the problems which arise in grinding.

Samans [5.2] has produced a use-related apparatus scheme for the chemical and pharmaceutical industries, an extract from which is shown in Table 5.2.

The following discussion of the individual types of grinder begins with apparatus for coarse grinding, unlike Samans' listing, which begins with grinders for obtaining very fine material and ends with coarse grinders.

5.2.2.1. Hammer mills

Hammer mills are robust coarse grinders for shattering coarse, brittle or friable substances with rotating hammers. The 'hammers' are percussion devices fitted to the rotor of the grinder and generally hang freely on four carrier bolts. These grinders are therefore sometimes also called pendulum hammer mills. Figure 5.7 shows the principle of a hammer mill.

The 'hammers' are flung out radially by their rotation and strike the

Table 5.2. Principle, milling action, rotation speed and areas of use of various types of grinders

Machine type	System	Milling action	Rotation speed (m/s)	Examples of application in drug material comminution (overlappings common, depending on purpose)
Pin mill		Impact	80–100	*Strophanthus* seeds, ergot, dried extracts
Blower mill (micronizer)		Impact	40–110	Pulverization of leaf, bark and root drug materials
Cross beater mill (disintegrator)		Impact and shearing	50–70	Leaf, bark and root drug materials
Hammer basket mill		Impact	70–90	Tobacco, leaf, root and herbal drug materials
Toothed disc mill		Friction and shearing	5–16	Dried extracts, freeze dried products, fruits and seeds
Knife or cutter mill		Cutting	5–18	Leaf, herb and root drug materials and also barks for subsequent percolation
Hammer mills		Impact	40–50 (100)	Coarse comminution of brittle, friable substance occurring in large pieces, such as root drug materials

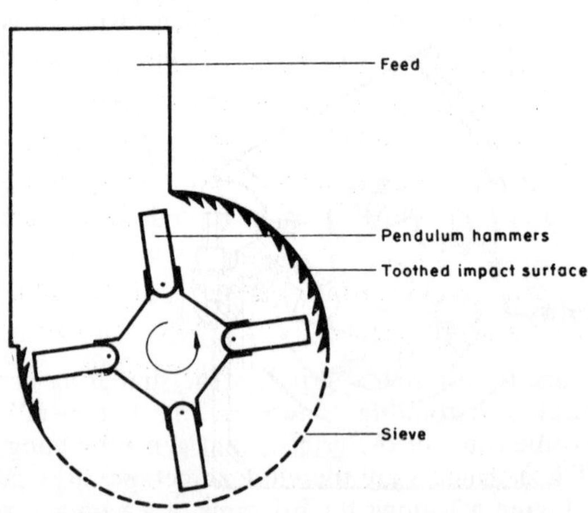

Figure 5.7 Principle of the hammer mill.

material fed in from above. The fineness of the ground material is governed by the sieve attachment. Grinding can also be modified for particular requirements by the selective use of various types of hammer, such as plate steel, ribbon steel or beam hammers. Plate and ribbon steel hammers differ only in the strength of the pendulum striker, whereas beam hammers consist of two penduli connected by a cross-beam which lie adjacent to one another on the rotor axle.

5.2.2.2. Cutting mills or knife mills

Cutting mills, sometimes also known as cutting granulators, were originally developed for grinding scrap plastic. Their mechanism consists of the cutting action between rotating blades and fixed blades mounted in the casing of the machine. Cutting mills have vertical and horizontal axles. Figure 5.8 [5.3] shows an example of a horizontal design in longitudinal section.

The illustrated grinders have fixed quadrilateral blades of which all four edges can be used, thus guaranteeing a long useful life. Figure 5.9 [5.3] shows a vertical design in cross-section.

The fineness of the ground material produced by cutting mills depends on the particular sieve used. The resulting proportion of dust, particularly in the grinding of drugs or medicinal plants, is influenced by the speed of the rotor. A high rotor or circumference speed favours the production of dust. The peripheral speeds, which are normally 10–18 m/s, are therefore reduced to 4–6 m/s for drug grinding.

The principal application of cutting mills in drug plant grinding is the low-dust preparation of leaves, bark and roots for subsequent percolation or maceration.

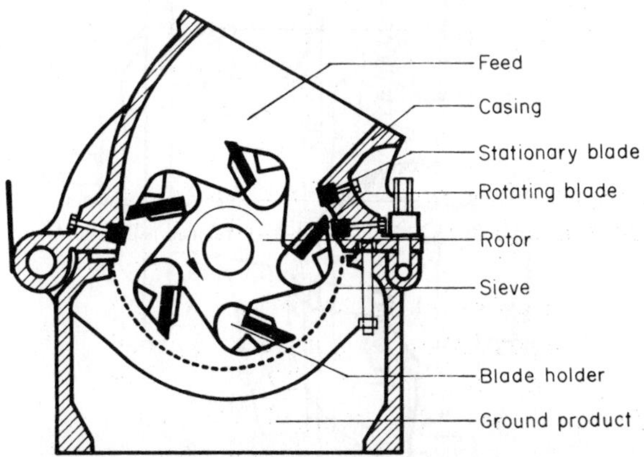

Figure 5.8   Section through a horizontal cutting mill.

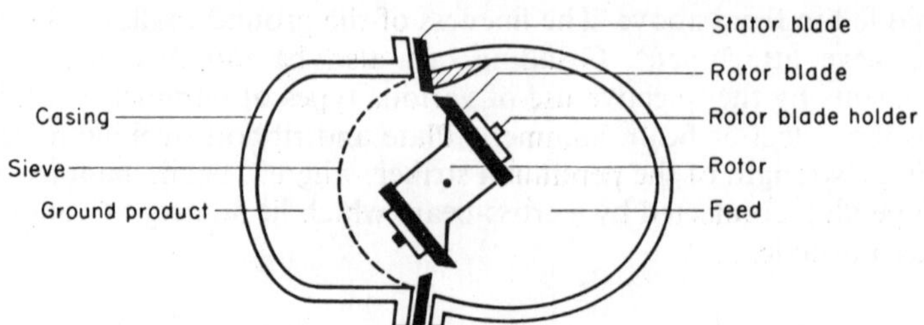

Figure 5.9   Section through a vertical cutting mill.

5.2.2.3. Toothed disc mills

Toothed disc mills are likewise used for low dust grinding of drug plants, in this case by friction and shearing. They are available in vertical and horizontal design. The principle of a mill with a horizontal axle, which is the type most frequently used for grinding drug plants, is shown in Figure 5.10 [5.4].

The grinding component of the mill consists of a fixed and a rotating toothed milling disc cast in one piece and bearing teeth arranged in concentric circles. The interlocking teeth of both discs diminish in size from

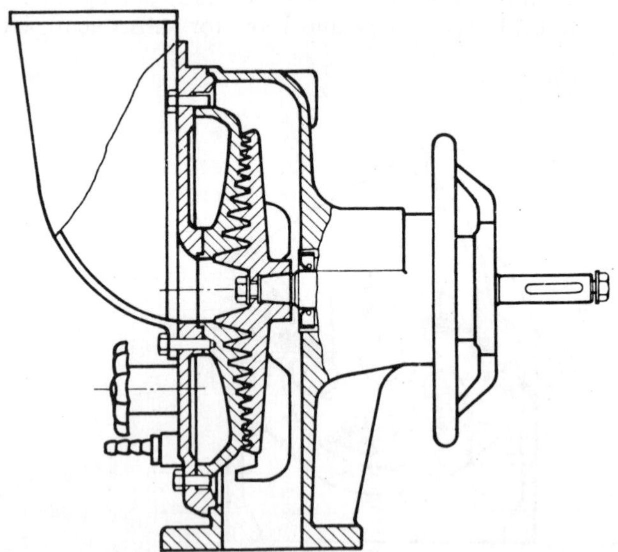

Figure 5.10   Cross-sectional view of a toothed disc mill with horizontal axle.

the centre to the periphery. The material to be ground is fed in at the centre and progressively ground up as it flows between the discs. Two process parameters can be altered at will here, namely the speed of the rotating disc and the distance between the discs.

The fineness of the ground material is governed mainly by the distance between the milling discs. The wider the gap, the coarser will be the resultant ground product.

The amount of dust produced depends on the speed of the rotor. The greater the speed, the greater will be the amount of dust. Very narrow particle size ranges with a very low proportion of fine material (dust) can be obtained with optimum adjustment of the mills and hence toothed disc mills are mainly used in drug plant grinding prior to extraction. The peripheral speed of the rotating disc here is about 16 m/s. The stationary grinding disc can be cooled.

5.2.2.4. Hammer basket mills

The hammer basket mill is a centrifugal mill. The material to be ground is fed into its centre. It is not to be confused with the hammer mill, into which the material is fed peripherally and in which the entire flow of the material being ground takes place peripherally. Figure 5.11 [5.4] shows the principle of a hammer basket mill.

The hammers hang freely on the periphery of the rotor and are flung out radially as the mill rotates. The material to be ground is fed into the centre of the mill and struck by these hammers against the circular sieve basket surrounding the rotor and thereby pulverized. The fineness of the ground material is governed by the particular sieve used. Grinding is achieved purely by an impact action.

Hammer basket mills are used particularly for pulverizing tough, fibrous material such as medicinal leaves, roots, weeds and herbs.

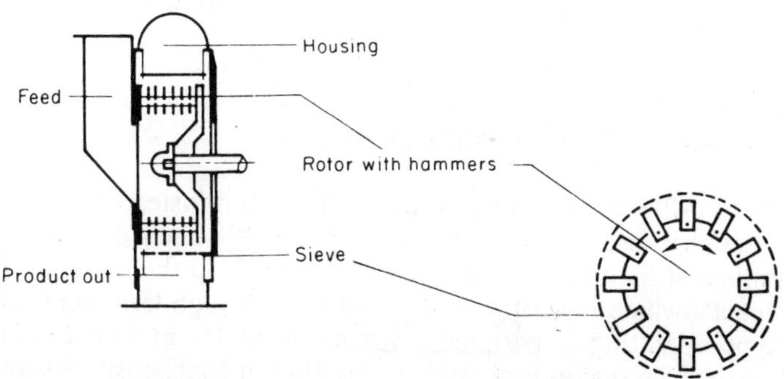

Figure 5.11   Principle of the hammer basket mill.

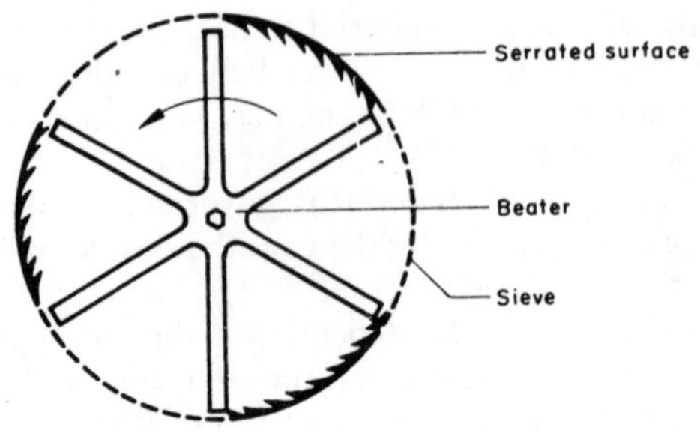

Figure 5.12   Cross beater mill.

5.2.2.5. Cross beater mills or disintegrators

These grinders, which are also widely used in the laboratory, have a rotating grinding action against a sieve basket fitted around its circumference. They thus show a certain resemblance to hammer basket mills, though they achieve a finer grinding than the latter. Figure 5.12 shows the principle in front view.

Instead of a continuous annular sieve surrounding the mill, this annular track can be divided up into alternating sections of serrated friction graters and sieves, as shown in the illustration. The sieves are thus situated in strips between the friction graters. This increases the impact and shearing action and thus makes it possible to obtain finer products, though the smaller sieve area causes the material to remain longer in the grinding zone, which is a disadvantage with material sensitive to heat produced by the friction. The factor determining the fineness of the ground material is the mesh-width of the sieve.

Cross beater mills are used for pulverizing medicinal leaves, bark and roots. Their robust construction makes them less susceptible to damage by foreign bodies than pin mills, though the latter grind the material more finely.

5.2.2.6. Blower mills or micronizers

Blower mills differ from hammer cross mills in that instead of a 5–6-armed cross beater they use a rotor with $\sim$ 30–40 firmly attached hammers. This large number of hammers has the effect of subjecting the material to many more impacts, which in turn provides a more thorough treatment. Grinders of this type operating at peripheral speeds of 40–110 m/s have a high air-throughput, which makes it possible to grind even heat-sensitive substances, such as hard fats, at room temperature. The name 'blower mill' is attribu-

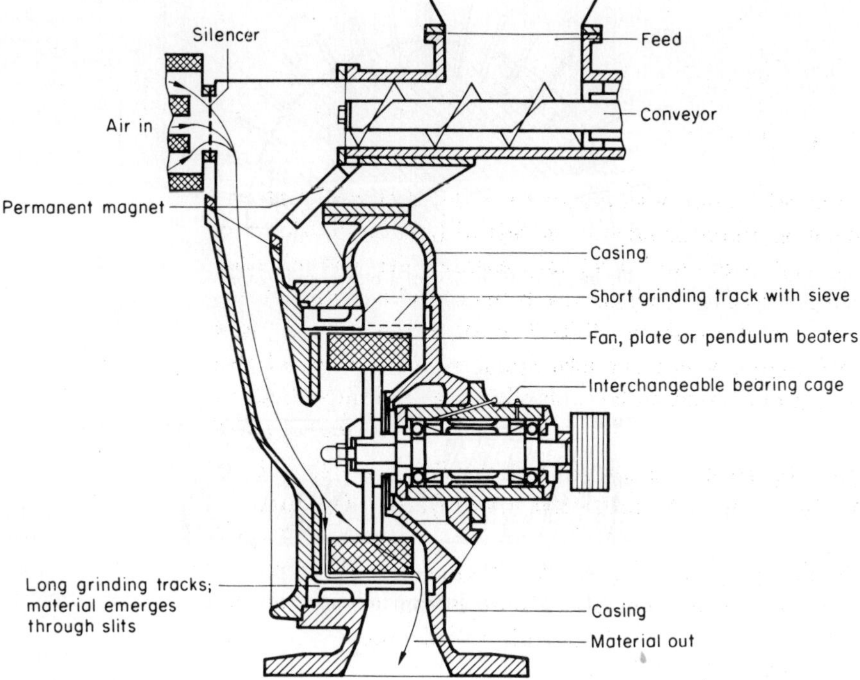

Figure 5.13 Longitudinal cross-section of a blower mill with screw conveyor and various grinding surfaces.

table to this high air-throughput. Figure 5.13 shows the principle in longitudinal section [5.5].

The material is transported by the conveyor screw to the centre of the rotor and is then flung outwards by the hammers against the annular sieve or the impact tracks. Figure 5.14 [5.5] shows a front view of the rotor with hammers.

In addition to ribbed impact tracks with a slit (Fig. 5.15a), combined tracks with short impact ribs and sieves can be used (Fig. 5.15b).

The fineness of the ground material can be regulated both by the construction and revolution speed of the rotor and by incorporation of various sieve and ribbed impact tracks. Blower mills are suitable for pulverizing medicinal leaves, bark and roots and other tough, fibrous materials. The fineness of the ground material obtained is such that its particle size is much smaller than the mesh size of the sieve and hence the final fineness is determined not by the sieve but by the powerful impact action produced by the high speeds.

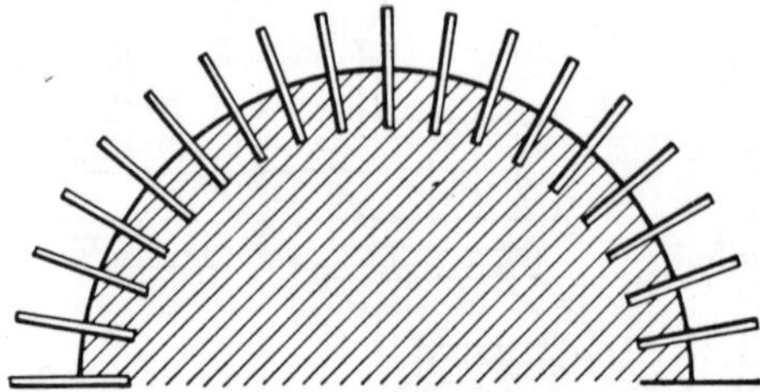

Figure 5.14   Plan view of a blower mill rotor.

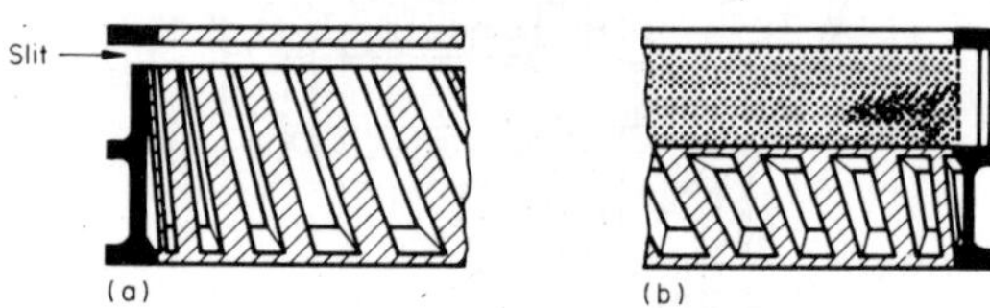

Figure 5.15   Types of grinding track in a blower mill.

5.2.2.7. Pin mills

Pin mills are high-speed grinders for fine and extremely fine milling by the principle of impact pulverization. The milling takes place between two discs fitted with concentric circles of pins, the tips of which interlock.

Figure 5.16 shows the principle of a mill with a rotating and a stationary pin disc in longitudinal section (side view).

Grinders with two rotating pin discs are especially used for milling drug plant materials to powders. These can run unidirectionally at different speeds, or in contra-rotation, depending on the product. Finer products are generally to be expected with contra-rotation, though the tendency to clogging and sticking of the product sets limits here. As grinding is brought about purely by an impact action from pin to pin, pin mills operate without sieves.

The degree of grinding is influenced by the following factors:

Arrangement of the pins. Widely spaced pins give a coarser end-product, though this reduces the risk of the product sticking to the pins. Wide spacing is therefore indispensable for certain products.

Ratio of mill rotation speed to diameter of disc. The smaller the diameter of the disc, the greater must be the rotation speed of the mill

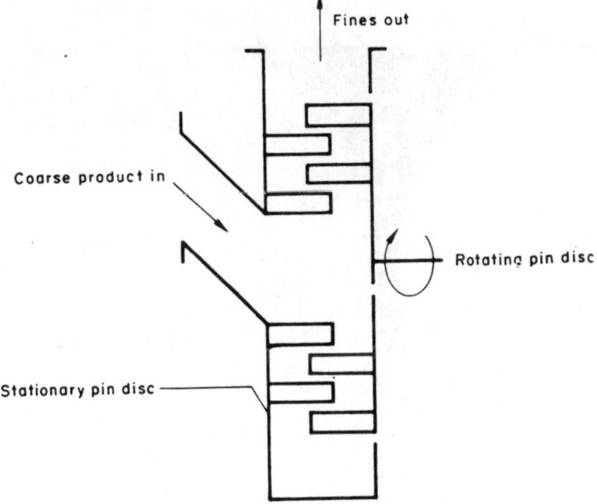

Figure 5.16 Schematic view of a pin mill.

to attain the same ultimate fineness. The rotation speed of the outer row of pins is decisive for the fineness. In grinders with one rotating disc its maximum is about 160 m/s, whereas relative speeds of up to 240 m/s have been attained in grinders with two contra-rotating discs. This corresponds to rotor speeds of up to 22 000 rev/min.

Supply of grinding material. Reduction of the supply increases the ultimate fineness of the ground material and vice versa. The coarse material must always be fed in evenly if a uniformly ground material is to be obtained.

The contra-rotating pin mills mostly used for grinding drug plant materials have a wide milling chamber casing to prevent rapid clogging. Under favourable milling conditions ultimate finenesses of the order of 20 μm can be attained, depending on the product, in grinders with one rotating disc, and of less than 5 μm with contra-rotating discs.

Pin mills are therefore used for the pulverization of materials where very fine end products are required. This is particularly the case with grinding of dried extracts.

5.2.2.8. Fluted roller grinders

These are used in comminution processes where a narrow particle size range and the minimum amount of fine material and dust are desired. They operate on the principle of shearing comminution and thus resemble cutter mills (Fig.

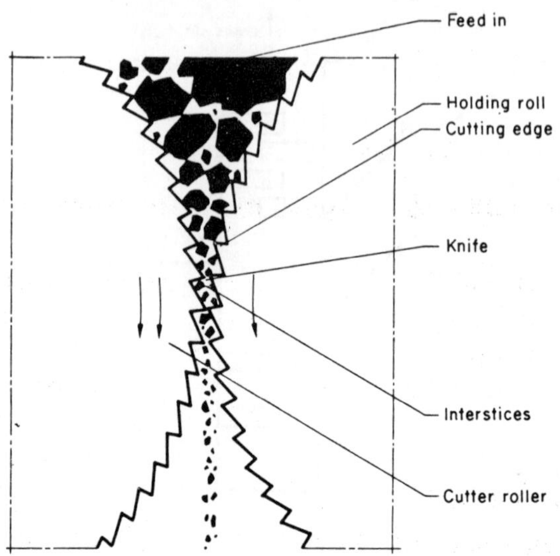

Figure 5.17 Comminution principle of the fluted roller grinder.

5.17). The fluted roller grinder consists of a slowly rotating holding roller (shown on the right in the illustration) and a more rapidly rotating cutting roller (on the left in the illustration).

To make them interchangeable, both rollers have the same number of equally large cutting teeth. The holding roller is designed so that in the working space the cutting surface points upward and there is a clear angle beneath it. The cutting roller is arranged the other way round. When running in the same direction, but at different speeds, the rollers produce friction. The edges of the holding roller act as cutting edges here and those of the more rapidly revolving roller as the actual cutting blades or knives. The material is poured between the rollers from above and comminuted by the 'knives' and 'cutting edges'.

The cavities formed between the blades and cutting edges of the rollers cause only those lumps of the product which are larger than the cavities to be comminuted. This minimizes the amount of dust produced. The process is governed by the following parameters:

 Size and number of the cutting edges.
 Revolution speed of the rollers.
 Distance between the rollers.

It has the following advantages:

 Short period of stress on the material.
 Little heating.
 Small amount of dust.

One disadvantage is that only products with a maximum hardness of 4 mohs, i.e., relatively soft substances, can be comminuted. Fluted roller mills are used for the comminution of spices, aromatics and other drug plant materials containing ethereal oils, as the latter are conserved during comminution with this apparatus. The smallest attainable particle size, depending on the product, lies between 100–500 µm.

5.2.2.9. Cold milling plants

These are grinders fitted with devices for (1) cooling the grinder, (2) cooling the air or an inert gas passing through the grinder, or (3) cooling the air passing through the grinder in order to render the starting material brittle and friable.

Nitrogen, air or carbon dioxide are used as cooling media. Nitrogen and carbon dioxide are generally preferred over air as with these there is less risk of explosion of dust. The simplest method is to cool the grinder by forcing the liquefied gas into the grinding zone through a jet as illustrated in Fig. 5.18. Only the grinder is cooled in this system. The cooling process is controlled by

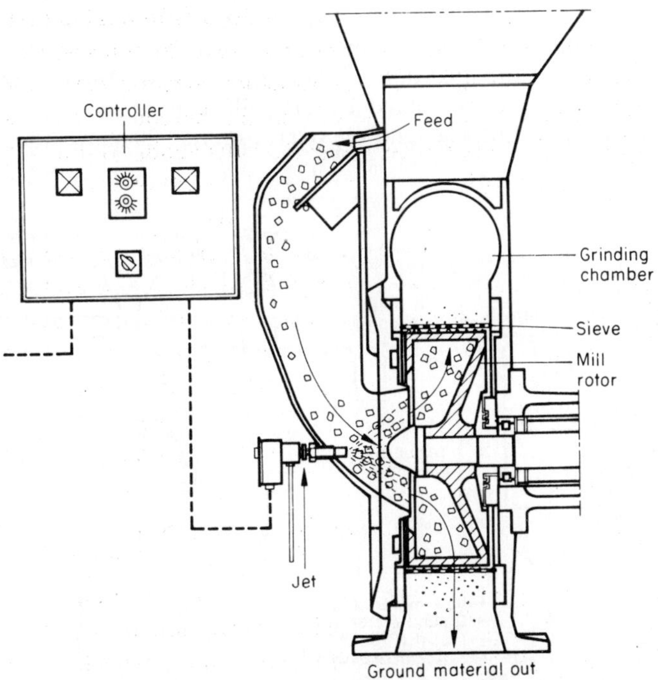

Figure 5.18 Cold milling using liquefied gas.

thermoelements placed in the grinder. The cooling medium goes to waste after having passed through the grinder. This accessory equipment is easily installed, but has the disadvantage of high operating costs due to the great losses of cooling agent.

Cold milling plants in which the cooling medium is recycled are therefore more advantageous and their running costs are increased only by unavoidable losses due to vaporization. Figure 5.19 [5.6] shows the principle of such plants. In system 1 the temperature of the grinder is regulated, whereas in system 2 the material is also rendered friable before being milled. The controlling parameter in both cases is the temperature at the mill exit. The closed circulation necessitates discharge of the product through a bucket-wheel sluice. The temperature in cold milling can fall to below $-20°C$. However, for reasons of economy the product is cooled only until it can obviously be ground, which as a rule is at temperatures between $+10$ and $-5°C$. The consumption of nitrogen or carbon dioxide, depending on the product, then amounts to 0.02–0.6 kg per kg of milling material.

5.2.2.10. Sifter or classifier mills

Sifter or classifier mills combine a grinder with a sifter or classifier in one apparatus. The material is put in the mill, whereupon the sifter wheel throws out the fine material. The coarse material remains in the milling chamber until it has reached the final desired degree of fineness. Impurities such as sand, which are not comminuted by the mill, can be removed from the milling chamber by a helical sand remover. Figure 5.20 shows the principle [5.7].

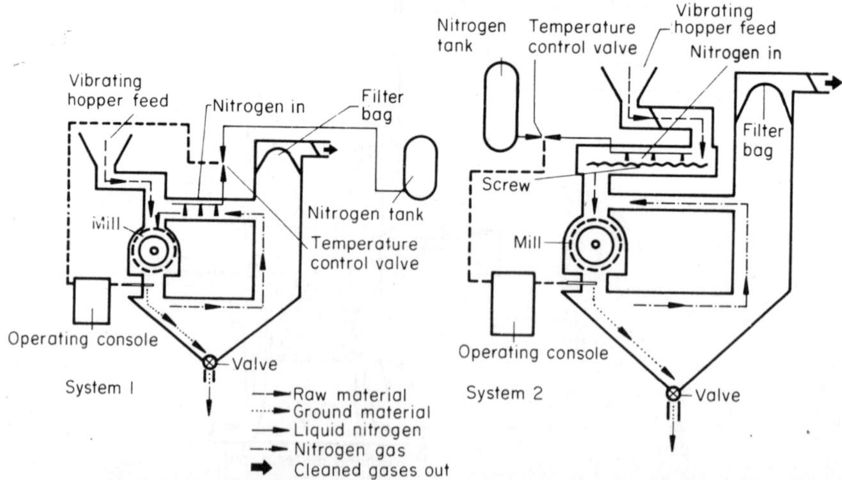

Figure 5.19 Cold milling plants with different circulation of coolant.

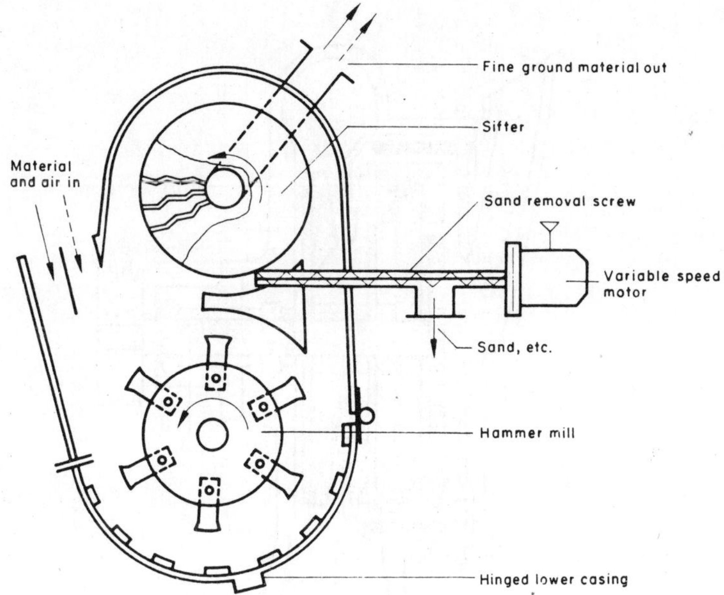

Figure 5.20 Schematic view of a Duoplex sifter mill with sand removal screw.

The sifter wheel here is situated above the grinder. The limiting particle size of the fine material is governed by the speed of rotation of the sifter wheel. In another form of this type of mill the sifter wheel lies behind the mill, as shown in Fig. 5.21 [5.8]. This arrangement permits a more compact design.

Sifter mills are used for the manufacture of powders with a sharply defined range of large particle sizes.

5.2.3. EXAMPLES OF USE

The choice of a mill for the comminution of a drug plant material is determined both by the properties of the starting material and by the purpose for which the milled product is subsequently to be used. Table 5.3 gives examples of the use of various types of mills in the comminution of drug plant materials and spices. Hammer mills are mainly used here for precomminution; no definite statements on fineness are made for this. Examples of materials which undergo this precomminution are alginates, *Curcuma* or turmeric roots, dried figs, gelatins, hops, ginger, cloves, tea and cinnamon bark.

When examining the data in Table 5.3 it must be remembered that other degrees of fineness can sometimes be attained by varying the fittings and rotation speeds of the mills.

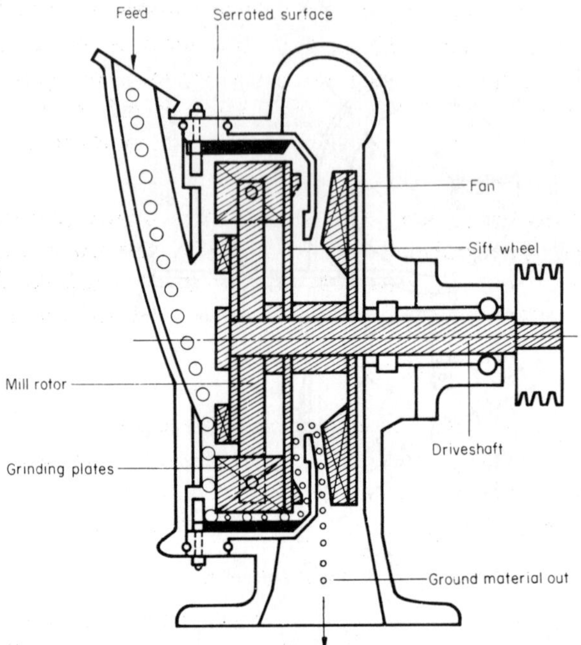

Figure 5.21 Cross-section through a Biplex crossflow sifter mill.

5.3. Machinery for classifying drug materials

5.3.1. AIR CLASSIFIERS

Lauer [5.9] divides air classifiers into the following groups:

> Gravity countercurrent classifiers
> Gravity crossflow classifiers
> Centrifugal classifiers
> Scatter blast classifiers

The principal and distinguishing characteristics of these classifiers are shown in Table 5.4. The particle size range in which sifting can be carried out extends from about 2 µm to several mm; in such a wide range sieving is superior to air classification as it is more economical. As the density of the particles affects the separation, the data on separating range and throughput given for the techniques are generally based on a material with a density of $\sim 2.5\,\mathrm{g/cm^3}$

Table 5.3 Degree of comminution attained by various processes for milling drug materials

Drug	Comminution achieved by				
	Cutter mill	Cross beater mill	Blower mill	Pin mill	Cold milling
Agar-Agar			90% < 200 μm	95% < 1000 μm	
Aniseed			99.5% < 500 μm		99% < 500 μm[a]
Birch leaves					99.5% < 200 μm[a]
Fenugreek seeds			98% < 500 μm		
Turmeric root		90–97% < 200 μm	97% < 160 μm		
Fennel		98% < 800 μm		90% < 600 μm	
Gelatin		99% < 300 μm	99.5% < 500 μm		
Gum arabic				99% < 200 μm	
Rose hips	+[b]	+[b]	90% < 200 μm		husks
					99.5% < 200 μm[a]
					98% < 400 μm[a]
Ginger		90% < 400 μm	99.9% < 500 μm		98% < 300 μm[c]
Cardamom		90% < 300 μm		90% < 300 μm	98% < 750 μm[c]
Coriander					95% < 750 μm[c]
Caraway		90% < 700 μm		90% < 400 μm	99.5% < 400 μm[a]
Bay leaves			97% < 200 μm		
Marjoram			99.9% < 300 μm		
Mallow flowers	35% < 1000 μm				
	85–90% < 2000 μm				
Mace					97% < 400 μm[a]
Nutmeg				90% < 1000 μm[d]	99% < 1000 μm[c]
				90% < 500 μm[e]	
				90% < 400 μm	
Clove		90% < 700 μm			
Pectin			98% < 300 μm		
Pepper, black		90% < 300 μm	97.5% < 300 μm		
Pepper, white		90% < 300 μm			99.9% < 400 μm[c]
Peppermint leaves	20–30% < 500 μm				
	98% < 2000 μm				
Rosemary					97% < 160 μm[a]
Celandine					98.5% < 215 μm[a]
Mustard seeds				90% < 500 μm	
Thyme			99.6% < 200 μm		
Juniper berries				90% < 2000 μm	
Cinnamon bark		90% < 230 μm	90% < 150 μm		99.7% < 300 μm[c]

[a] Cold milling with nitrogen. [b] No fineness data. [c] Cold milling with carbon dioxide. [d] Two rotors, same direction of rotation.
[e] Two rotors, counterrotating.

Table 5.4 Air classifiers

Type of sifter/classifier	Separating force	Range of use
Gravity countercurrent classifier	Gravity	$>200\ \mu m$
Gravity crossflow classifier	Gravity	$>200\ \mu m$
Centrifugal classifier	Centrifugal force	$2–100\ \mu m$
Scatter blast classifier	Gravity and centrifugal force	$50–200\ \mu m$

5.3.1.1. Gravity countercurrent classifiers

The simplest and oldest gravity countercurrent classifier, which is particularly widely used in the analysis of clay minerals, is the Gonell sifter, shown in Fig. 5.22 [5.10].

The Gonell sifter consists of a long vertical tube (a) with a conical lower part (b) and a glass attachment (c) at the lower end. Air enters through a thin tube (d) and whirls up the weighed material being processed in the tube. The fine material is carried up through the separator tube and is deflected at the cone (e) and then deposited in the upper glass bell jar (f). The coarse material falls back into the glass attachment (c).

The principle of the Gonell sifter has also been used, with certain modifications, in other analytical sifters.

Preparative gravity sifting into three different fractions is possible with an apparatus such as is shown in Fig. 5.23 (taken from photograph and illustrations [5.9]). The material to be processed is conveyed into the axis of the sifter. Air is forced in through valves or flaps at the bottom of the sifter and then upwards through a ring fissure between the outer wall and various fittings. The material being processed first goes on to a conical deflector and from there into the above-mentioned ring fissure. The airflow speed here is selected so that the coarsest particles in the material are separated by falling

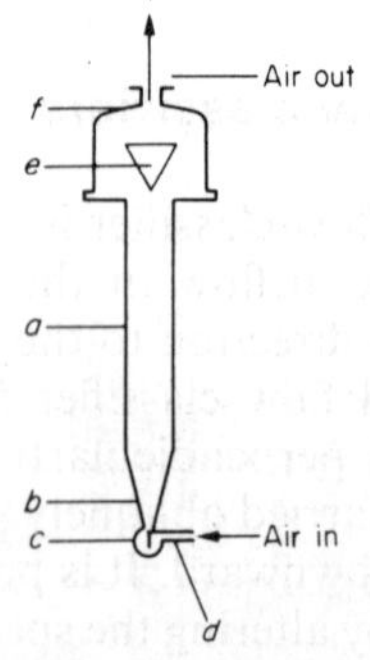

Figure 5.22 Gonell sifter.

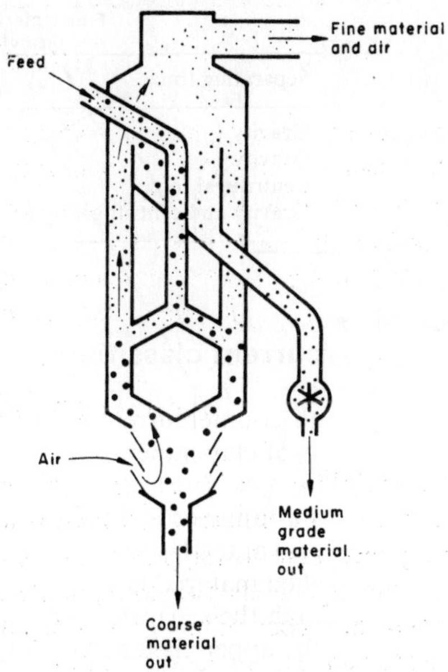

Figure 5.23 Preparative gravitational sifting into three fractions.

downwards. Medium sized and fine materials continue to be carried upwards by the airstream in the ring fissure. The flowspeed is so greatly reduced by the widening at the upper end of the apparatus that at this point the medium-sized material is also deposited. It leaves the sifter via a laterally attached tube closed with a chamber sluice. The fine material is conveyed with the air to the upper part and deposited in an attached cyclone. A separate blower is necessary for the air supply. The areas of use of this apparatus are the purification of plant products of all kinds (e.g. grains, drugs) and the removal of dust from synthetic granulates.

5.3.1.2. Gravity crossflow classifiers

The name gravitational crossflow classifier is derived from the construction of this classifier. Whereas the airflow in the gravitational countercurrent classifier flows in the opposite direction to the material being separated, the air enters the cross or lateral flow classifier from the side and meets the material supplied from above perpendicularly. Figure 5.24 [5.9] shows the principle. The fine material is carried obliquely upwards together with the air, and the coarse material falls downward. It is possible within certain limits to alter the limiting particle size by altering the speed of the airflow. Lateral flow classifiers of this simple design are especially suitable for removing dust, e.g.

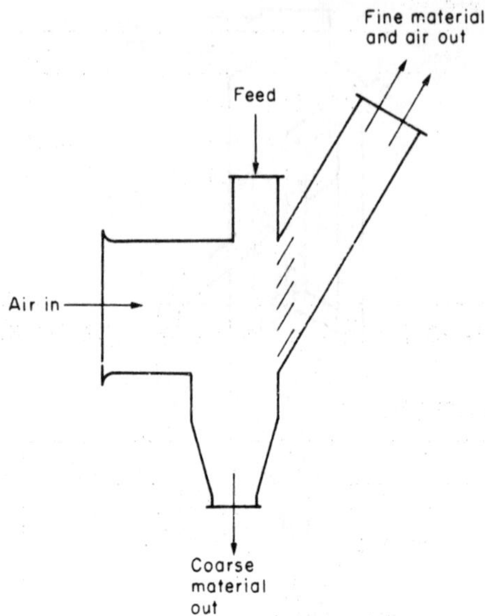

Figure 5.24　Principle of the crossflow sifter.

from coal. The zig-zag classifier, the principle of which is shown in Fig. 5.25 [see Ref. 5.9], is frequently used, though it is somewhat more complicated in design and mode of action. It consists of a vertical tube of rectangular cross-section and numerous sharp bends through which a stream of air flows upwards from the bottom and is turned at each bend. The material to be separated is fed in at one of these bends. The coarse material slides down to the 'lower' side, crossing the airstream at the next bend. This causes intensive whirling. The fine material is transported upwards by the air to the 'upper' side of the sifter. Figure 5.26 shows the conditions in detail.

The actual separation takes place in the region of the bend in which the stream of coarse and fine material cross the high-speed airstream. Although the separating effect of each individual stage is small, a very clean separation is eventually attained by the repetition of the process at each bend. The maximum output is ~ 1 ton/h. The particle size separation range lies between 100 μm and 10 mm. Larger throughputs are attained by using several sifter tubes in parallel. Figure 5.27 [5.11] shows two modifications of this arrangement [makers: Alpine AG, Augsburg].

In both types the air is led upwards from below. In type A the material to be separated is put on a sloping perforated plate through which air flows from below. The material is whirled up and goes into the separating channels; the coarse material falls back on to the perforated plate and leaves the classifier through the coarse material exit. The fine material is carried upwards under the impetus of the air.

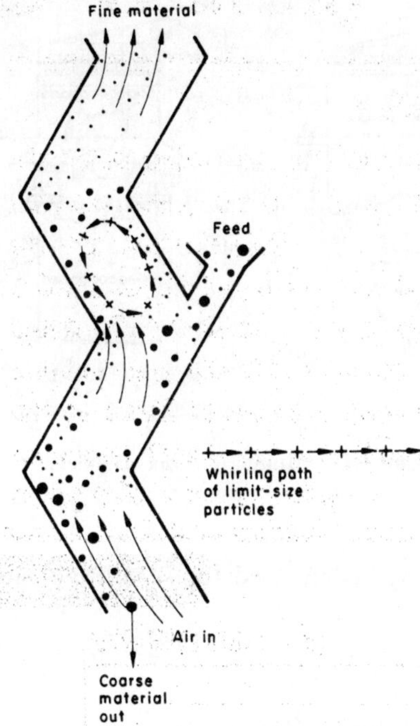

Figure 5.25 Principle of the zigzag sifter.

In type B the material is fed into the centre of the classifier by a distributor screw. An additional centrifugal roller distributes agglomerates and feeds the material into the classifier tubes.

Such arrangements are used in the preparation of drugs for the separation of sand and other impurities such as leaves and stalks from leaf drug

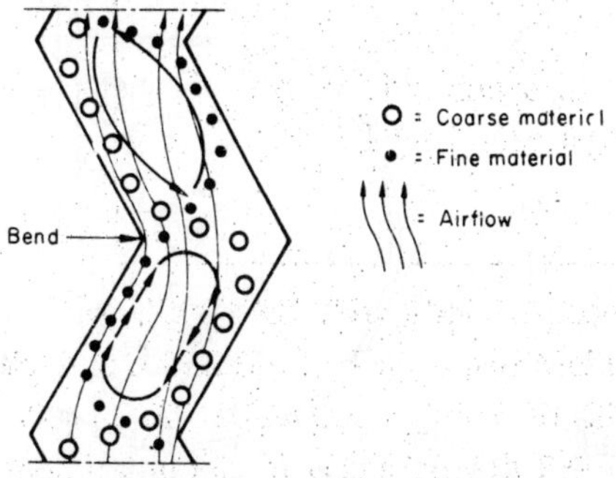

Figure 5.26 Particle activity in a zigzag classifier.

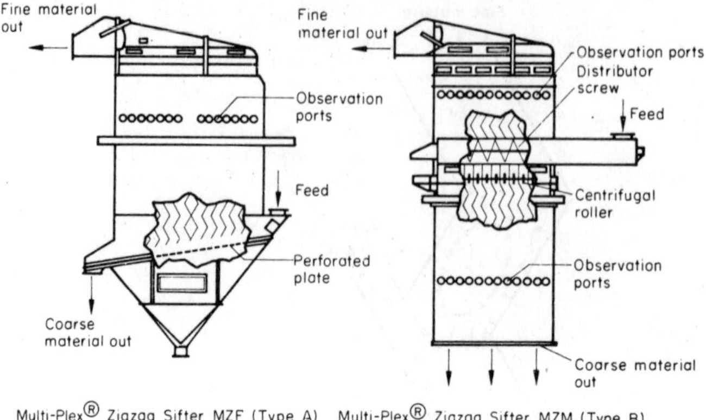

Figure 5.27   Multiplex zigzag classifiers.

materials; it is especially widely used for separating sand and stalks from shredded tobacco leaves.

5.3.1.3.   Centrifugal classifiers

The most familiar centrifugal classifier is the spiral air classifier. Figure 5.28 [5.9] shows one type with rotary walls surrounding the sifting space. The air sucked in through the sifting air inlet passes between the adjustable blades in the sifting space, which is in the form of a flat cylinder. The air flows spirally inwards and is turned in a vertical direction in the centre; it then leaves the

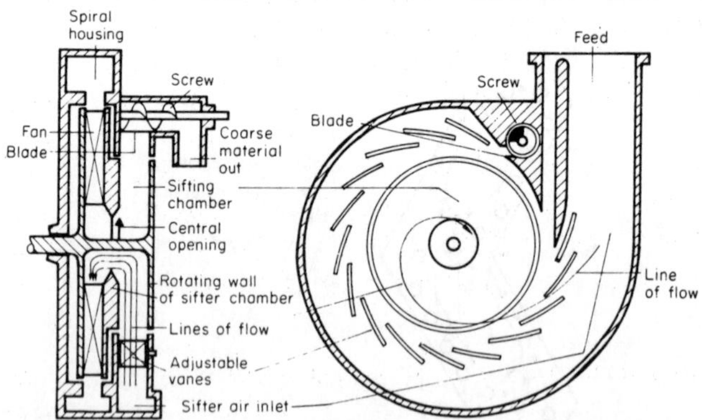

Figure 5.28   Centrifugal separator (Mikroplex Spiral Air Classifier, Alpine Co.).

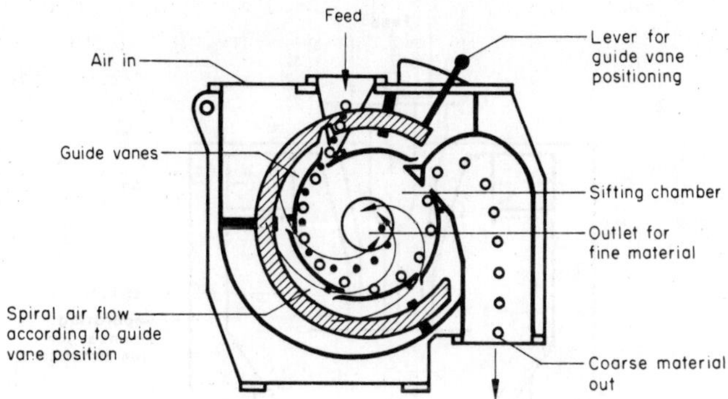

Figure 5.29 Spiral air classifier without rotors.

classifier through the ventilator. The material to be sifted comes down the feed shaft and is caught up by the sifting air. The outwardly directed centrifugal force and the inwardly directed forces of friction of the air act on the particles in the spiral track. Either one of the forces predominates, depending on the particle size. Fine particles are carried into the centre with the effluent air, whereas coarse particles collect at the periphery, and are scraped off by the blades and carried away by the screw. Depending on the size of the classifier, the particle size separation range can be set at between 3–30 µm or between 8–60 µm by adjusting the blades or by altering the rotation speed of the ventilator. The rotating sifter wall also sits on the axle of the ventilator. Figure 5.29 [5.12] shows an arrangement operating without rotors. The separation limit is set by adjusting the spaces between the baffle-blades. Centrifugal classifiers are used especially when clean separation of fine particles is required.

5.3.1.4. Scatter blast classifiers

These classifiers, which are used especially in heavy industry, such as the manufacture of cement, operate in the range 50–200 µm and thus fill the gap between centrifugal and gravitational classifiers. Their throughputs are up to several hundred tons per hour. In principle they represent a combination of crossflow and spiral air classification. Figure 5.30 [5.9] shows the principle. The incoming material falls through a vertical shaft on to the rotating scatter plate, which flings it into the upward-flowing and simultaneously spinning airstream (a) produced by the ventilator. A crossflow classification then takes place at point (c), whereupon the coarse material is centrifuged out and falls downwards and undergoes a further crossflow classification at point (d). The upward-flowing airstream laden with fine material is diverted outwards at (b). Processes similar to those in a spiral air classifier operate here. The fine

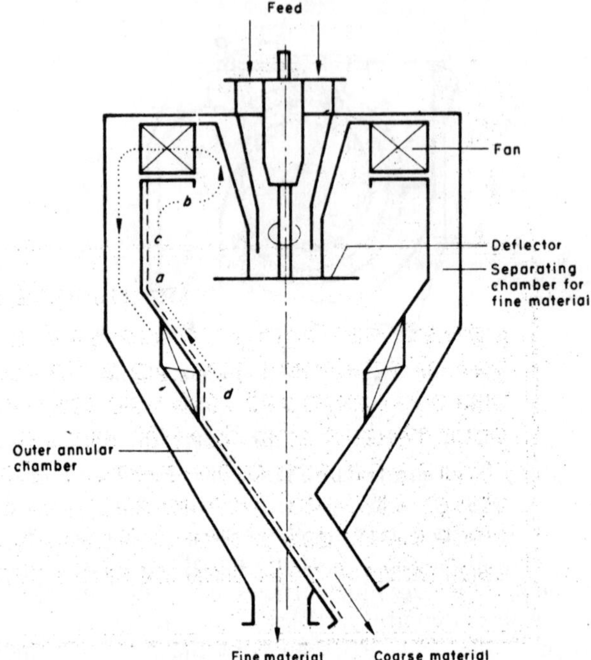

Figure 5.30 Schematic of a scatter blast classifier.

material flowing out through the ventilator is finally separated in the outer circular jacket. The air flows via (d) to the jacket and back into the separating space and thus completes the circuit.

5.3.2. SIEVES

Fine sieving is the most significant of the three ranges coarse (200–40 mm), medium (40–2 mm) and fine sieving (smaller than 2 mm) in the classification of medicinal plant materials. The medium sieving range also has a certain importance in the preparation of tea and of medicinal plant materials for extraction. A clear assignment of sieving machines to coarse, medium or fine sieving is not possible.

A fundamental distinction has to be made between simple classification through a sieve, e.g. for separation of a coarse or a fine fraction, and multiple classification for obtaining several fractions. The latter is, for example, the case with the separation of tea products into coarse cut, semi-fine and fine cut. A distinction must be made in multiple classification between:

Sieving with decreasing mesh width on sieves placed one beneath the other, the largest sieve thus being at the top.
Sieving with increasing mesh width, the sieves, beginning with the finest sieve, being arranged one after the other.

5.3.2.1. Sieving with decreasing mesh width

The coarsest sieve is in the highest position. The residue left on this sieve is the coarse fraction. The material which passes through this sieve goes on to the sieve below with the next smaller mesh width, and so on. Figure 5.31 shows two common arrangements.

In the tumbler or vibrating sieve the sieves are closely connected to each other as a set of sieves. The whole set is subjected to a combined rotary and vertical movement. The material being sieved is thrown up on the sieve, thus making circular hopping movements.

The advantage of this arrangement lies in the closed process and the small amount of space which it requires. Its disadvantage is that its throughput must always be governed by the quantitatively largest sieve fraction, as the sieve area is the same for all fractions and cannot be modified. The oscillating sieve is a similar arrangement. Its advantage lies in the possibility of using various sieves with different oscillation frequencies and adapting the sieve areas to the requirements. Its considerable height is a disadvantage. An advantage common to both arrangements is the fact that coarse particles do not come into contact with fine sieves and thus cannot damage these sensitive sieves.

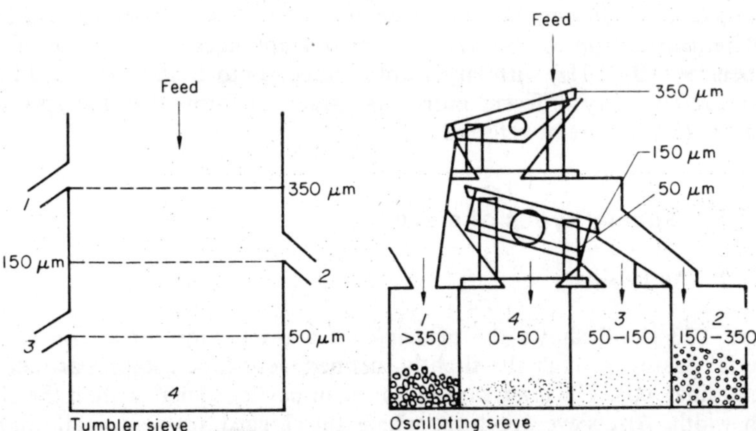

Figure 5.31 Sieving with decreasing mesh width. 1, Coarse material; 2, intermediate size material; 3, fine material; 4, dust.

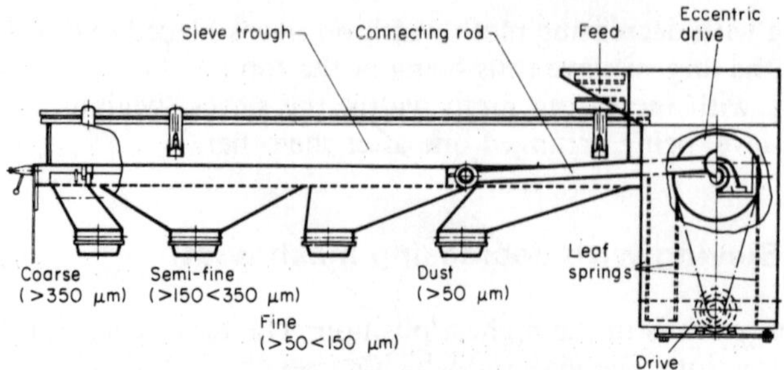

Figure 5.32   Sieving with increasing mesh width.

5.3.2.2. Sieving with increasing mesh width

The sieves are arranged one after the other by increasing mesh width, with only small differences in height between them. Figure 5.32 [5.13] shows the principle of this arrangement. The sieve trough in the freely oscillating sieving machine represented here is supported on oscillating leaf springs and is driven by an eccentric drive rod. The sieving machine oscillates horizontally. The inclination of the sieve trough can be adjusted to any angle between 0–4°. The material is fed into the sieve head situated before the finest sieve. The coarse fractions thus travel the furthest. The advantage of this arrangement is the possibility of adapting the sieve areas to the frequency of the individual fractions. Its disadvantages are the contamination of fine sieves with coarse material and the large amount of space taken up by the horizontal arrangement.

Oscillating magnets, as well as oscillating leaf springs and eccentric rods, are also used for driving this type of sieving machine. Combinations either of high frequencies (up to 100 Hz) with small amplitudes (~ 1 mm) or of low frequencies (10–25 Hz) with larger amplitudes (up to 10 mm) are used here. There are so many different individual practical forms that the specialist literature [5.14] must be referred to.

5.3.2.3. Special types of sieve

5.3.2.3.1. Air jet sieve

Figure 5.33 [5.14] shows the principle of a production air jet sieve. The material passes through the slightly inclined sieve-tube rotating around its longitudinal axis. The airstream removes all particles smaller than the sieve mesh width. Any sieve apertures which get clogged are continually blown clear. The longitudinal ribs built into the walls of the sieve tube ensure that the material is constantly stirred up. The air jet sieve is particularly suitable

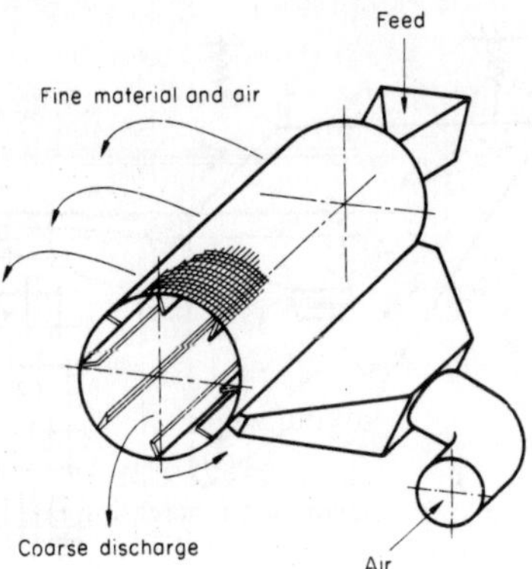

Figure 5.33   Air jet sieve.

for removing dust and fine material from shredded leaves of medicinal plants and is widely used in the preparation of tobacco. The separation limit is 40–100 µm.

5.3.2.3.2.   Mogensen sizer

In the sieving machines discussed thus far there is a fairly close connection between the mesh width of the sieve and the separation limit. The sieve retains particles larger than the mesh width and allows smaller particles to pass through. Particles of borderline size sometimes stick in the meshes of the sieve.

The Mogensen sizer lacks the strict assignment of separation limit to mesh width. The separation limit amounts to only a fraction of the clear mesh width. The resultant poorer selectivity is improved by passing the material repeatedly through several sieves placed one beneath the other. Such processes, which do not classify strictly according to the mesh width of the sieve, are also known as sieve like processes. Figure 5.34 [5.17] shows the principle of the apparatus developed by Mogensen [5.15 and 5.16].

The material is classified by several sieves placed one above the other and with increasing inclination and decreasing mesh width from top to bottom of the sieve stack. However, in contrast to thrower or shaker sieves, a very large sieve mesh width in comparison with the separation limit is used, i.e. the separation is carried out in the region of high probability of passage, which makes large inputs of material possible. The resultant poor sieving effect of the individual sieves is compensated for by repeating the separation process through further sieves. The separation is also sharpened by the increasing angle of inclination and the decreasing mesh width of the sieves from top to bottom of the sieve stack. The inclination of the sieve can be quickly altered by means of the springs on which the system depends. Multiple classification

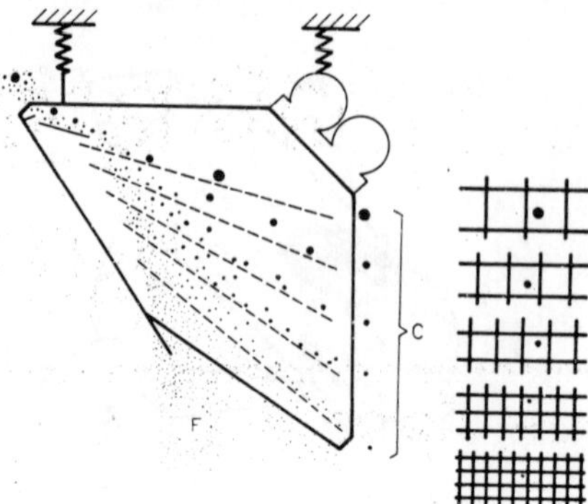

Figure 5.34 Principle of the Mogensen sizer. F, fine material; C, coarse material.

is possible, though in the example of Figure 5.34 a differentiation is made only between coarse material C and fine material F. A vibrator produces the necessary sieving movement.

The theory of Mogensen classification has been described in detail by Wessel [5.17]. The curves shown in Figure 5.35 [5.17] are obtained by considering a progressively increasing angle of inclination σ of the sieve, wire thickness δ of the sieve and passage rebound zone ψ. ψ is a factor introduced for particles which do not fall freely through a sieve.

The left hand part of the Figure illustrates the probability W of passage for a particle size α relative to the sieve mesh width l (corresponding to d/l) through four sieves placed at progressively increasing angles of inclination. The right hand side of Fig. 5.35 shows curves for selecting of the best angle of inclination for the separation of two different particle sizes. The number of throws for a 90% probability of passage is plotted on the ordinate against the inclination of the sieve (abscissa) for various *relative* particle sizes d/l. About 100 throws or oscillations are, for example, required for a 90% probability of particles of size $d = 0.8 \times l$ passing through a sieve inclined at 10°, whereas ~ 150 oscillations are required with a sieve inclination of 20°, in other words, the steeper the inclination of the sieve, the smaller is the probability of passage for large particles. This apparently trivial deduction can, however, be very useful for the separation of particles of different sizes. Whereas ~ 150 oscillations were required for a 90% probability of passage of particles of diameter $d = 0.8 \times l$ through a sieve inclined at 20°, only about 10 oscillations are necessary for a 90% probability of passage of particles of diameter $d = 0.4 \times l$ at this inclination of the sieve. The cited particle sizes can therefore be separated quickly and cleanly when the sieve is inclined at this angle. The figure also shows that an angle of $\sim 56°$ is a favourable sieve inclination for separating particles of diameter $0.4 \times l$ from those of diameter $0.2 \times l$.

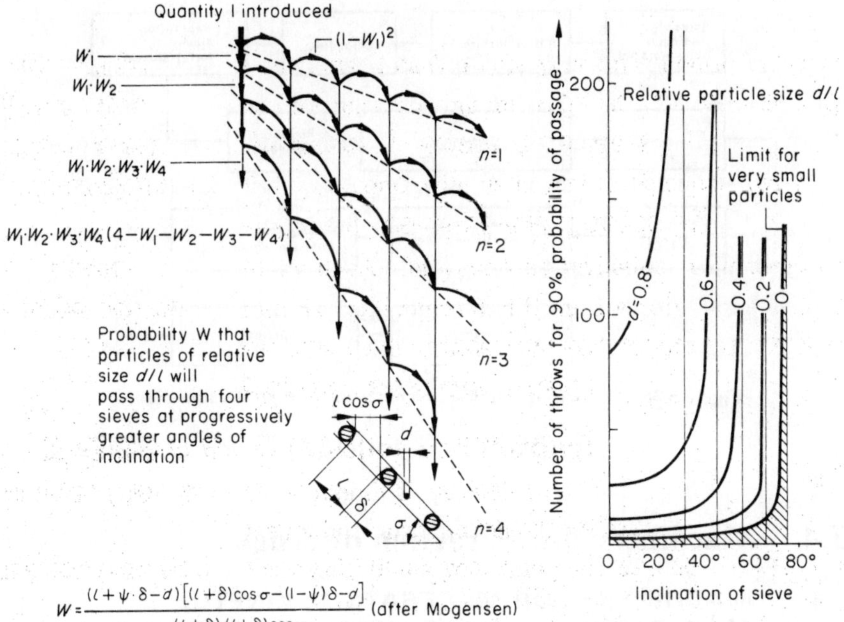

$$W = \frac{(l+\psi \cdot \delta - d)\left[(l+\delta)\cos \sigma - (1-\psi)\delta - d\right]}{(l+\delta)(l+\delta)\cos \sigma} \quad \text{(after Mogensen)}$$

Figure 5.35 *Left.* Probability W of passage of material through a stack of sieves of mesh wire thickness δ at various inclinations σ. A passage impact zone factor ψ is applied (after Mogensen). *Right.* Effect of angle of inclination of a sieve on the number of throws required to achieve 90% separation.

The Mogensen sizer is used mainly in the clay and minerals industries, although apparatus for preparing teas and phytopharmaceutical infusions is also already in use.

5.3.3. COMPLETE INFUSION PRODUCTION PLANT

Infusion mixtures are medicinal preparations which are widely prepared and used. The great demand for them has led to the development of complete herbal tea production lines where all the manufacturing processes from delivery of the bundles of herbs to the finished infusion mixture are carried out by total automation.

Figure 5.36 shows a scheme of such a production line. A moisturizing unit can be used to reduce the amount of dust in friable drug plant materials before these undergo the first sieving. Modern plants with capacities of up to 10 tons/h can be operated by one man from a control desk and can process practically any medicinal plant material.

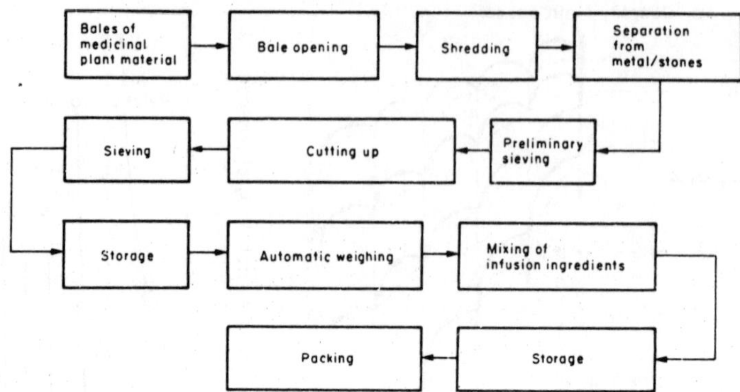

Figure 5.36   Block diagram of a complete infusion production plant.

5.4.   Machinery for extraction of drugs

5.4.1.   MACHINERY FOR MACERATION, KINETIC MACERATION, REMACERATION, DIGESTION AND SIMILAR PROCESSES

The preparation of extracts by maceration as per the pharmacopoeias, i.e. in sealed vessels left to stand for 5 days with occasional shaking, is now carried out only on a small scale.

Maceration is carried out on an industrial scale mostly as kinetic maceration, which can also include regular renewal of the solvent (remaceration) or maceration at high temperature (digestion). In this respect the definitions given in the chapter on Methods are not sharply delineated.

Two groups of machines are available for kinetic maceration:

Machines without built-in accessories (stirrers).
Machines with built-in accessories.

5.4.1.1.   Machinery without stirrers

These include all machines which bring about mixing of drug materials and solvent as free fall mixers. Figure 5.37 [5.18] summarizes the possible types of such machines. Most of these mixer types are available in various sizes and hence a machine suitable for any batch size can be found. A special form is feasible through the incorporation of sieves which fix the material to be extracted in one half or a corner of the mixer so that the solvent flows through the stationary drug material upon each revolution. The process is

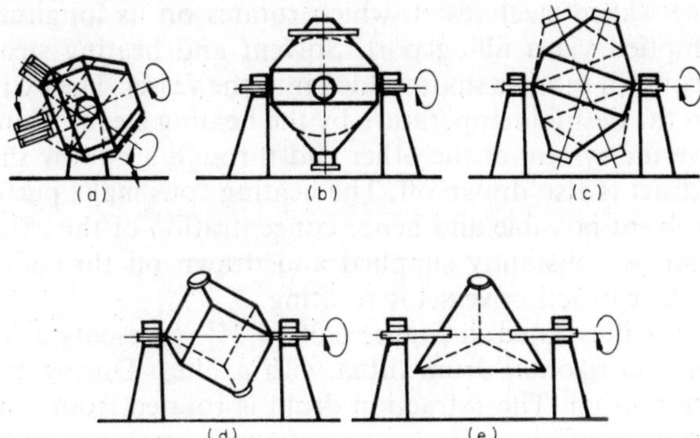

Figure 5.37 Various mixers for maceration: (a) mixing barrel, (b) twin cone mixer, (c) inclined twin cone mixer, (d) cubic mixer, (e) tetrahedral mixer.

also known as rotary maceration. The miscella is usually separated from the drug residue by emptying and expression.

An excess of solvent is normally used in simple kinetic maceration. A method in which the extraction vessel is charged fully with drug plant material but only partially with solvent was also proposed by Hartmann [5.19] for lignified (woody) drug plant material. In the rotating vessel the solvent then runs through the drug plant material and thus becomes enriched with crude extractive substances. After sufficient contact time the extract is drained off. The process can be repeated as many times as are needed until the drug material has been exhaustively extracted. An additional effect can be achieved by heating separate sections, preferably the lower part, of the extraction vessel. The vaporized solvent rises into the upper, cooler parts of the vessel, where it condenses and flows back into the warmer zones laden with extract. The rotary extractor illustrated in Figure 5.38 [5.20] operates on the principle of digestion.

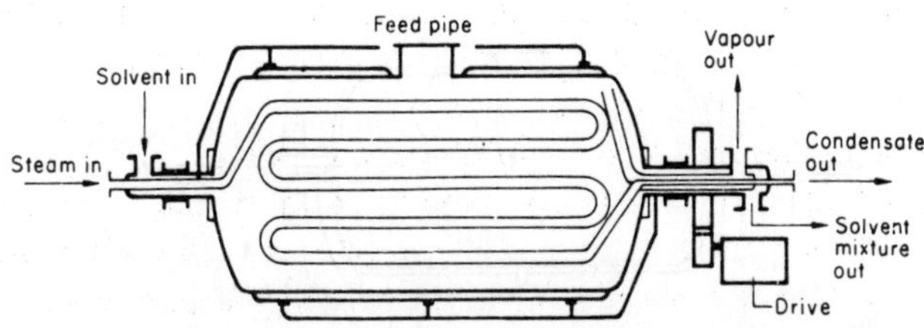

Figure 5.38 Double-walled (jacketed) rotary extractor.

The double walled steel vessel, which rotates on its longitudinal axis, is filled and emptied via a filling port. Solvent and heating steam enter the extractor through a tubular shaft welded into the vessel. The extraction space is brought to the desired temperature by the heating steam. Condensate and vapours leave the system at the other end through a hollow shaft, through which the extract is also drawn off. The heating coils make partial vaporization of the solvent possible and hence concentration of the extract. Liquids and gases can be constantly supplied and drawn off through the tubular shafts while the extraction vessel is rotating.

The extractor illustrated in Figure 5.39 [5.21] represents a special design consisting of a horizontal drum fitted with a filter. During extraction the filter is at the bottom. The extraction drum is rotated from time to time to prevent formation of channels in the extraction material and to improve contact between solvent and extraction material.

When a new filter cloth is put in, the drum is rotated 180° and the old clogged filter is cleaned by back-flushing. After completion of the extraction the drum can be used simply as a dryer for the extraction material, as the side walls, the jacket and the internal tube network can be heated. The extraction material is put in and removed from the drum through a porthole made for this in the jacket, and the supply and reflux of solvent fractions and heating agents take place through the central shafts. The great advantage of rotary extractors is that the entire process, including drying of the solid material, is carried out in just one apparatus.

Figure 5.40 [5.21] shows another version of a batch extractor which is emptied by tilting the apparatus through 180°.

The solvent and the steam for heating the jacket must be supplied through the hollow axle, through which the extract is also drawn off.

5.4.1.2. Machinery with stirrers

The efficiency of kinetic maceration can be improved by the use of machines with stirrers, which produce a more rapid equalization of concentration.

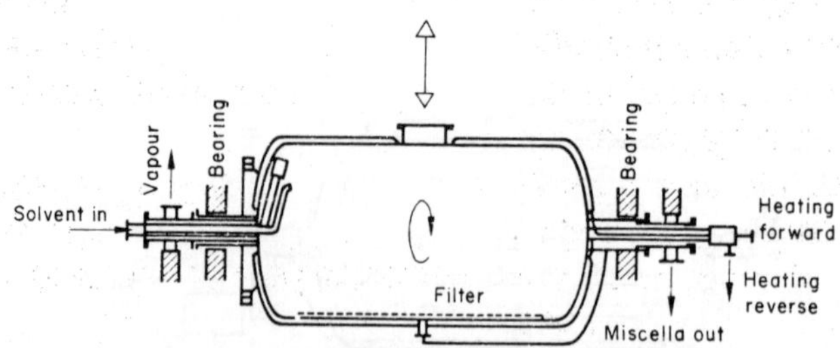

Figure 5.39 Horizontal drum extractor with filter.

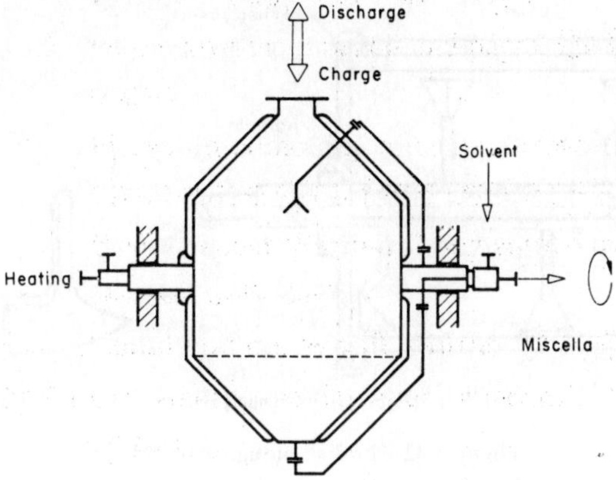

Figure 5.40   Batch extractor with tilting device.

In the simplest machine of this type the stirrer can be suspended in the extraction vessel. The stirrers should be of a design and dimensions such that they guarantee homogeneous mixing of the entire contents of the container. Examples of suitable stirrers are the Z-blade or cross-beam and the Intermig stirrers illustrated in Figure 5.41 [5.22].

However, complete mixers are much more often used for kinetic maceration. One example of such a high-speed machine is the Lödige ploughshare mixer illustrated in Fig. 5.42 [5.23].

Another principle encountered in extraction is that of the Nauta mixer shown in Fig. 5.43 [5.24]. This apparatus consists of a conical jacketed container which can be heated and cooled. A mixing screw lying against the wall conveys the material upwards from below. As the centrally pivoted swivelling arm simultaneously takes the screw along the wall it produces a combined mixing movement such as is shown in Fig. 5.44 [5.24].

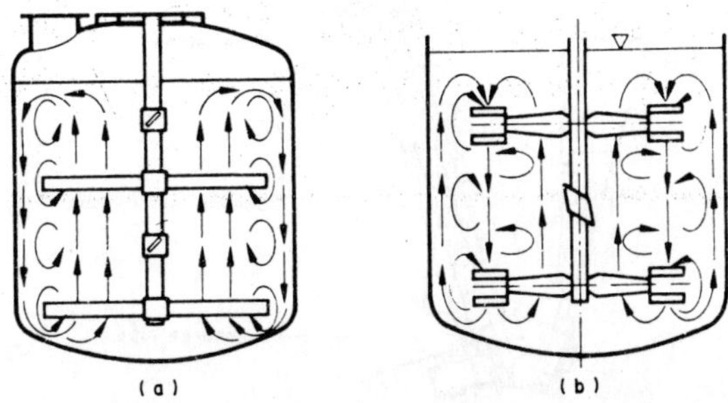

Figure 5.41   Stirrers for extraction: (a) cross-beam stirrer, (b) Intermig stirrer.

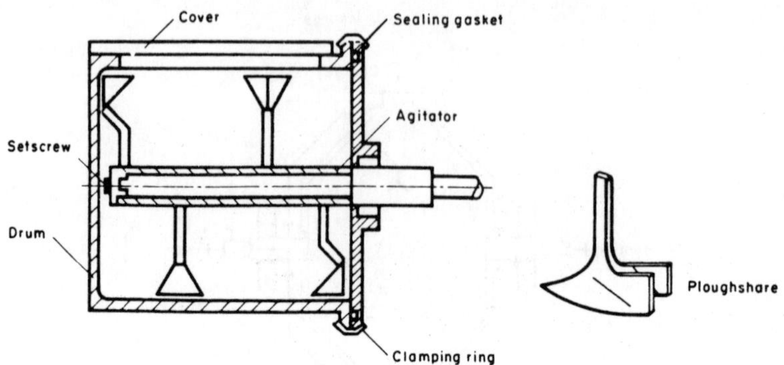

Figure 5.42   Lödige ploughshare mixer.

The mode of action of the Nauta mixer represents an efficient kinetic maceration which is also suitable for drug plant materials containing mucilage and swelling substances.

The drug plant material is stressed to a substantially greater extent in extraction with rotorstat immersion homogenizers of the Ultra-Turrax type [5.25]. In this method of maceration, which is of the high-speed vortical type of extraction, the cells are disintegrated by powerful shearing forces. Figure 5.45 shows the principle of this machine.

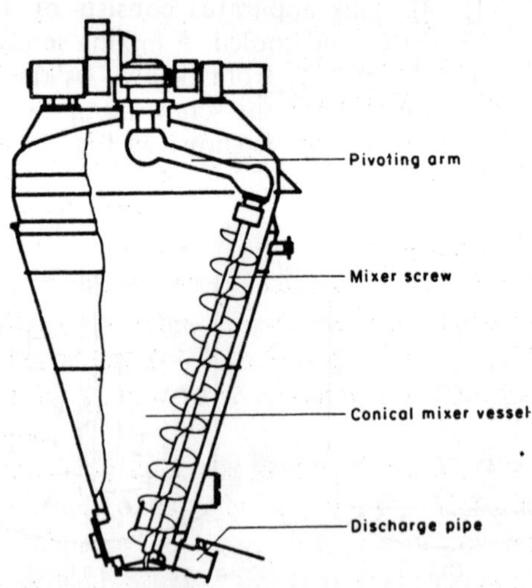

Figure 5.43   Nauta mixer.

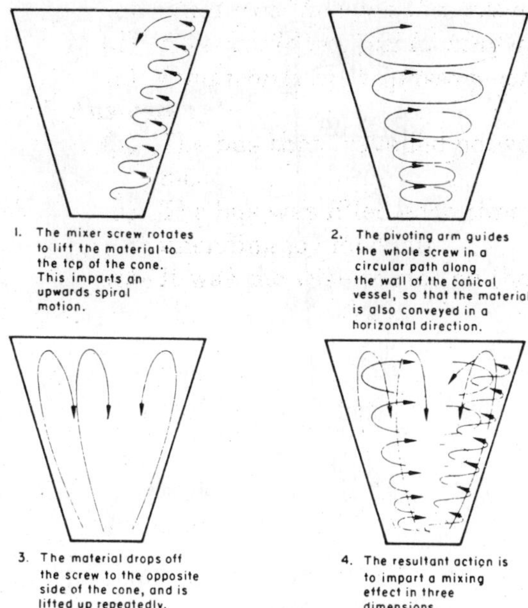

Figure 5.44 Combined mixing movement in the Nauta mixer.

The high-speed machine comminutes the drug material during extraction, which results in a faster washout of the extractive substances but on the other hand makes separation of fine skeletal material of the plant difficult. High-speed homogenizers must therefore be combined with filters or, better, centrifuges when being used for extraction.

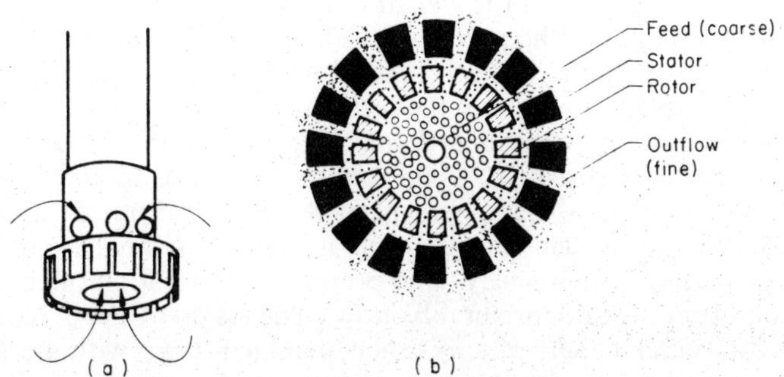

Figure 5.45 Side view (a) and plan view (b) of the action of the Ultra-Turrax.

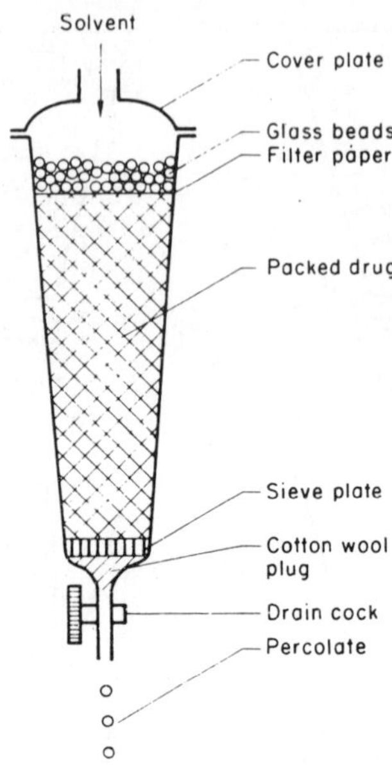

Figure 5.46 Percolator conforming to DAB 8.

5.4.2. MACHINERY FOR PERCOLATION

DAB 8 generally permits percolation as well as maceration for the preparation of liquid phytopharmaceuticals. The percolators used for this are cylindrical or conical vessels at the bottom of which is a stopcock for regulating the outflow rate. Figure 5.46 shows the usual type of percolator made of glass as per *DAB 8*. The Pharmacopoeia demands that the depth of drug material in the percolator should be 5 times the average diameter of the percolator. The comminuted material is first soaked with a quantity of menstruum weighing 30% of the weight of the drug material for 2 h to make it swell before it is put in the percolator. The degree of comminution of the drug plant material should be such that a sufficient area of contact with the solvent is available, though on the other hand the material should be substantially free of powder to prevent clogging and the production of a cloudy miscella. The degree of comminution depends on the drug plant material, though for most of these a particle size of 1–3 mm is suitable for percolation. The preswollen drug material is then pressed lightly and evenly into the percolator in such a way as to prevent the formation of channels in which the solvent would preferably flow. The cake of drug material is covered with a filter paper, which is then weighted down with glass beads. The solvent is poured in with the stopcock open to force the air contained in the cake of drug material out at the bottom. As soon as the liquid begins to

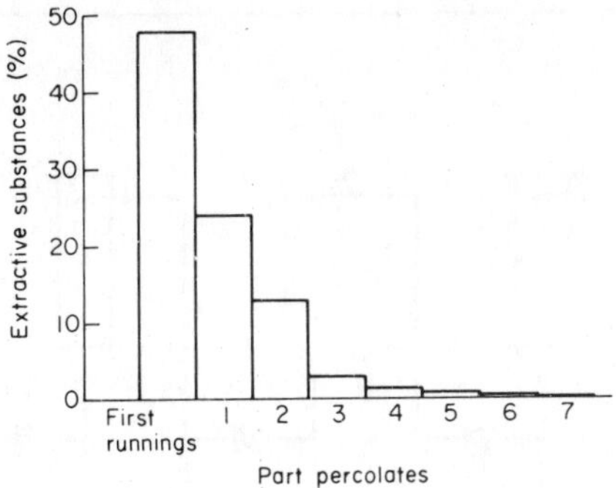

Figure 5.47   Extractive substance (%) for each part percolate stage.

drip from the percolator the stopcock is closed and the drug material left to macerate in the solvent for 24 h. The extract is then allowed to drip out of the percolator while menstruum is continually added at the top. According to *DAB 8* the percolation rate should be 4–6 drops per minute per 100 g of drug material.

The first run of miscella has the highest extract content and is known as the first runnings or extraction head. It is followed by the part percolates. The weight of the preliminary runnings and of each part percolate is about 80% of the weight of the drug material. Figure 5.47 shows a graph of the course of extraction [5.18]. The simple percolation described here uses a large quantity of solvent. The closer the percolation approaches completion, the smaller the content of extractive substances in the part percolates. One obvious way of achieving a more economical use of the solvent is therefore to use low-content part percolates as solvent for fresh drug material. Miscellae rich in extractive substances are thus obtained with a saving of solvent. This has led to the invention of percolator batteries which are now widely used in the industrial preparation of extracts. Figure 5.48 [5.21] shows a diagrammatic representation of such a battery.

Semi-continuous operation is possible by running the fresh solvent through one extractor after another so that it acquires an increasing concentration of extract. As soon as the desired degree of extraction has been attained, the particular extractor which worked last with fresh solvent is taken out of circulation, allowed to drain, emptied and refilled with fresh material. The first extraction in this freshly filled extractor is then carried out with the most concentrated miscella. After emptying and recharging of the next extractor (say C), the extractor B is charged with a miscella fraction which has run through A and contains little extract, and so on. The advantage of these standing extractors is that it is easy to switch to the next fraction and that they can operate under vacuum or under pressure. Their disadvantage lies in the numerous filling and emptying operations and the rather complicated connections between the filling and emptying silos and the extractors required for these operations.

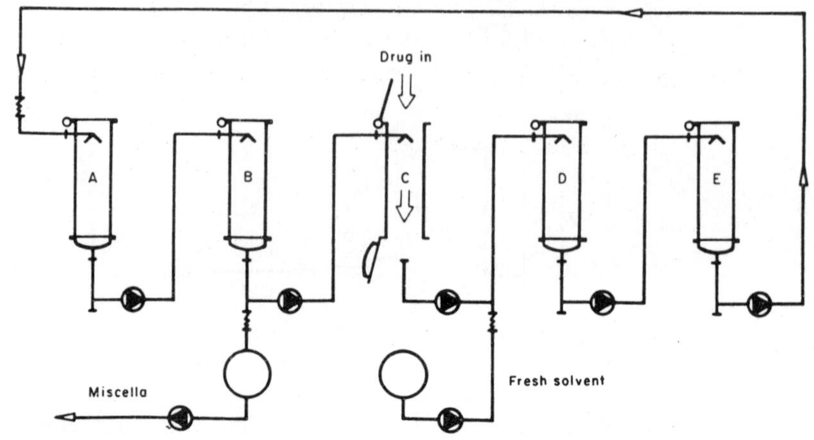

Figure 5.48   Percolator battery.

Extraction units of this type are used, among other things, for the recovery of tannins and medicinal plant extracts. Filling and emptying can be simplified by pneumatic or hydraulic drive.

This mode of operation, which produces relatively highly concentrated extracts, has the advantage of economizing on solvent when used on an industrial scale. Fluid extracts can be obtained directly with skilled operation of the process. However, the process has certain disadvantages, particularly when carried out on a large industrial scale. It is uneconomical to soak and swell the material before it is put in the percolator. A considerable increase in pressure in the percolator must therefore be expected with strongly swelling plant drug materials, especially those containing a large amount of mucilage; the latter therefore cannot be percolated.

Despite careful packing of the percolator, 'nest formation', i.e. the formation of dry areas within the cake of drug plant material which are not reached by the solvent, cannot always be avoided. The drug material in the 'nests' is not extracted and represents a real loss. Finally, charging and emptying of the percolators are labour intensive processes. Swelling of the drug material in the percolator and the associated formation of a tightly-fitting cake of drug material can cause additional problems. This can be partially counteracted by the design of the percolators. Figure 5.49 shows two possible practical forms of these.

Whereas the cylindrical drug cake of type A in Fig. 5.49 can tend to jam, the conical shape of type B, which is wider at the bottom, is better for emptying. The drug cake can be made to drop out simply by sliding the filter out of the whole outlet. However, despite these technical improvements, percolator batteries are being increasingly replaced by continuous counter current extraction methods, especially where the quantities are sufficiently large.

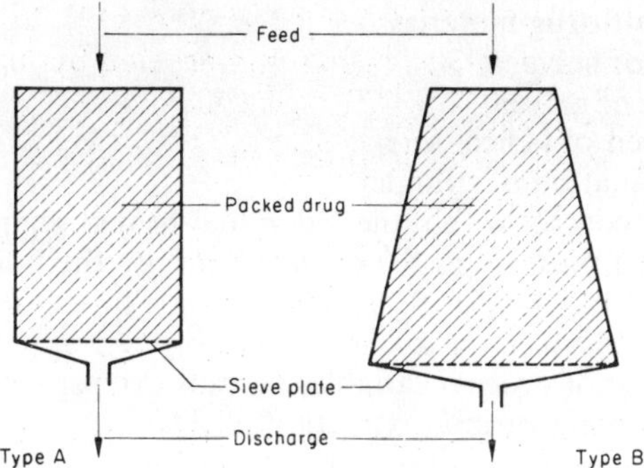

Figure 5.49 Designs for percolators.

5.4.3. MACHINERY FOR COUNTERCURRENT EXTRACTION

5.4.3.1. Countercurrent screw extractor

The countercurrent screw extractor or helical extractor operates on the principle of continuous absolute countercurrent extraction, in which the drug material and the solvent are kept in continual motion in opposite directions to each other. Figure 5.50 [5.26] shows a diagram of the apparatus.

The apparatus consists of a pressure resistant cylinder, sections of which can be heated or cooled, and inside which is a conveyor-screw. The drug material is fed into the extractor through a hopper and is conveyed by the screw, with mixing and partial compaction, towards the discharge end of the cylinder. The solvent flows against the drug material from the discharge end. The following process variables are involved in this arrangement:

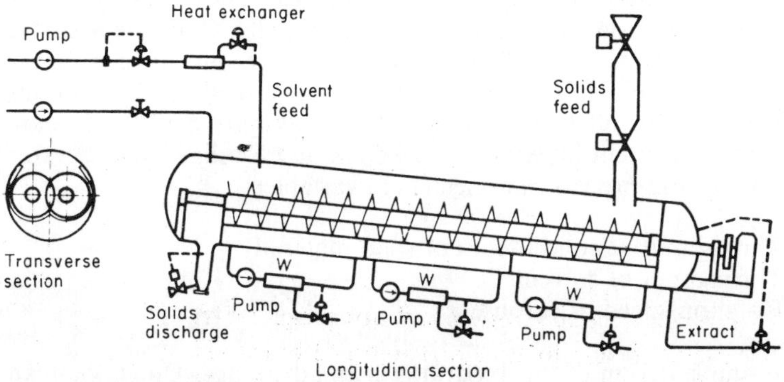

Figure 5.50 Countercurrent screw extractor.

Throughput of drug material.
Throughput of solvent.
Temperature of solvent.
Rotation speed of helical screw.
Angle of inclination of cylinder.
Temperature control within the individual heated segments.
Increase in the pressure in the cylinder to raise the boiling-points of
the solvents.

The large number of process variables permits extensive adaptation of the
apparatus for resolving various extraction problems.

Drug materials with high contents of mucilaginous substances or pectins
cause difficulties when aqueous solvents are used. When the proportion of
water is too high the swelling processes in the cylinder cause a build-up of
pressure, which impedes the straightforward conveyance of the drug
material. Mucilaginous substances also have a lubricant effect which causes
slipping of the screw, thus stopping the extraction. The countercurrent screw
extractor is therefore particularly suitable for extraction with solvents which
do not swell the drug material. It can be used to advantage for hard barks
and seeds, and for ginger.

5.4.3.2. Carousel extractor

The carousel extractor works on the principle of continuous relative countercurrent extraction. It may also be regarded as a countercurrent operated
percolation battery in which each chamber of the extractor represents a
percolator. Figure 5.51 [5.27] shows the principle.

The clockwise rotating carousel is loaded chamber by chamber. Fresh
solvent is normally put in the last chamber. However, if the number of
separation stages permits it, the part miscella of the third stage is led directly
on to the freshly supplied solid material to reduce the turbidity. This prevents
the naturally large quantity of substances causing turbidity in the first
extraction stage from getting into the final miscella. Once the first stage has
been passed, this part miscella is led back to the second stage, where it gives
up again the turbid substances which it has absorbed. In this case the final
miscella is drawn off from the second stage. Figure 5.52 [5.27] shows the
course of the content of extract in solid and in miscella in relation to route.

The following process parameters are variable:

Throughput of solid (depth of solid material).
Throughput of solvent.
Rotation-speed of carousel.

The simple design of the apparatus is an advantage. Difficulties can arise
through swelling of the drug material when aqueous solvents are used. The

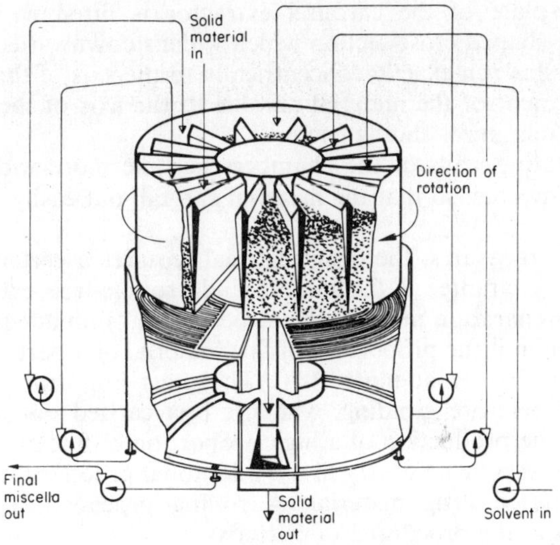

Figure 5.51 Principle of the carousel extractor.

problem of nest formation must also be watched, as a pure percolation process takes place inside the chambers.

An even drainage of the miscella is achieved by the following constructional measures:

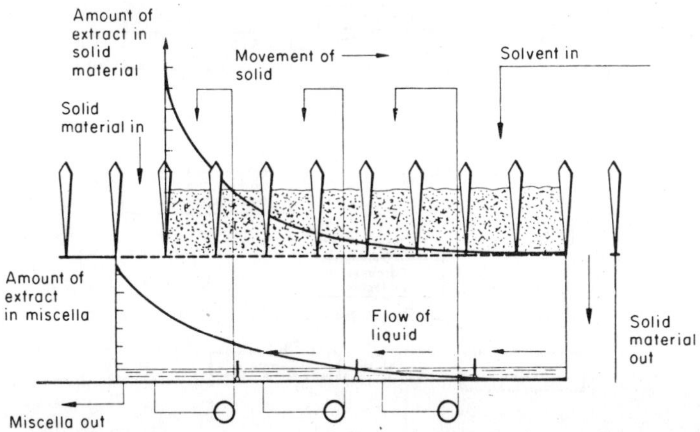

Figure 5.52 Changes in concentration during carousel extraction.

The sieve-plate of the carousel extractor is fitted with wires of trapezium-shaped cross-section which widens downwards.

The sieve-slits run exactly concentrically to the axis of the rotor.

The movement of the material parallel to the axis of the sieve thus produced minimizes sliding friction.

The separating-walls of the chambers become more widely spaced apart downwards, so that the material can fall out easily.

Despite these measures, the drug material requires a certain amount of preparation to guarantee a frictionless and trouble-free extraction. The object of this preparation must be the achievement of an adequate percolation rate and to limit the proportion of fine material of a particle size of less than 0.5 mm to 5% or at most 10%. The most important stage of the preparation is therefore grinding, which is best carried out with knife or cutting mills if the production of a high proportion of fine material is to be avoided. It may also be necessary to use additional processes such as rolling of the comminuted drug material. A rolling process is (for example) frequently used in the processing of oilseeds.

The extent to which other methods of preparation suggested in the literature, i.e. granulation, pelleting and extrusion, have actually found practical application in preparation for extraction is not known.

Figure 5.53 shows a flow chart of the preparation and extraction of drug plant material. Depending on the type of drug plant material, it may be possible to take the material directly from the mill to the extractor. However, intermediate stages leading to the extractor via alkalization in a helical

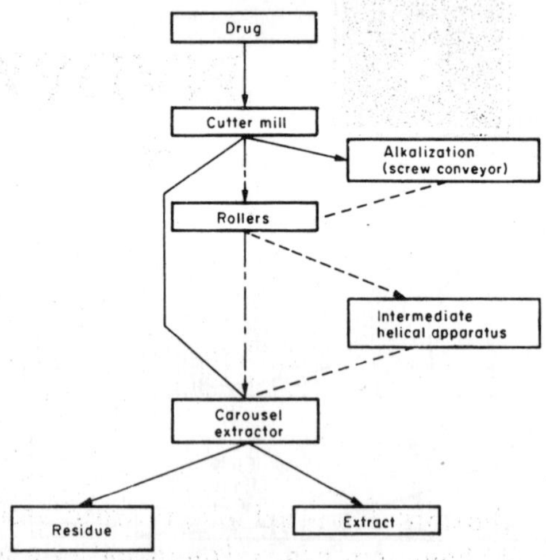

Figure 5.53 Flow diagram showing the passage of plant drug material through a carousel extractor.

apparatus to disintegrate the drug material, rolling and possibly some time in a helical apparatus to preswell the drug material may also be necessary. An adaptable production plant should have all these possibilities and be able to resort to them as options. The resultant drug residue still contains 'bound solution'. The processing of this residue is described in Section 4.2.2.6.

The carousel extractor has become widely used in the last few years, particularly in the pharmaceutical industry for the extraction of active substances from drug plant materials.

5.4.3.3. Circulation extractor (U-extractor)

The circulation extractor [5.28], like the carousel extractor, operates on the principle of continuous relative counter-current extraction. In both cases the solvent flows against the drug material contained in fixed chambers. The difference lies in the arrangement of the chambers; whereas in the carousel extractor the chambers are moved horizontally, in the circulation extractor the movement is in a vertical direction. Figure 5.54 shows a schematic drawing of the circulation extractor.

The apparatus consists of a U-shaped tube in which sieves are suspended on a conveyor belt. The drug material falls from the conveyor-screw on to the sieves and is transported through the U-tube. The solvent flows in the opposite direction, thereby passing over the cake of drug material lying on

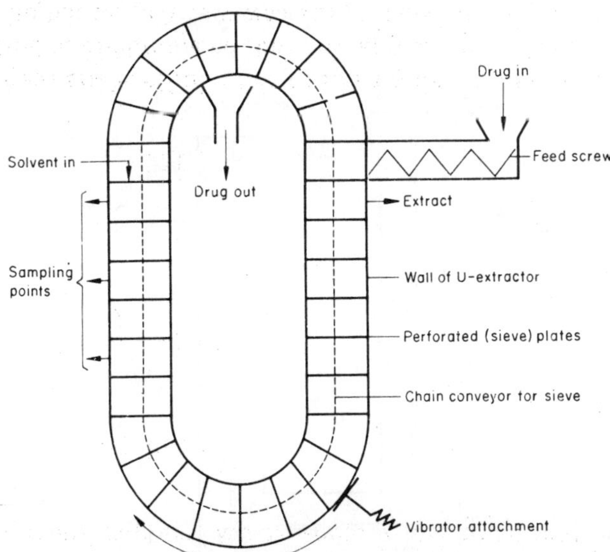

Figure 5.54   Circulation extractor (U-extractor).

each sieve and through the perforated plate on to the next cake of drug material. A vibrator in the lower part of the U-tube prevents formation of nests and channels. The spiralling of the drug material during transport through the U-tube serves the same purpose. The solvents principally used are water, ethanol, methanol and mixtures of these. The points provided in the left-hand half of the U-tube for taking samples facilitate determination of completion of the extraction and make possible an optimal adaptation of the apparatus to the particular extraction problem in question while the plant is in operation. The ratio of solvent to drug material is 2–2.5:1. The hourly capacity with leaf and herbal material is about 100 kg.

5.4.3.4. The Hildebrandt conveyor-screw extractor

This apparatus operates on the same principle as the U-extractor. The constructional difference is that helical screws instead of conveyor baskets transport the drug material in the U-tube. Figure 5.55 [5.20] shows a diagram of this apparatus.

The mode of operation is basically similar to that of the U-extractor.

5.4.3.5. Belt box/tank extractor

The belt box or tank extractor likewise operates on the principle of counter-current percolation. Figure 5.56 shows it in diagrammatic form [5.20].

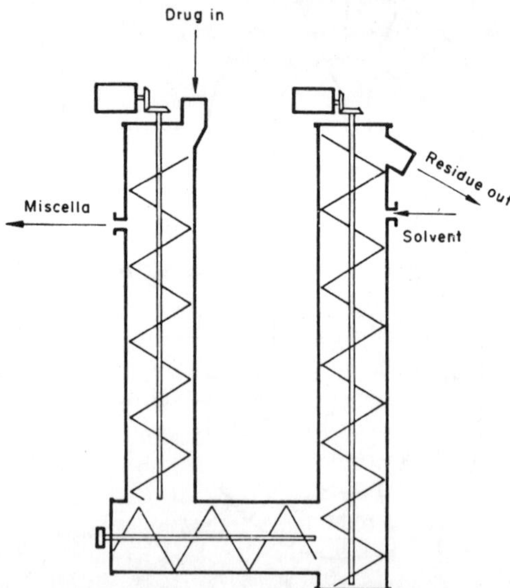

Figure 5.55 Hildebrandt conveyor-screw extractor.

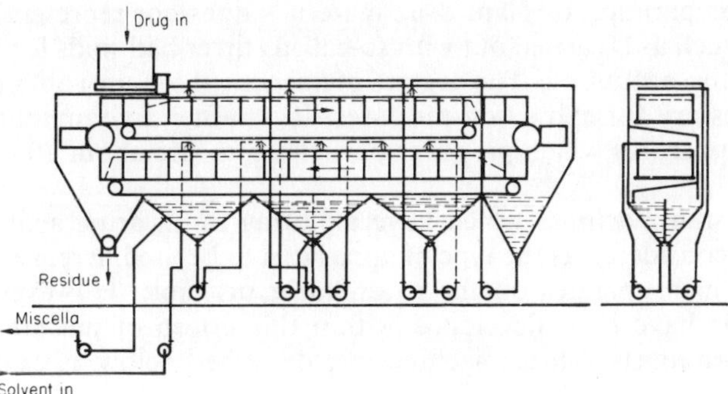

Figure 5.56   Belt box (tank) extractor.

Part miscella fractions are also collected here and sprayed on to fresh drug material. Its special constructional peculiarity is that the boxes containing the wet drug material run in two rows, one above the other, and the part miscella fractions are collected separately from the upper and the lower row, as shown in the side view.

5.4.3.6.   Extraction centrifuges

Although the combinations of apparatus discussed here are summarized under the heading extraction centrifuges, a distinction must be made between the apparatus used for disintegrating the drug material and those which bring about the actual extractive separation. Both types of apparatus are necessary for a complete extraction plant. The drug material can be broken down dry by grinding on a hammer mill to produce the very fine powders required for extraction processes in a centrifuge; however, it can also be broken down by wet comminution. Dry comminution is described in Sections 4.1 and 5.2. Only the apparatus for wet comminution will be discussed below, where it should first be stated that wet comminution must always be preceded by a coarse comminution of the drug material.

5.4.3.6.1.   Machinery for wet preparation of drug materials

Wet comminution of plant drug material as a preparatory stage for extraction has the advantage over dry milling that the material which is to be extracted is brought into contact with the solvent right from the start. The result of this is that the comminuted cells are already filled with solvent and the extraction process actually begins here. Whereas the fine powder particles produced in dry milling are surrounded by a layer of air, which in certain circumstances makes processing with the subsequent addition of solvent difficult, these difficulties do not arise in wet preparation.

The wet preparation of plant drug materials does not represent the very fine milling such as is carried out with so-called stirrer ball-mills for lacquers, plant protective agents, etc. The degrees of fineness (0.5–5 μm) obtained there are not necessary for extraction purposes. It is rather a comminution with dispersion apparatus which produces particles of a size about 10–100 times larger.

All dispersion machines which have a cutting or shearing action can in principle be considered as the type of apparatus to be used here, i.e. as a rule apparatus which operates on the rotor/stator principle. However, certain special forms have been developed within this group of machines which operate continuously. Three machines are described below as examples of this group.

5.4.3.6.1.2. Dispax reactor

In addition to the Ultra-Turrax, which is also obtainable for in-line operation (see Section 5.4.1; Machinery for maceration), machinery is available in the form of the Dispax reactor which makes the three-stage wet preparation of plant drug material possible [5.25]. Figure 5.57 shows the principle. The slurry of drug material containing 15–35% solid is first fed in for the very coarse comminution stage and then passes through stages 2 and 3, in which

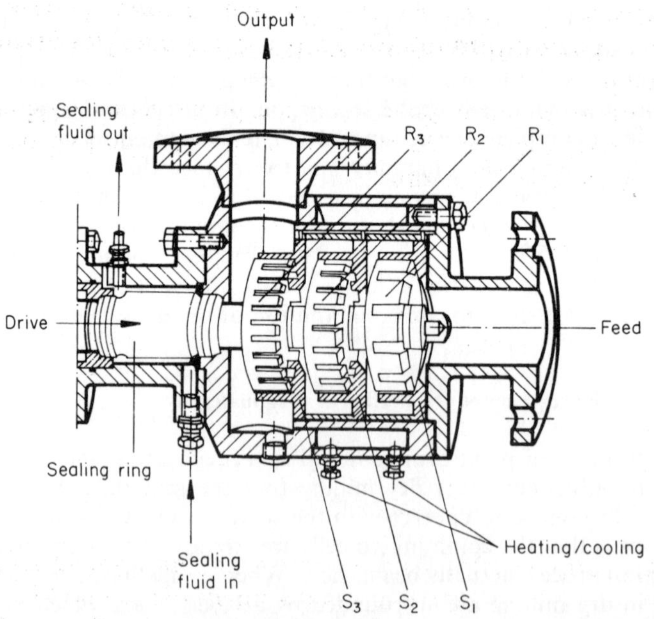

Figure 5.57 Dispax reactor.

high shearing forces predominate. Here the cell membranes are torn open and the extractive substances released. The machine produces its own driving pressure of ~ 2 atmospheres and hence the centrifuges can be loaded for the separation of the solid material without having to be connected to pumps.

5.4.3.6.1.2. *Gorator*

This is, in principle, an inclined disc pump. Figure 5.58 shows the design [5.29]. The inclined disc is attached to the rotor axle at a certain angle of inclination and is serrated in the axial direction. It is enclosed in a cylindrical stator made up of three likewise serrated individual segments, each of which curves through 90°. The remaining 90° is filled by a sieve which lets the comminuted material out. Each tooth of the serrated inclined disc fits into a groove in the stator, producing strong shearing forces. The stator can also contain an additional groove running in the axial direction which intensifies the shearing forces. The width of the space between rotor and stator is constant. The machine combines mixing, pumping and comminuting actions and has been used successfully for disintegrating drug plant materials.

5.4.3.6.1.3. *Supraton*

This continuously operating machine for the wet preparation of drugs is in principle a rotor/stator apparatus. However, whereas in the Dispax reactor

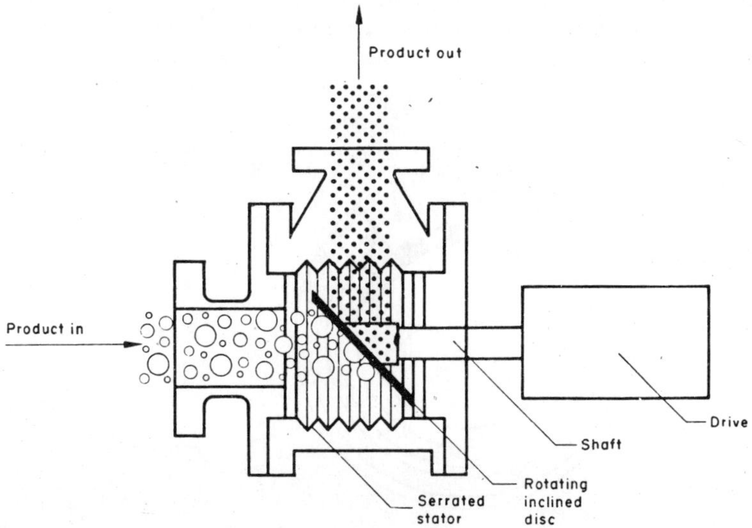

Figure 5.58   Gorator (longitudinal section).

three stages lie one behind the other, in the Supraton system the stages are arranged radially on a disc, as shown in Fig. 5.59 [5.30]. The teeth of the rotor and stator interlock in a manner similar to that in the serrated disc mill. The movement produces cavitation zones with intense stress on the material being extracted, as well as 'resting zones'.

This alternation accelerates an already rapid and complete extraction. The manufacturer states that this also produces less fine material which might cause problems in the separation of the extract. One advantage of the Supraton machine is that it can operate under pressure (up to 300 atmospheres). It has been used successfully for extracting spices and medicinal plant materials such as liquorice roots, etc.

5.4.3.6.2.   Centrifuges for extractive clarification

The centrifugation of extracts is, strictly, a purification process. The centrifuges are therefore primarily dealt with in Section 5.5.2. At this point we shall just briefly look at the apparatus used in solid–liquid extraction.

The most important criteria for selecting a centrifugal extractor are: (1) the percentage content of centrifugable solid matter, and (2) the specific gravity of the solvent (lighter or heavier than water).

According to Brunner [5.31], so-called extraction decanters best fulfil the task of solid–liquid separation with the solid content under discussion. Automatically emptying clarifying separators can also be used for reclarification.

In a further review paper, Theiler and Paschedag [5.32] give selection criteria for extractors for solid–liquid and liquid–liquid extraction. Figure 5.61 shows the check list which they have drawn up. The list is regarded by the authors as an aid in the preselection of apparatus. It is certainly

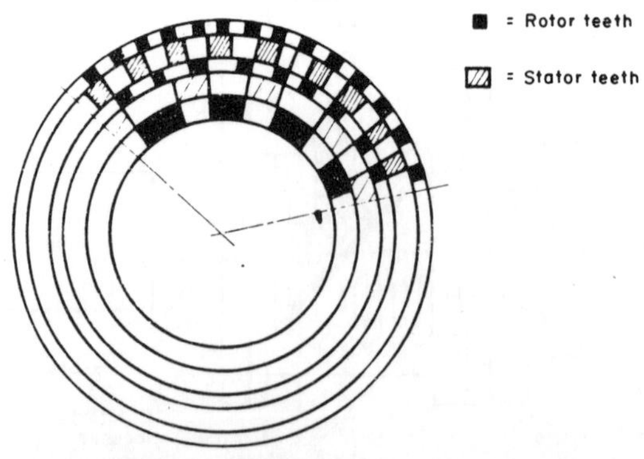

Figure 5.59 Supraton (cross-section).

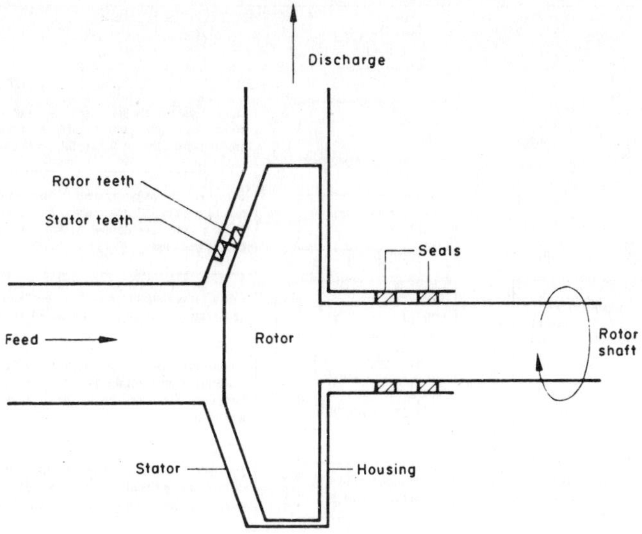

Figure 5.60 Supraton (longitudinal section).

influenced by the apparatus offered by the manufacturers, but nevertheless gives a good overall view of the criteria to be considered.

Extraction decanters are used almost exclusively in the first stage of separation of drug residues from extract by the principle of centrifugal extraction [5.31]. The principle and the mode of action of a decanter are described in Section 5.5.2.4.

Automatically emptying, clarifying separators or nozzle separators (see Sections 5.5.2.7 and 5.5.2.8) are used for reclarification and hence for removal of small remnants of solid. The choice of these machines is governed by the solid content of the liquid to be clarified and by the type of sediment produced.

5.4.3.6.3. Combination of individual elements to form a production plant

Wet preparation of the plant drug material, extraction in the decanter and clarification of the extract in the clarification separator form a coherent sequence for centrifugal extraction. Figure 5.62 [5.31] shows a section relating to this from such a plant for extracting morphine from poppy straw. Only the section of this plant up to the point where the clarified extract is obtained has been shown, as the subsequent preparative stages such as liquid–liquid extraction, distillation and precipitation serve for recovery of the pure morphine.

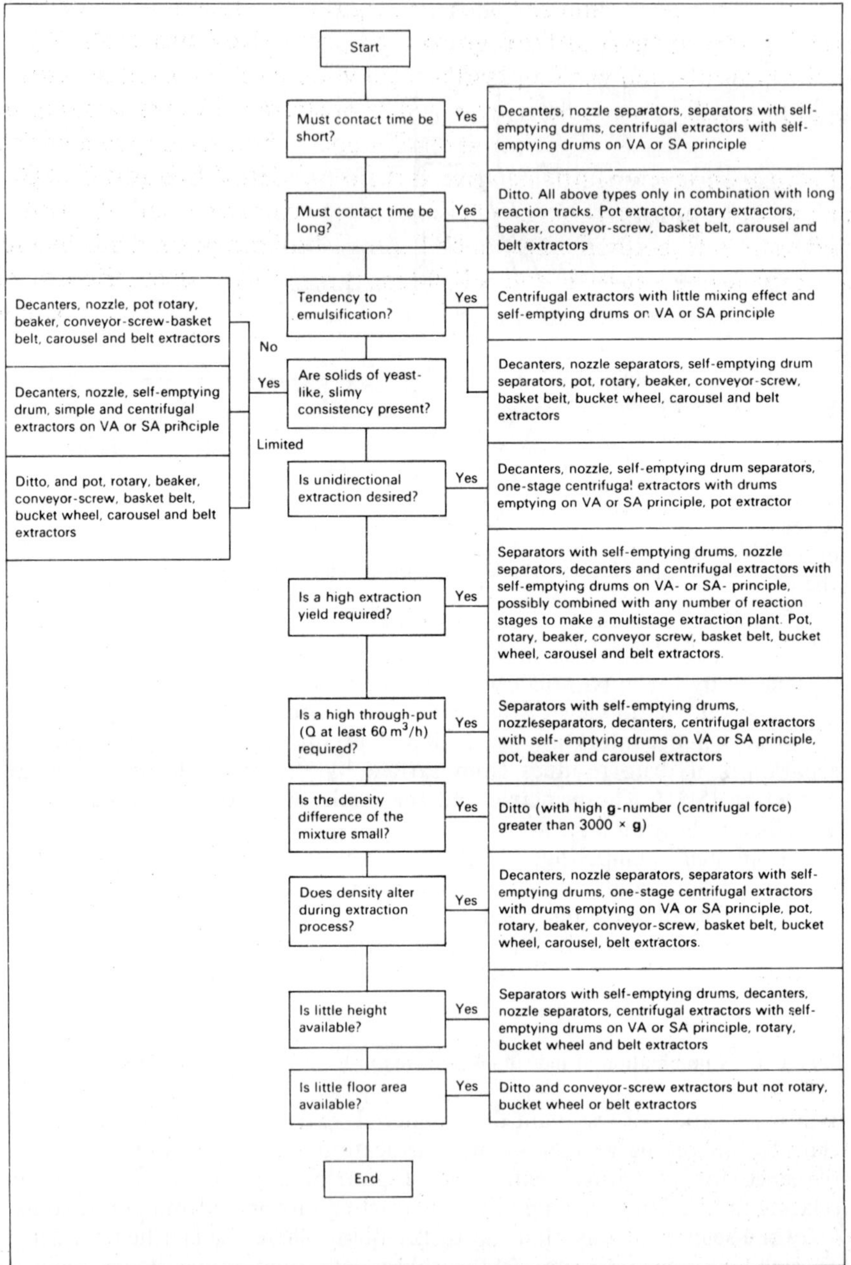

Figure 5.61 Choice of an extractor according to process criteria for solid–liquid extraction processes.

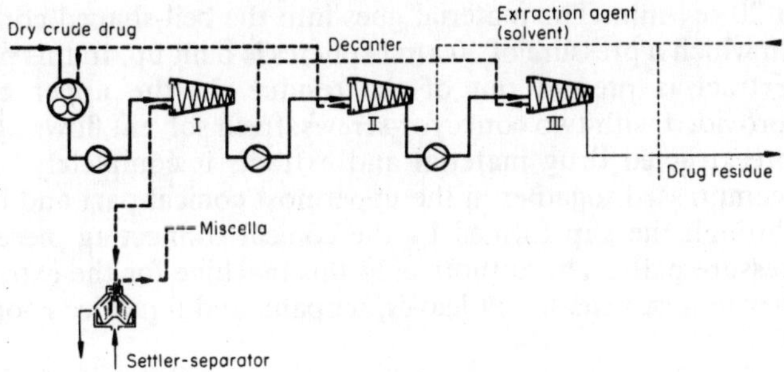

Figure 5.62 Extraction plant for processing medicinal plant materials.

5.4.3.7. Continuous extraction press

This continuously operating machine described by Minina *et al.* [5.33] could also be described under Section 5.4.5 because of its use of pressure. However, it seems more logical to deal with it here as solvent is supplied during the extraction process with the object of recovering an extract. Figure 5.63 shows the principle of the apparatus.

The material to be extracted goes into a mixing chamber through an inlet and is then caught and pushed vertically upwards by a conveyor-screw

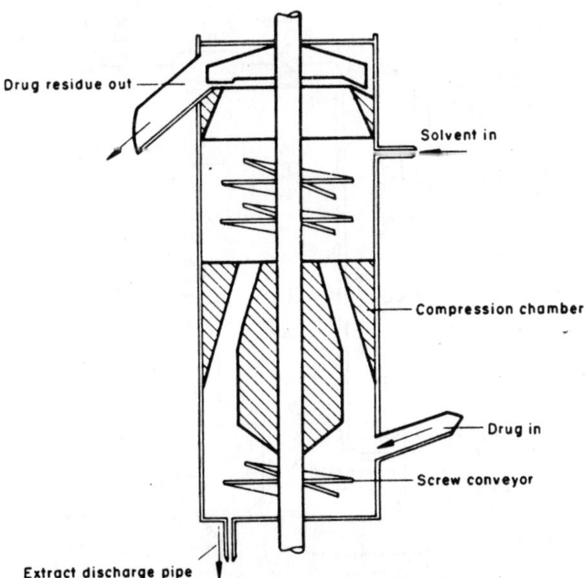

Figure 5.63 Continuous extraction press.

rotating at 20 rev/min. The material goes into the bell-shaped compression chamber, in which a pressure of 30 atmospheres is built up. In this part of the machine extract is pressed out of the residue. In the upper extraction chamber, provided with two conveyor-screws, fresh solvent flows against the partially preextracted drug material and extracts it completely. The drug residue is compressed together in the uppermost conical part and leaves the machine through the gap formed by the conical connecting piece and the counter-pressure plate. The authors used this machine for the extraction of, among other things, belladonna leaves, scopalia and liquorice roots.

5.4.3.8. Pulsation column

The pulsation column operates on the principle of continuous absolute counter-current extraction. Figure 5.64 [5.34] demonstrates this. The comminuted drug material is fed into the top of the column through a funnel and a chute. Solvent fed in via a measuring-pump flows upwards against it. Plates and baffles in the column ensure intensive mixing of solid and liquid. The liquid is also made to pulsate by rhythmic compression and expansion of a gas reservoir in a pulsation apparatus. The solid particles arriving at the bottom of the column are conveyed by the circulation pump into the solid–liquid separator, where the solid is separated off. The solvent is recycled to

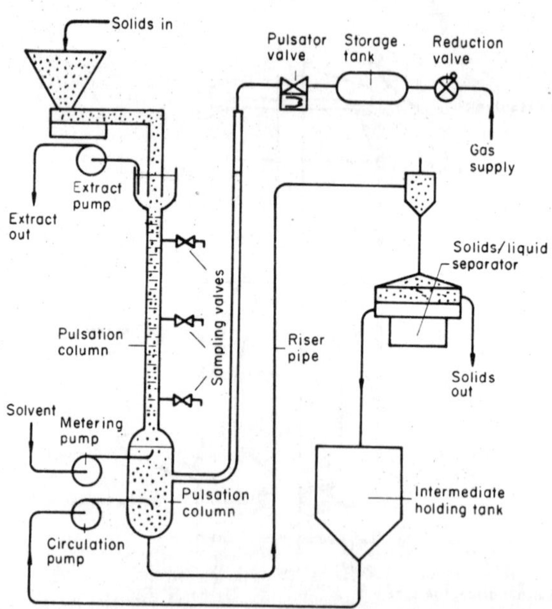

Figure 5.64 Pulsation column.

the circulation pump via a temporary storage tank. The extract is drawn off at the upper end of the column by the extract pump.

The machine permits the regulation of the following parameters:

Contact time.
Particle size of the solid.
Differences in density of the phases (solid–liquid).
Height of the column.
Type and quantity of the solvent.
Pulsation.
Temperature.

Adaptation for solving many extraction problems is thus possible. The machine is available in various sizes permitting throughputs from 10 kg up to several tons per hour. It is used for the extraction of leaves, roots, barks and seeds.

5.4.4. MACHINERY FOR EXTRACTION WITH SUPERCRITICAL GASES

Extraction with supercritical gases is such a new method that there are still very few machines for it on the market. The machines shall therefore be discussed in order of the individual makers. We shall also briefly discuss the machines specially designed by Stahl, with which most of the pharmaceutically relevant investigations were carried out. Two Patent Applications are referred to as examples of continuously operating machines. The following commercial machines are now available on the market:

The Nova-Swiss high-pressure extractor (Nova Werke AG, Vogelsangstrasse 24, CH-8307 Effretikon)
HDA High-pressure extractor (HDA Anlagenbau, Bücklestrasse 7, D-7750 Konstanz).
High-pressure extractor (Friedrich Krupp Industrie- und Stahlbau, Hamburg-Harburg).

5.4.4.1. Nova-Swiss high-pressure extractor

This is a laboratory machine with a 4 L extraction vessel and a 2 L separating vessel and is therefore the simplest conceivable form of a high-pressure extractor. Figure 5.65 [5.35] shows a flow chart of the machine. The drug material which is to be extracted is put into the 4 L pressure extraction vessel. The gas goes into the extraction vessel via a gas inlet valve, the heat exchanger, the compressor and the flowmeter. The extractive substance-laden gas then leaves the extraction vessel under pressure at the top and

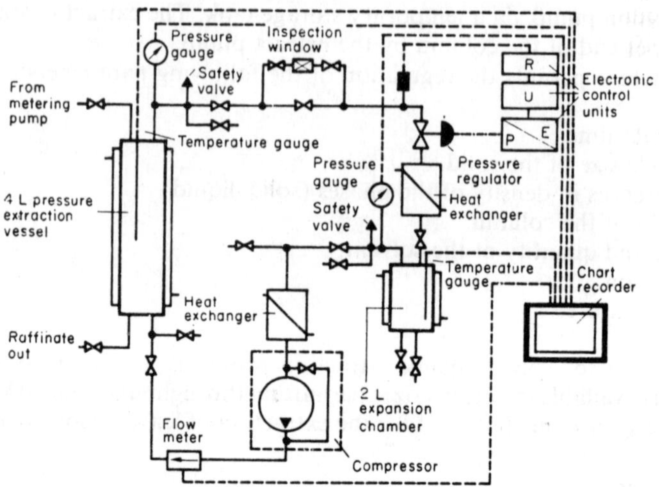

Figure 5.65 Nova-Swiss high-pressure extractor.

passes through a pressure regulating valve into the expansion chamber (the 2 L pressure vessel). The pressure regulating valve opens only when the preselected extraction pressure has been attained. The separation temperature is regulated by the heat exchanger. After completion of the starting phase the gas inlet valve is closed and the gas valve between the expansion chamber and the compressor is opened. The gas in the expansion chamber is thus led to the compressor, where it is compressed and then recycled to the extraction vessel. The material is thus extracted with circulating gas.

The pressure in the machine is regulated with extensometer strips. The valves for attachment of a dosing pump and for drawing off the refined liquid make continuous liquid–liquid extraction possible. Larger machines are available.

5.4.4.2. HDA high-pressure extractor [5.36]

The design is, in principle, the same as in the machine discussed above. The extraction volume of 4 L is also the same. Machines for multistage extraction and separation and likewise technical and production plants are also on the market, though their supporting documents contain no information on maximum permissible pressure and temperature.

5.4.4.3. Krupp high-pressure extractor

The plant described by Coenen *et al.* [5.37] has two extraction vessels and five

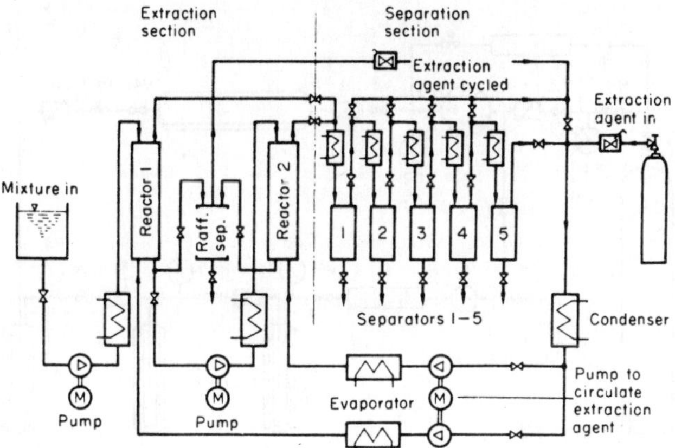

Figure 5.66   Schematic of an experimental plant employing two-stage extraction and five-stage separation.

separation stages. It is shown diagramatically in Fig. 5.66. The machine is fitted with pumps for compressing the gases and has two 7.5 L extraction vessels. The five separator units each have a volume of 1 L. This machine can meet heavy demands, the whole of it being made of V4A-standard industrial material so that it can withstand temperatures of up to 350°C and pressures of up to 350 atmospheres. It can operate discontinuously or continuously.

5.4.4.4.   Stahl high-pressure extractor [5.38]

This is shown in Fig. 5.67. The gas is obtained from a commercial pressurized cylinder (G) via a pressure-reducing valve (RV1) and compressed to the extraction pressure with a membrane compressor (C). The extraction pressure is regulated with the pressure valve (RV2). The hot compressed gas is then brought to the desired extraction temperature in a heat-exchanger and passed through the extraction vessel (E), where it becomes charged with the drug constituents soluble in it. The flow of the loaded compressed gas is controlled with the fine control valve (RV3). The dissolved substances are separated from the extraction gas in the separator part of the machine by reducing the density and pressure of the gas. Fractionation of the extract is possible because of the controlled conditions (pressure and temperature) in the separators. The gas thus freed of extract then passes through the flowmeter (S) and is then recycled to the compressor (C). The extraction gas is thus circulated in a manner similar to that in a Soxhlet extraction, so that the gas used for the extraction is recovered almost in its entirety. The machine is designed as a laboratory apparatus and operates with quantities of drug material of the order of 100 g.

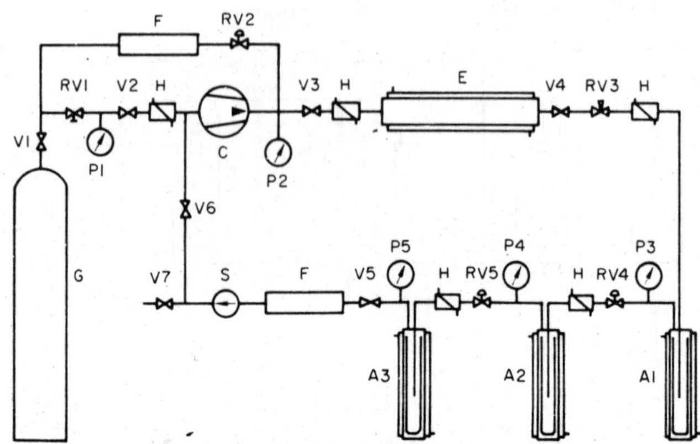

Figure 5.67   Schematic of the "Stahl" high-pressure extractor. V1–V7, stop valves; RV1–RV5, regulatory valves; P1–P5, pressure gauges; G, gas reservoir; H, heat exchangers; C, membrane compressor; E, extraction autoclave; A1–A3, extract separators; F, activated charcoal–blue gel filter; S, flow meter.

5.4.4.5. Plant for deodorization of fats and oils as per West German Patent 2332038

The plant used for deodorizing fats and oils, which operates with supercritical carbon dioxide at temperatures of 150–250°C and pressures of 101.3–253.3 atmospheres, consists of a column 15 m long with an internal diameter of 6 cm filled with glass beads and which is widened at the lower end. Figure 5.68 [5.39] shows a schematic drawing of the plant. The column is enclosed in a heating jacket and is heated to 200°C. The oil is fed in from the top at a rate of 5 kg/h and the carbon dioxide is led in from below.

5.4.4.6. Decaffeination of coffee as per West German Patent Disclosure 2905078 [5.40]

The continuous mode of operation resembles that of the previously described plant. The height of the extraction column is 6 m and its diameter 6 cm. Aqueous coffee extract is extracted with carbon dioxide at 50–51°C and 200–202 atmospheres. The rate of flow of the aqueous coffee extract is 17 cm^3/min and of the carbon dioxide 800 g/min. The depressurized conditions are 0.5–1 atmosphere at 80–90°C. Caffeine is principally removed in a yield of 98%.

5.4.5. MACHINERY FOR TREATMENT OF THE DRUG RESIDUE

The oldest methods for pressing out liquids in batches use basket presses or

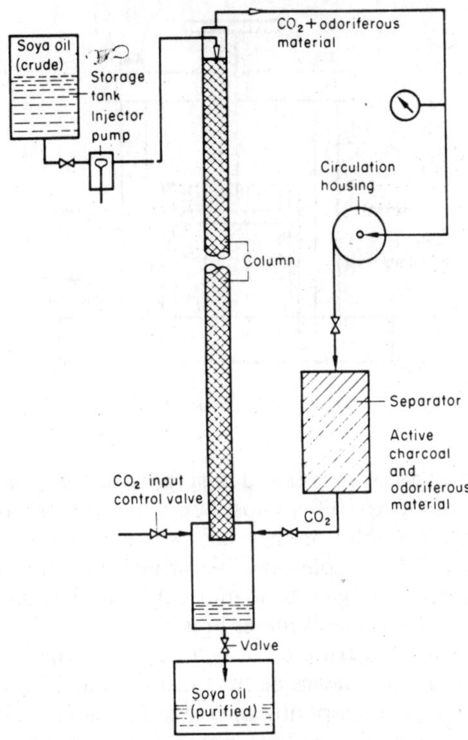

Figure 5.68 High-pressure extraction plant based on West German Patent Application 2332038.

wine presses, the principles of which are shown in Fig. 5.69 [5.41]. The material to be pressed is put in a sieve basket normally constructed of wooden stakes standing vertically alongside each other. The press piston is moved downwards on the spindle with a handwheel, thereby exerting a pressure on the cake of pressed material. The liquid squeezed out flows down between the wooden stakes into a collecting channel. Pressures of up to ~15 atmospheres can be attained with basket presses.

Filter or strainer presses, in which steel plates 10–12 mm thick provided with boreholes widening towards the exterior are arranged in a jacket to form a cubical pressing chamber, operate very similarly. The plates are strengthened by ribs, between which the liquid can flow. The pressure is produced by a hydraulic piston. Pressures of up to 350 atmospheres are possible and hence filter presses can also be used for the cold pressing of fatty oils out of seeds and fruits.

Pack presses also operate in a similar manner. Here the material which is to be pressed is wrapped in filter cloths, and the resultant packets are stacked in the press. Between 10–25 packets with an individual height of 7–12 cm are

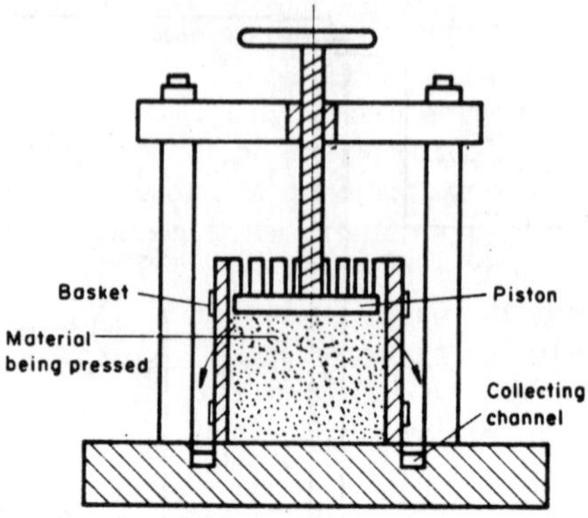

Figure 5.69   Basket press.

generally put in. The pressure is raised in stages in order to preserve the filter cloths. Pressures of up to 100 atmospheres are eventually applied. Pack presses are likewise suitable for the cold pressing of oils out of rapeseed, peanuts and linseed, for example. Another principle is utilized in the Willmes presser used for the recovery of fruit juices. Figure 5.70 shows a schematic drawing of the apparatus. The Willmes presser consists of a horizontal drum with sieve like boreholes distributed over the entire surfaces of the long sides of the drum. A nozzle, which can be turned downwards to empty the drum, serves as both a filling and emptying opening. The sides of the apparatus are sealed, curved walls to which the hose-like, inflatable rubber bag is fastened.

The apparatus is first filled with the material which is to be pressed. The rubber bag is slowly inflated, pressing on all sides evenly on the press material. As the pressing continues the rubber bag is inflated further, continuously reducing the thickness of the press cake. After completion of the pressing the pressure is released, whereupon the press cake collapses. The

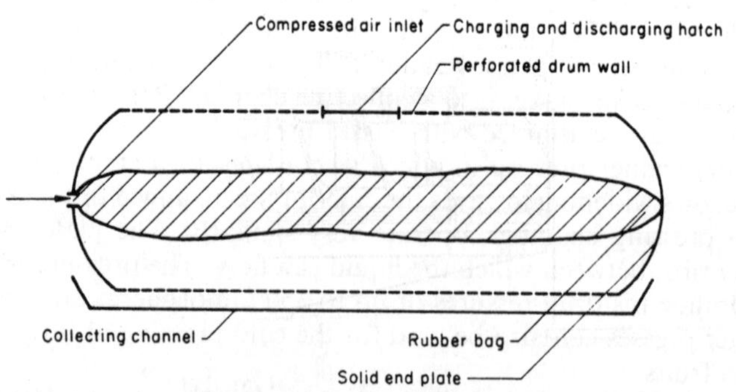

Figure 570   The Willmes presser.

apparatus is rotated, bringing the filling and emptying aperture underneath so that the press cake falls out.

Whereas in the basket press the available pressure area substantially corresponds to the area of the piston, i.e. the basket wall has practically no pressure effect, the pressure area in the Willmes presser is considerably larger. The thickness of the layer of material is reduced by spreading it over a larger area and this in turn accelerates the expression process. A further advantage is that, in contrast to the basket press, the press and run-off areas are not reduced during the pressing process. The press cake is partly loosened up by the run-off of the juice. An even run-off of juice under constant pressure conditions thus takes place throughout the entire pressing process. The flexible rubber bag ensures even pressure over the whole area of the sieve.

All of the presses described so far operate batchwise. Screw presses and sieve belt presses are used for continuous operation. Figure 5.71 [5.41] illustrates a screw press, which consists of a screw-shaft which revolves in a strainer made of longitudinally arranged rods. The slits between the rods can be adjusted to the material being pressed. The diameter of the screw increases in stages in the direction of conveyance of the material, thus compacting the material which is being pressed. The degree of compaction is controlled by a conical ring at the point where the pressed material is discharged. A perforated sieve can also be used instead of the ring with materials which can be pressed easily. The machines are designed for pressures of 100–300 atmospheres. The screw turns at a speed of 10 rev/min. The pitch of the screw differs in the individual pressing sections. A steeper pitch in the first zone ensures that the material being pressed is drawn into the next longer pressing zone with a gentler pitch. Only a little liquid is expressed in the first section. Most of the pressing takes place in the actual pressure zone. Scrapers fitted between the strainer and the screw prevent the press cake from being turned with the screw. This turning can also be prevented by the positioning of the strainer rods.

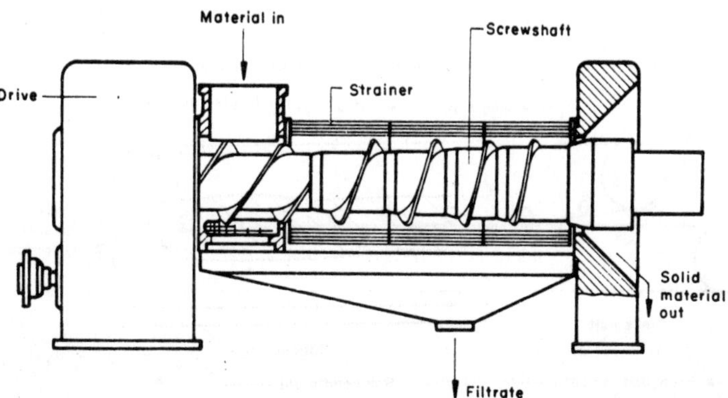

Figure 5.71 Screw press.

Screw presses are often used in combination with continuously operating extraction plants which have suitable throughput capacities. They are widely used for the expression of shredded sugar beet in the manufacture of sugar. Here the water content of the shreds is reduced from 94 to 14–15%.

Sieve belt pressing is an economical process. Figure 5.72 shows the functional principle of a sieve belt press [5.41]. The material which is to be expressed is put in a thin layer on an endless sieve belt consisting of metallic, synthetic (plastic) or mixed material with mesh widths of 0.2–1.5 mm. The mesh width must be larger than that of the particles of the expressed product to prevent the meshes getting clogged. The compaction of the material into a cake prevents individual particles getting into the expressed juice in any noticeable quantity. The pressing belt runs under pressure rollers which press it from above on to the sieve belt, which is supported by a number of supporting rollers; the material to be expressed is compacted to a thin layer by the pressing belt in a predehydrating zone. The thinness of the layer makes high compaction pressures unnecessary. Most of the pressing takes place in the next pressing zone where the rollers exerting a higher pressure are situated. This is followed by the shearing zone, in which the sieve belt is led in wave form between the pressure and supporting rollers. This produces displacements within the material being pressed, which facilitates the expression process. Sieve belt presses cannot achieve such a high degree of dehydration as that attained with screw presses. It is generally possible to reduce the amount of water contained in a material from 90–95 to 65–75% by pressing with a sieve belt press.

Gromova *et al.* describe an apparatus for recovering expressed juices from soaked and swollen plant drug materials [5.42] (Fig. 5.73). The drug material is taken by the press arm in the horizontal pressing chamber and pressed out. The authors investigated, among other things, the effect on expression of the preswelling time in valerian roots with solvent and of certain additives such as Tween 80 or alcohol. The best results were obtained with stage-wise expression of a drug plant material soaked with water.

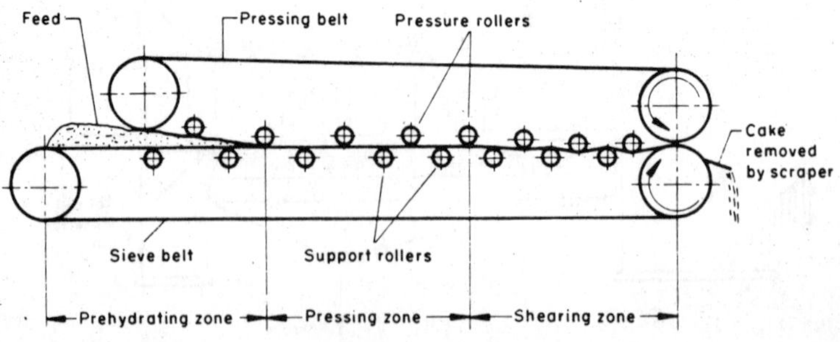

Figure 5.72   Sieve belt press.

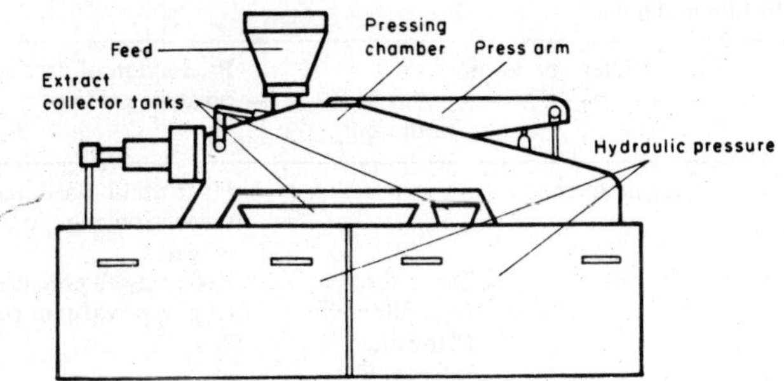

Figure 5.73 Horizontal extractor press.

5.5. Machinery for the purification of extracts

The purification of extracts mainly involves clarification and reduction of the
bacterial count. Accordingly the following types of apparatus are used:

> Filtration apparatus.
> Separators and decanters.
> Apparatus for pasteurization and ultrahigh heating.

5.5.1. FILTRATION APPARATUS

Filtration apparatus is constructed with a great variety of designs. Table 5.5.
gives a classification of these [5.43] by:

> Type of filtering material (poured, flexible or rigid).
> Mode of operation (continuous, batchwise).
> Type of driving force (pressure or vacuum).

The filtration of extracts is a clear filtration, i.e. the desired product is the
filtrate. All apparatus such as vacuum drum filters, pressure drum filters and
vacuum belt filters used for recovering the residue are therefore excluded.
Membrane filters are also unsuitable as they quickly become clogged during
filtration of extracts, with the result that the filtration rate is also rapidly
reduced. The only presses suitable for the clear filtration of extracts thus
remaining in the Table are the frame and chamber filter presses. Both operate
discontinuously.

Frame filter presses form the filter cakes in hollow frames 20–150 mm deep
placed between two filter plates covered with filter cloths. Figure 5.74 shows
a schematic representation of the arrangement [5.44]. The filter plates are
hung alternately with the frames vertically on tensioned rods, with a filter
cloth placed between the frame and the plate. The stopcocks of the plates are

Table 5.5 Filtration equipment

Type of filter	Type of filter operation		Production of driving pressure gradient
	Batchwise	Continuous	
Aggregate, pouring	Aggregate filter	—	Hydrostatic pressure of suspension column (feed height)
Filter cloth (flexible)	Suction filter	Drum filter Disc filter Plane filter Internal filter Belt filter	Reduced gas pressure by water-jet pump (vacuum pumps)
	Frame and chamber filter press	Drum filter (push–pull filter)	Fluid excess pressure of suspension by pressure pumps
Filtering material (rigid layer)		Candle filter Plate filter	

opened. The solution which is to be filtered enters an upward-sloping pipe and is led into the chambers through apertures in the frames.

The filtrate is pressed through the filter cloths and flows via guide-grooves in the plates into the collecting channels. The filter cakes build up slowly on the frame side of the filter cloths and eventually completely fill the frames, with the result that the filtration rate falls towards zero.

The liquid used to wash the filter cakes (Fig. 5.74(b)) flows into a separate channel which also runs in the upper part of the plates and frames. This only opens into every second plate. The stopcocks of these plates are closed. The washing liquid penetrates through the entire filter cake and flows down over the adjacent plate (run-off plate). When the plates have been drawn apart the filter cakes can be removed, after having been blown dry with compressed air.

Chamber filter presses have square or circular filter plates with raised edges. Fig. 5.75 illustrates their design and mode of operation [5.44]. Pairs of

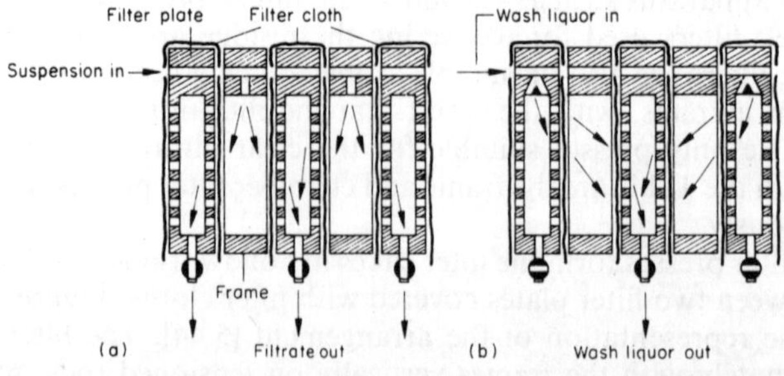

Figure 5.74 Frame filter press showing (a) filtration and (b) washing.

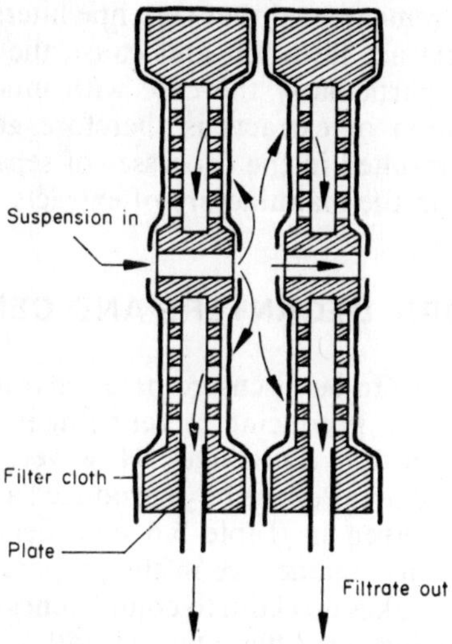

Figure 5.75 Filtration process in a chamber filter press.

filter plates covered with filter cloths form chambers for collecting the filter cakes. The capacity of these chambers is generally smaller than in frame filter presses. The chambers are sealed by the filter cloths being pressed against each other. The suspension is fed in at a point half-way down the apparatus. The filtration and washing processes do not differ essentially from those in frame filter presses. The filtration pressures in both systems are 3–15 atmospheres.

The choice of filter plates or filter cloths depends on the desired degree of filtration. Preliminary and fine filtration can be combined in one apparatus by a combination of various filters, using a revolving chamber as shown in Fig. 5.76 [5.18].

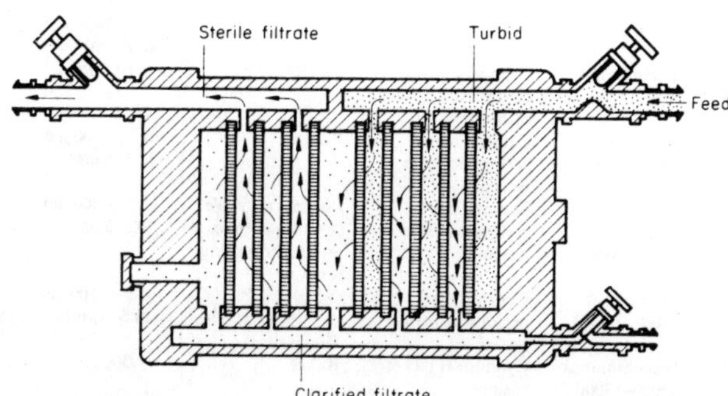

Figure 5.76 Schematic diagram of a Seitz filter bed with a revolving chamber.

It must of course be remembered that very fine filters also become clogged very quickly during extract filtration and cause the filtration rate to fall towards zero. This is particularly the case with mucilaginous drug plant material. Sterile filtration of extracts is therefore generally not possible. These problems have resulted in the processes of separation and decanting becoming widely used in the manufacture of extracts.

5.5.2. SEPARATORS, DECANTERS AND CENTRIFUGES

All the types of apparatus to be discussed here separate solid–liquid or even liquid–liquid mixtures by producing a centrifugal acceleration and can therefore also be described as centrifuges. However, their varying designs, operating conditions and areas of use have produced a great variety of names for them. The survey given in Table 5.6 considers only those types of centrifuge which have any significance in the preparation of phytopharmaceuticals and therefore makes no claim to completeness. We have deliberately made no attempt at a further differentiation within the individual groups

Table 5.6 Centrifuges, decanters and separators

Type	Mode of operation, acceleration factor	Separating element	Method of discharging solid materials	Particle-size range	Solid concentration
Sieve centrifuges	Discontinuous or continuous 200–1500	Conical or cylindrical perforated sieve drum, vertical or horizontal	Rakes, screw or vibrator	5 to 10000 μm	5–60%
Scraper centrifuges	Continuous 300–3000	Cylindrical perforated sieve drum	Scraper blades	5 to 10000 μm	5–60%
Shearer centrifuges	Continuous 200–2000	Cylindrical perforated sieve drum, horizontal, single- or multi-stage	Oscillating sliding base	100 to ca 80000 μm (40000 μm)[a]	20–75% (10–40%)[a]
Decanters	Continuous 1000–4500	Cylindrical/conical jacket, horizontal	Screw	3 to 20000 μm (200 μm)[a]	2–40%
Ring chamber separators	Discontinuous 6000–11500	Cylinders placed concentrically inside each other	Rakes	< 1000 μm	< 2%
Separators with settling drums	Discontinuous 4000–13000	Conical, non-perforated plate system	Rakes	0.5–~500 μm (0.5–5 μm)[a]	0–3%
Separators with self-emptying drums	Partly continuous 4000–13000	Conical, non-perforated plate system	Pistons, movable centrifugal bases or internally turning screw	0.5–~500 μm (0.5–5 μm)[a]	2–10%
Separators with nozzle drums or conveyor screw	Continuous 4000–13000	Conical, non-perforated plate system	Nozzles	0.5–~500 μm (0.5–5 μm)[a]	5–25%
Tube centrifuges	Discontinuous 13000–17000	Cylindrical jacket casing	Rakes	< 300 μm	< 1–2%

[a] The numerical data vary according to source.

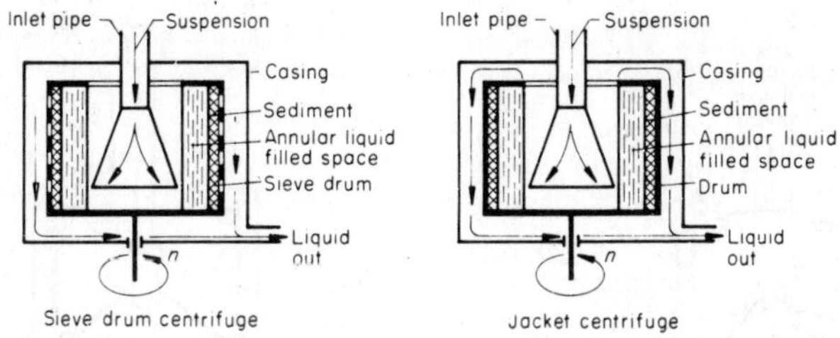

Figure 5.77 Differences between sieve and jacketed centrifuges.

such as has been undertaken in the standard text books on technical chemistry.

A difference exists in principle between sieve and filter centrifuges, and jacketed centrifuges. In the former the solid material being separated off is retained by sieves or filters, and the liquid runs down to the outside through the sieve cloth or the filter bed, which during operation is covered with a 'filter cake'. In jacketed centrifuges the solid collects on the outside on a sealed centrifuge wall, from which it must be removed from time to time by hand (discontinuous operation) or automatically carried away by devices such as conveyor screws or push-plates. Figure 5.77 [5.43] shows the various modes of operation of these types of centrifuge.

Decanters and plate separators, in particular the automatically emptying separators, have now become the most important of all the types of centrifuge listed in Table 5.4. Other models, such as ring chamber separators, have been suppressed by these.

5.5.2.1. Sieve centrifuges

Sieve centrifuges are simple, relatively slow-running machinery. The basic component is a vertical or horizontal sieve drum. A distinction is made between discontinuously and continuously operating types. In both cases clarified liquid runs down continuously, the difference being only in relation to the way in which the solid substance is discharged. Figure 5.78 shows two discontinuously operating simple sieve centrifuges [5.45].

Although these models appear somewhat outmoded, they are numerically still very important. They are used in the chemical and pharmaceutical industries and in the preparation of fruit juices and foods. Emptying can be facilitated by filter inserts in the form of so-called 'lift-out' bags. The conical shape of the sieve jacket as shown in Fig. 5.79 [5.45] makes it possible for the solid which has been centrifuged off to slide to the larger diameter as a result of the components of centrifugal force directed parallel to the sieve insert. These machines, which are also known as sliding centrifuges, thus operate continuously.

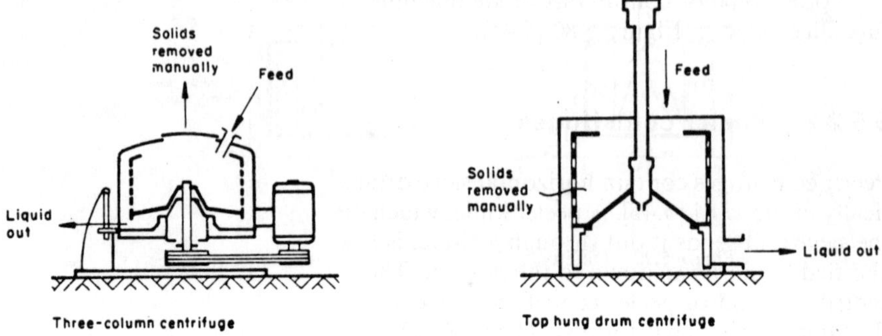

Figure 5.78   Simple sieve centrifuges for batch operation.

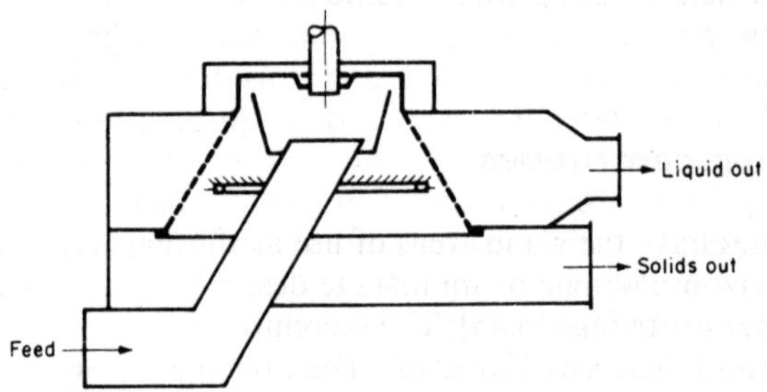

Figure 5.79   Conical sieve centrifuge.

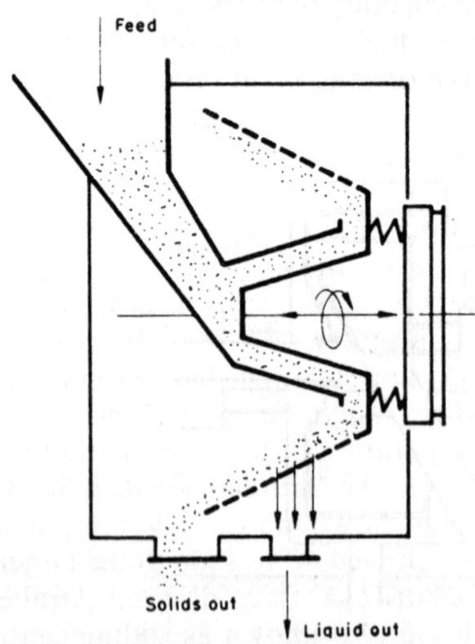

Figure 5.80   Oscillating sieve centrifuge.

Another type of continuous mode machine is the oscillating sieve centrifuge illustrated in Figure 5.80 [5.45].

5.5.2.2. Peeler centrifuges

Peeler centrifuges contain horizontal sieve drums and are illustrated schematically in Fig. 5.81 [5.45]. A peeler knife, which peels off sediment build-up on the sieve and sends it out through a chute, is fixed inside the centrifuge. The clarified liquid flows down to the bottom. The capacity both of simple sieve centrifuges and of peeler centrifuges depends on the drainage properties of the filter cake which is formed. The permeability of the filter cake depends on the granule structure and the density of the cake. A stiff cake, such as builds up in discontinuously operating centrifuges, is therefore less permeable than a filter cake which is continually removed in continuously operating machines.

5.5.2.3. Pusher centrifuges

Pusher centrifuges have the same areas of use as sieve and peeler centrifuges and are of a horizontal single or multistage design. Figure 5.82 illustrates a three-stage pusher centrifuge [5.45]. The incoming material is led to the first, innermost stage and then centrifuged off. The cake formed on the sieve wall is pushed forward from time to time by pusher plates and thus passes into the next centrifugation stage. In multistage models the circular collar of a drum is used as the pusher plate for the next drum towards the outside.

Pusher centrifuges are encountered very frequently in the potash, salt and fertilizer industries. Their high capacity for solid materials makes them suitable for the separation of drug plant residues after extraction processes.

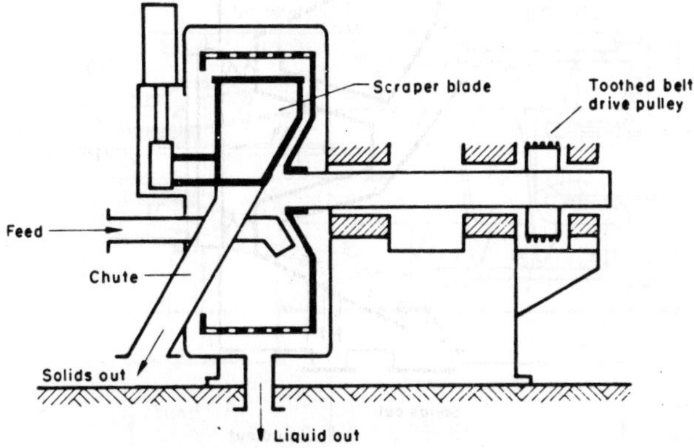

Figure 5.81 Peeler centrifuge.

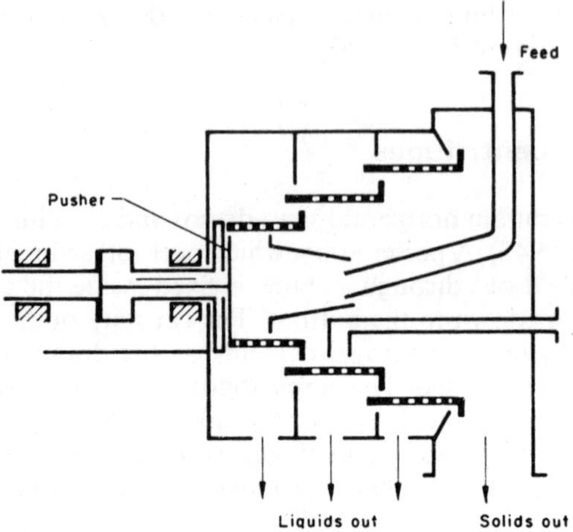

Figure 5.82   Three-stage pusher centrifuge.

5.5.2.4. Decanters

Completely enclosed centrifuges with a conveyor screw for removing sediment are known as decanters. They can be used where a solid sediment suitable for removal by a conveyor-screw is formed. Figure 5.83 shows the principle of a decanter [5.43]. The mixture flows continuously through the inlet tube into the distributor head. It is delivered at the end of the conical part of the jacketed drum by the shaft of the conveyor-screw transporting the solid. The height to which the cylindrical part of the drum is filled is regulated

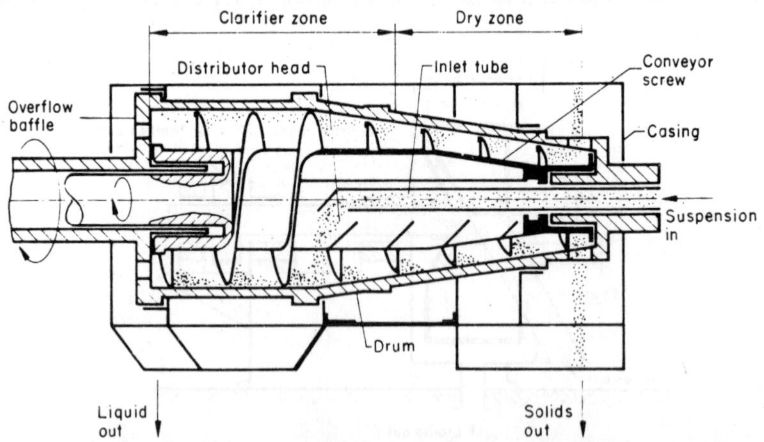

Figure 5.83   Schematic diagram of a decanter.

by the adjustable overflow baffle at the end. The centrifugal force deposits the solid on the surface of the jacket. The conveyor-screw, which revolves faster than the enclosed drum, pushes the solid out of the liquid zone into the dry zone, where it is transported by the tapering screw to be discharged at the conical end.

The liquid flows in the opposite direction to the cylindrical end of the clarifying drum. The finest particles are separated off in this area, which is also termed the clarifier zone. The liquid leaves the machine through the overflow baffle. The separating action of a jacketed centrifuge increases with increasing acceleration force and is limited only by the tensile strength of the rotor. The greater the acceleration factor, the smaller must be the diameter of the drum at any specified tensile strength, and hence the smaller will be the throughput, which is determined from the basic equation of continuous sedimentation, the sedimentation equilibrium:

$$\dot{V}_1 = A v_c = 2 r_m \pi L v_c.$$

Here $\dot{V}_1$ is the centrifugate throughput (m^3/s), L is the length of the drum (m), v_c is the centrifugal sedimentation rate in (m/s), and r_m is the mean radius of the circle of liquid (m) as per Fig. 5.84.

$$r_m = \frac{r_a + r_i}{2}$$

r_i is produced from the thickness s of the liquid ring, which can be adjusted by the overflow baffle.

$$r_i = r_a - s.$$

Decanters are commonly used in the pharmaceutical and foodstuffs industries where separators can no longer be used because the content of solid is too high or where the solid can be separated with low acceleration factors.

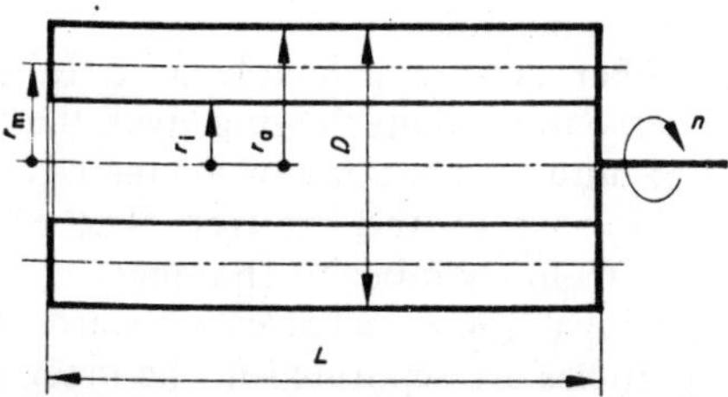

Figure 5.84   Jacketed centrifuge in section, showing principal dimensions for drum throughput calculations. D is the drum diameter and n is the speed of revolution.

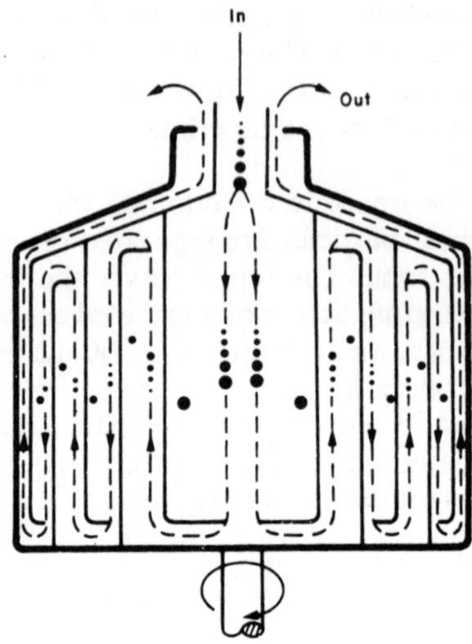

Figure 5.85   Ring chamber separator.

Decanters are very often used in combination with separators. Here the decanter does the preliminary clarification, and the fine clarification takes place in the separator.

5.5.2.5.   Ring chamber separators

The ring chamber separator is an expansion of the tube separator (tube centrifuge). Here the surface area available for clarification is increased by putting several concentric cylinders one inside the other, as shown in Fig. 5.85.

The separation of the solid particles by size from the inside outwards shows the classifying action of the apparatus. The liquid flows through the chambers outwards from the inside, thus lengthening the time for which it undergoes separation in the machine.

Each individual chamber acts in principle as a tube centrifuge. The separation takes place in the thin axially flowing layer, the thickness of which decreases from the inside outwards because of the increasing diameter of the chambers the nearer they are to the exterior. The effective length for deposition thus decreases from chamber to chamber.

As the effective centrifugal force increases outwards from chamber to chamber, the coarser particles are separated in the inner chambers and the finer particles in the outer chambers. Ring chamber separators therefore have a certain classifying action. Figure 5.86 shows the flow conditions in a ring chamber [5.45].

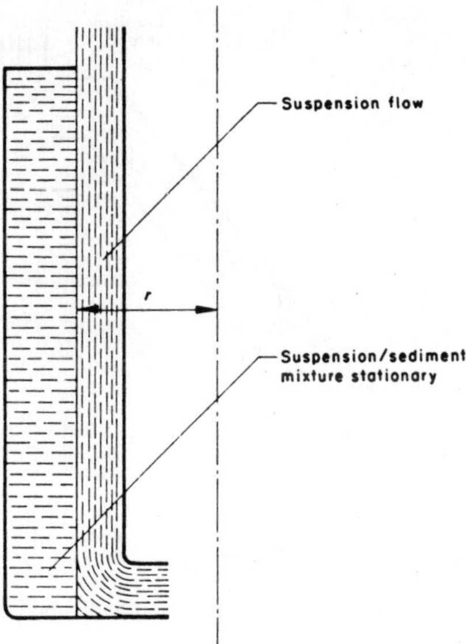

Figure 5.86 Flow conditions in a chamber of a ring chamber separator.

The advantages of these separator drums are a good clarifying effect (until the chambers are filled with solids) and a large capacity for solids. They have the disadvantages that they can only be used batchwise as clarifiers for separating solids from a suspension; they cannot be used as separators for fractionating a liquid mixture into its various liquid components; and when the chambers are filled with solid substance the separator is stopped and cleaned manually. The chambers must be dismantled individually for this.

Ring chamber separators are used, for example, for clarification in the soft drinks industry and wherever solids are to be recovered in compact form together with the liquid. Continuously operating separators have now replaced many ring chamber separators.

5.5.2.6. Separators with clarifying drums

The name 'separator' has become popular for a group of centrifugal separators which should be more correctly called plate centrifuges or plate separators. Here the clarifying area is enlarged by the incorporation of a large number of conical plates 0.5–0.75 mm thick placed 0.3–2 mm apart. The angle of inclination of the plate is generally 30–40°. The stream of liquid is split up into many thin layers by the large number of individual separation spaces connected in parallel, producing very short deposition paths [5.45].

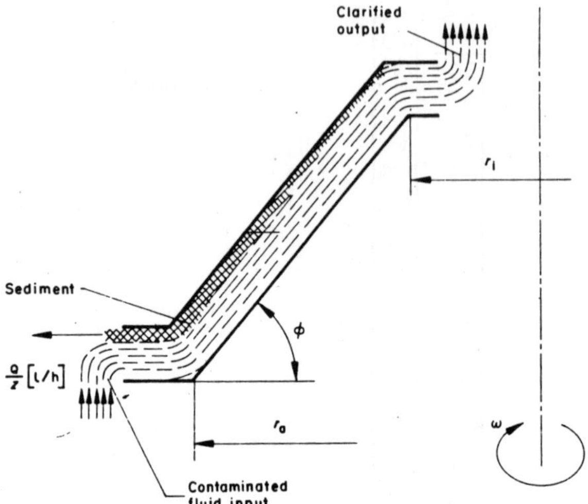

Figure 5.87   Flow between two plates in an individual separating space.

Figure 5.87 shows an individual separation space, which is a space in the form of a conical ring. The number of these individual separation spaces of a plate drum is z. The quantity Q/z flows through an individual separation space upon even distribution of the incoming hourly throughput quantity Q (which may be assumed on the basis of the prevailing pressure conditions). A particle of solid is considered to have been separated from the liquid when it has reached the upper conical surface of this individual separation space. The deposited solid particles slide continuously as a coherent layer on this surface into the solid-space of the drum. The clarified liquid flows inward and leaves the plate insert via its inner edge.

The advantages of these plate drums are:

> A good clarifying effect.
> Automatic cleaning of the plates by correct setting of the angle and distances of the plates.
> The possibility of universal application in discontinuously and continuously operating separators.

If the throughput equation is worked out with a few simplifying postulates, then

$$Q = \underbrace{\frac{d^2 \Delta\rho}{18\eta} g}_{v} \times \underbrace{\frac{2\pi}{3\,g}\omega^2 (\tan\varphi)\, z (r_a^3 - r_i^3)}_{\Sigma_T}$$

$$Q = v\Sigma_T$$

The equivalent clarifying area is thus

$$\Sigma_T = \frac{2\pi}{3g}\omega^2(\tan\varphi)z(r_a^3 - r_i^3)$$

where Q = throughput rate (kg/h), d = particle size of solid (m), $\Delta\rho$ = difference in density between solid and liquid phase (kg/m³), η = viscosity of liquid phase (Pa/s), φ = angle of inclination of plate (30–40°), ω = angular velocity of drum (rad/s), r_a = outer radius of separator compartment (m), r_i = internal radius of separator compartment (m), g = acceleration due to gravity, z = number of individual separation spaces, v = Stokes's sedimentation rate (m/s), and T = comparison number for the clarifying action, derived from the constructional data of the separator drum (Σ_T is the equivalent clarifying area).

The formula for the equivalent clarifying area contains only values relating to the apparatus. It represents a practical rule of thumb or rough formula yielding comparative values for various separators.

The following factors influence the output of a plate separator:

1. Product controlled factors:
    (a) Density difference—the greater $\Delta\varphi$, the greater the output.
    (b) Viscosity of the carrier liquid—the lower η, the greater the output.
    (c) Particle-size—the greater d, the greater is the output.
    (d) Consistency of the solid material which is discharged (e.g. matt, pasty, etc.) is important in self-emptying separators.

2. Construction controlled factors:
    (a) Revolution speed of drum.
    (b) External radius of plate or separation radius.
    (c) Angle of plate.
    (d) Separation path or distance or number of plates.

An increase in the constructional factors increases the output, though material and technical limits are set on this increase. Hence the permitted tension of the drum material, for example, limits the revolution speed and the external diameter of the drum. Figure 5.88 [5.45] shows the principle of a separator with a clarifying drum. The solid deposited on the outer wall remains in the drum and has to be removed manually. Separators with clarifying drums are therefore only suitable for clarifying liquids with very low solid contents (0–3%). They are being increasingly replaced by separators with an emptying mechanism, although they are of a simpler construction and permit higher revolution speeds because of the absence of an opening mechanism.

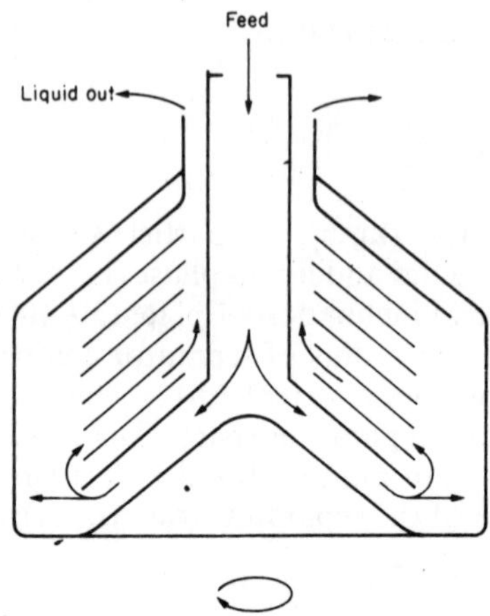

Figure 5.88   Separator with clarifying drum.

5.5.2.7.   Separators with self-emptying drums

The separators previously described operate continuously as regards the supply of material and discharge of the liquid, but discontinuously as regards the solid remaining in the drum, which has to be removed at intervals. Two variants have been developed for automatically discharging the solid:

>   Partially continuous discharge with self-emptying separators.
>   Continuous discharge from the separator with a nozzle drum or conveyor screw.

The drum of a self-emptying separator has a double conical space for the solid in which the sediment collects. When the solid space is full, a circular aperture is opened hydraulically and the solid is thrown out when the drum is revolving at full speed.

Two construction principles are used for partly continuous automatic emptying:

>   Emptying with movable floors of the centrifuge chamber.
>   Emptying with internal pistons.

5.5.2.7.1.   Emptying with movable centrifuge chamber floors

Hemfort [5.45] has described emptying with movable centrifuge chamber bases in detail. The base of the centrifuge chamber (Fig. 5.89) limits the separation space including the space for solid in the drum. It also forms,

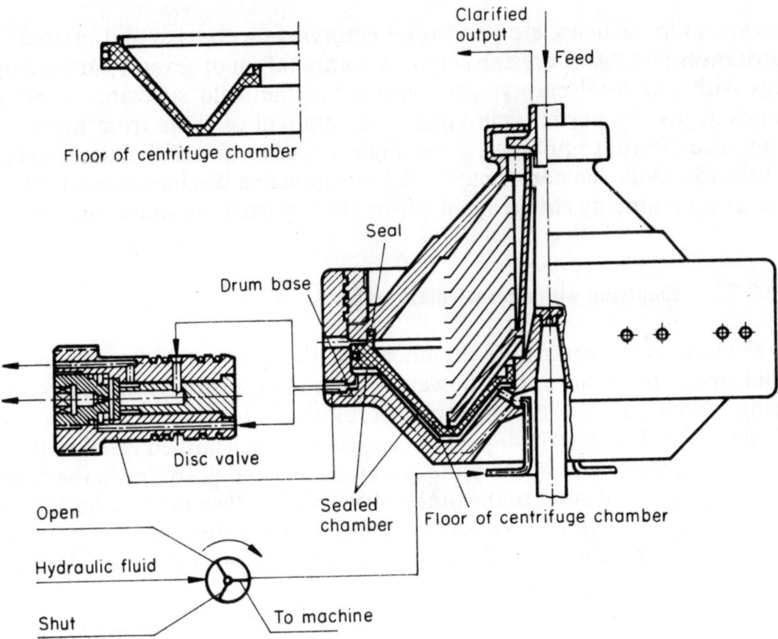

Figure 5.89 Self-emptying separator drum with movable centrifuge chamber floor.

together with the lower part of the drum, the closed chamber. When liquid is let into this closed chamber a 'sealing pressure' is built up by the rotation speed. This produces a force in the axial direction, pressing the movable base of the centrifuge chamber (the push-piston) against the main cover seal of the drum, thus sealing the separating chamber. The pressure in the separating chamber of the drum acts in the opposite direction on the piston, though the force from the separating chamber is smaller than the sealing force, as the radius of the sealed chamber is greater than that of the separating space.

When the space for solids is filled with sediments and the sealing fluid is drained out of the sealed chamber by means of hydraulically controlled disc valves, ring valves or slide valves, the pressure in the separating space displaces the movable base of the centrifuge chamber axially and opens the drum. The liquid level in the sealed chamber recedes radially to the outside. The sliding piston moves downward as soon as the sealing force under the piston is less than the opening force above it.

Emptying of the separating space commences. If the sealed chamber is refilled while emptying is still in progress, the piston returns to the closed position. This is a partial emptying. If the sealed chamber is emptied completely, the separating space also empties completely (total emptying). Partial emptyings are carried out with solid substances of a plastic nature (yeasts, pulp particles, etc.) and total emptyings with hard and matted solid substances (catalysts, crystals, etc.). A combination of several partial emptyings with one total emptying is used when the solid substance cannot be removed just by partial emptyings, e.g. removal of slime from must (fresh fruit juice). Partial emptyings give a high concentration of the discarded solid substance. Only the compacted solid substance is discharged and rejected. The drum is already closed again before the supernatant liquid can come out.

5.5.2.7.2. Emptying with internal sliding pistons

In the case of the centrifuge chamber base the pressure of the accumulating solid opens the solid outlet, however, the sliding piston is moved by water being admitted into the opening or closing chamber. The principle is illustrated in Fig. 5.90 [5.46]. The suspension to be clarified is fed in through the inlet and clarified in the plate system. The fixed grab draws the clarified liquid up and conveys it to the outlet. The solids collect in the solid substance chamber. The ejection of the solid is controlled via the controlling device by means of water, which is led into the external and internal closed chamber or opening chamber.

During the normal clarifying operation the sliding piston is in the closed position, i.e. the closed chambers are filled with water. The water pressure is sufficient to keep the sliding piston in the closed position (left half of Fig.

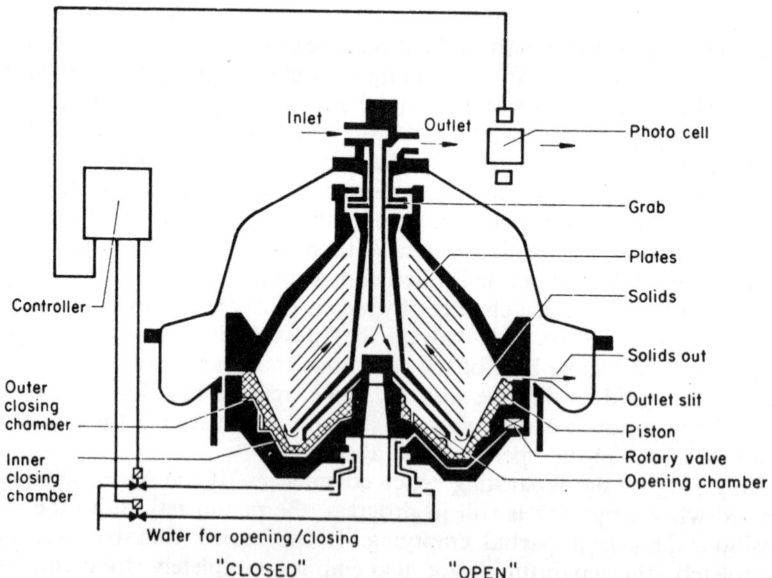

Figure 5.90 Separator with internal sliding piston.

5.90). Only the outer closed chamber is emptied via the drum-valve for partial emptying. This opens the outlet slit between the piston and the upper wall of the container so that the solids are discharged (right half of Fig. 5.90) until the counter pressure in the inner closed chamber is sufficient to bring the piston back to the closed position. Constant quantities of solid are thus discharged at every partial emptying.

For a total emptying the water is discharged from both closed chambers. The additional water pressure of the opening chamber causes a complete opening of the discharge slit. Even solids which are difficult to expel by centrifugul force are discharged in this position. The piston is closed again by filling the closing chamber with water.

Fully automatic opening and closing can be controlled either with a time switch, by monitoring the discharging liquid with a photocell (see Fig. 5.90) or by contact scanning of the level of the solid in the drum by means of feeler fluid.

Self-emptying separators are widely used in both the foodstuffs and pharmaceutical industries, particularly in the clarification of extracts.

5.5.2.8. Separators with nozzle drums or conveyor screws

These are used for the fully continuous discharge of solid material from the separating chamber. Figure 5.91 shows the design and operating principle of a nozzle separator. The incoming suspension is separated into a clear liquid phase and the solid. The solid is continuously discharged through nozzles.

The possibility of continuous discharge of solid with a conveyor screw is

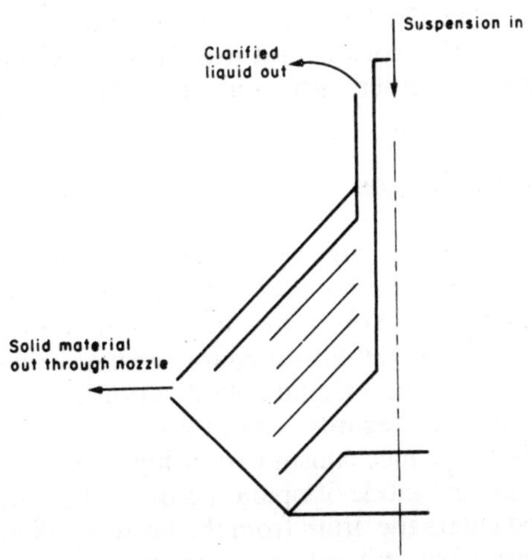

Figure 5.91   Construction principle of a nozzle separator.

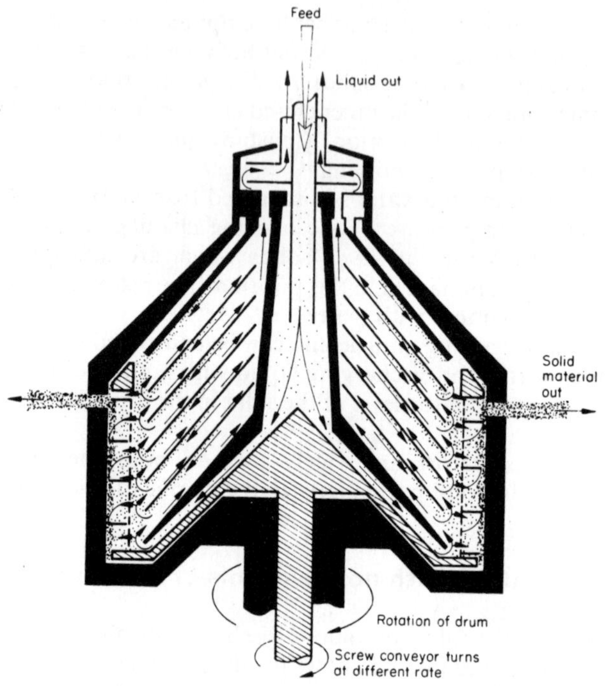

Figure 5.92   Separator using a conveyor screw for the continuous discharge of solid material.

illustrated in Fig. 5.92 [5.47]. The screw scrapes off the solid deposited on the inside of the drum and carries it upwards into the vicinity of the nozzle, where it is discharged together with some of the liquid phase.

5.5.2.9.   Tube centrifuges

Tube centrifuges attain the highest acceleration factors of all and are therefore suitable for the separation of the finest particles. The drums must have a small diameter because of the limited strength of the material of which they are made. The axial height must therefore be many times the diameter of the drum in order to obtain adequate clarification areas. Tube centrifuges operate discontinuously as regards separation of the solid. The design and the absence of a discharge mechanism permit high revolution speeds. Figure 5.93 shows the design and mode of operation of a tube centrifuge [5.48]. The liquid to be clarified enters the drum from the bottom, filling the free space in the drum and stationary still relatively to the drum. Only the central cylindrical layer of liquid flows upwards from the bottom and leaves the

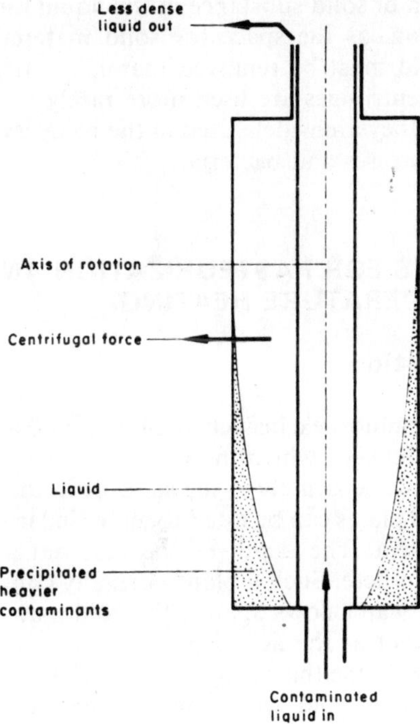

Figure 5.93 Operation of a tube centrifuge.

drum through the overflow. The maximum external diameter of this layer is equal to the internal diameter of the overflow.

The advantages of this construction are:

Fairly constant clarifying effect, until the sediments have attained the diameter of the overflow.
Simple dismantling of the drum.
Convenient and easy cleaning.

Its disadvantages are:

Small equivalent clarifying area.
Because of the almost cylindrical axial flow only the area corresponding to the diameter of the overflow is effective for separation and hence the effective centrifugal acceleration is substantially less than that acting on the outer wall.
Small sediment space.

A low concentration of solid substance in the liquid mixtures is necessary for economic operation, as the space for solid material is small and the deposited cake of solid must be removed manually after the machine has been stopped. Tube centrifuges are used more rarely in the clarification of drug extracts, though they are widely used in the recovery of vaccines and in the centrifugation of viruses and bacteria.

5.5.3. APPARATUS FOR PASTEURIZATION AND ULTRA-HIGH-TEMPERATURE HEATING

5.5.3.1. Pasteurization

This is now carried out almost exclusively in plate heat-exchangers consisting of shaped steel plates between which the material to be pasteurized and the heating medium circulate separately in the same or in the opposite direction to each other. The steel plates can be fitted together in large numbers to form a plate packet or system. The depth of the material and of the heating medium is only 3–6 mm here. Such a plant normally consists of three parts. In the first part warm water flows against the cold material. In the second stage the material is kept at the actual pasteurization temperature for the requisite length of time. In the third stage the material is cooled again. In the plates of this third stage, cold water or cold extract flows against the pasteurized extract, thereby becoming prewarmed but rapidly cooling the material. Figure 5.94 shows the principle of a plate heat exchanger.

The exchanger is set up here as a counter-current exchanger. Figure 5.95 shows the arrangement of the plates in a packet with open exchanger frames.

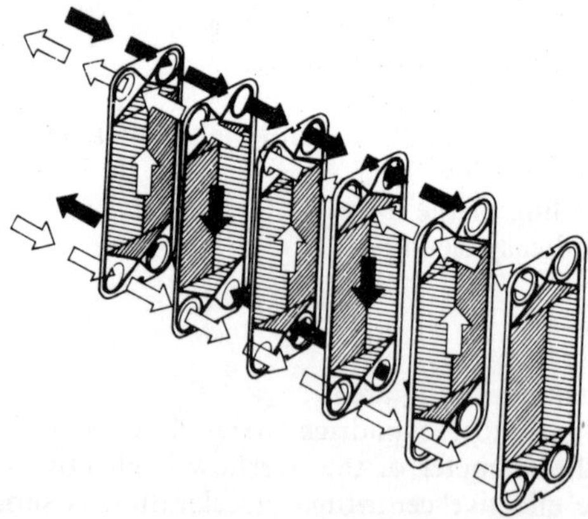

Figure 5.94   Circulation in a plate heat exchanger.

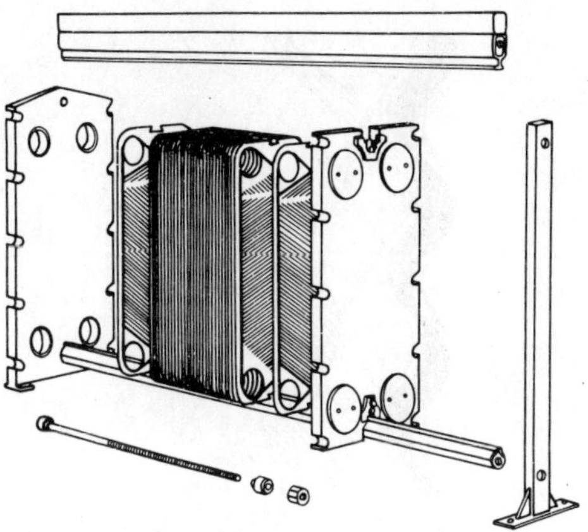

Figure 5.95   Plate packet.

Turbulent flow along the plates can be produced in various ways. Two familiar systems are the washboard and the herring bone patterns.

In the washboard pattern turbulence is achieved by constantly changing direction and speed of flow, whilst with the herring bone pattern a channel is formed whose geometry confers a whirling motion on the liquid (Figs 5.96, 5.97 and 5.98).

5.5.3.2. Ultra-high-temperature heating

In this apparatus product is brought to a temperature of $\sim 150°C$ for 2–4 s by injection of steam. Figure 5.99 [5.49] shows a schematic representation of such a plant.

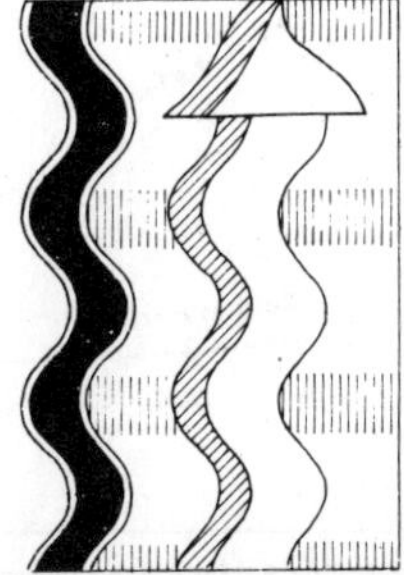

Figure 5.96   Washboard pattern plate.

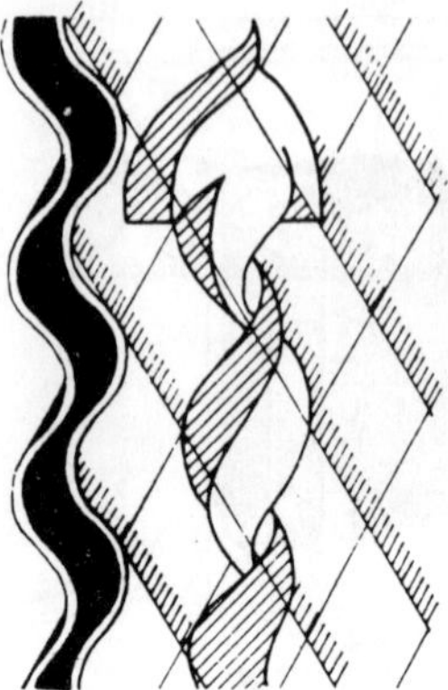

Figure 5.97   Herring bone pattern plate.

Saturated steam is injected into the uperizer, where it condenses in the product and raises its temperature to 150°C, at which it is kept for 2–4 s to kill all bacteria and spores. In the subsequent reduction of the pressure *in vacuo* the water vaporizes again, with sudden cooling of the material. To prevent dilution or concentration of the uperized product the supplied quantity of fresh steam must be regulated so that it is exactly equal to the quantity removed by the expansion *in vacuo*.

5.6.   Apparatus for concentrating extracts

The quantity of material to be processed per time unit and the properties of the solution being concentrated are the critical parameters for the choice of a

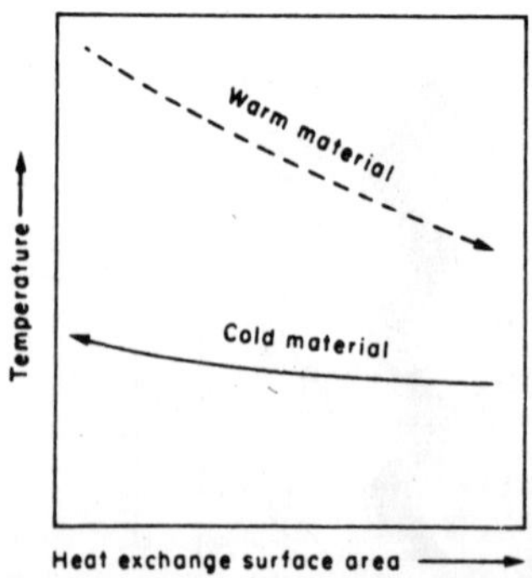

Figure 5.98   Temperature plot of a counter-current plate heat exchanger.

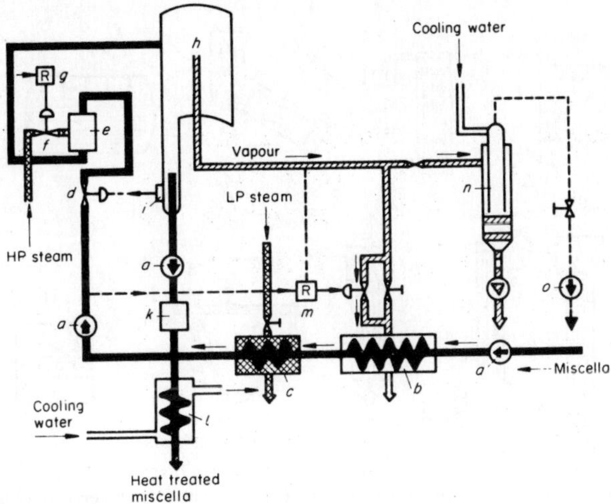

Figure 5.99 APV "uperization" process: a, pump; b, preheater; c, additive preheater; d, pneumatic flow-regulating valve; e, high temperature heater (uperizer); f, pneumatic steam-regulating valve; g, high temperature regulator; h, expansion chamber; i, level tube with measuring cell; k, homogenizer; l, cooler; m, regulator for specific material; n, injector condenser; o, vacuum pump; high pressure (HP/steam at 180°C/10 bar; low pressure (LP) steam at 120°C/2 bar.

vaporizer. Stable solutions which do not tend to form crusts and are non-corrosive can be boiled down in moderately priced apparatus. Solutions of heat-sensitive active substances which can only be exposed to heat for a short time require more expensive apparatus, which sometimes operate under vacuum. The various requirements have resulted in the development of a series of basic types of vaporizers which have been summarized in a review by Billet [5.50] (Fig. 5.100).

Whereas the same operating principle is used, and the vaporization process is combined with circulation of the liquid, in types (a), (b) and (c), substantial differences in the functional principle exist, both in comparison with (a) to (c) and in types (d) to (i).

In addition to the oldest types, i.e. tube vaporizers with and without forced circulation, we shall only discuss those types of vaporizer which have acquired some significance in the pharmaceutical industry.

5.6.1. TUBE VAPORIZERS WITH AUTO-CIRCULATION

These represent the oldest type of vaporizer. Figure 5.101 shows the principle of the Robert vaporizer [5.50].

It consists of a bundle of narrow vaporizer tubes arranged concentrically around the wider central fall-tube. The vaporizer tubes are heated externally

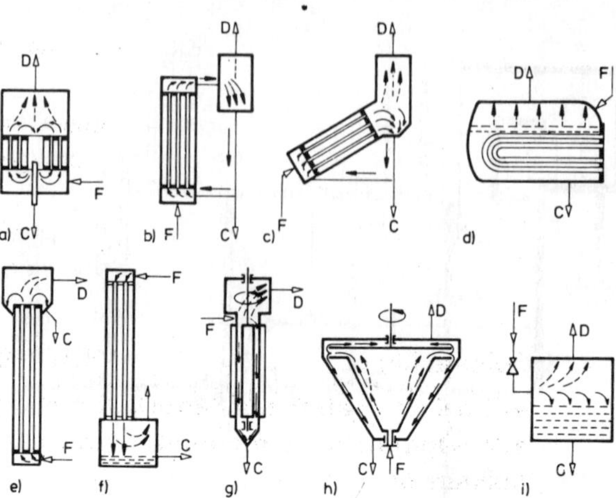

Figure 5.100 Basic practical forms of vaporizers. (a) Apparatus with vertical, externally heated vaporizer tubes and central fall-tube with auto-circulation of liquid. (b) Apparatus with vertical, externally heated vaporizer tubes in the external heater and with a fall-tube attached to the outside of this for auto- or forced circulation of the liquid. (c) Apparatus with inclined and, in the limiting case, horizontal, externally heated vaporizer tubes inside the external heater and with a fall-tube attached to the outside for auto- or forced circulation of the liquid. (d) Apparatus with horizontal, internally heated vaporizer tubes immersed in the liquid. (e) Apparatus with vertical externally heated vaporizer tubes for single throughflow of liquid from bottom to top. (f) Apparatus with vertical externally heated vaporizer tubes for throughflow of liquid from top to bottom by the falling film principle. (g) Apparatus with rotating wipers for even spreading of the liquid flowing as a thin film from top to bottom on the inside of the externally heated surface of the vaporizer. (h) Apparatus with rotating, externally heated conical vaporizer surface with throughflow of liquid as a thin film from bottom to top. (i) Apparatus without vaporizer heater for vaporization of preheated liquid by pressure reduction. F, Fresh solution; D, vapours; C, concentrated solution.

with steam. In the vapour chamber above the tubes there is a liquid separator, which prevents droplets of concentrated solution from being removed with the steam. The fresh solution (crude extract) flows upwards in the narrow heated tubes, vaporizing as it does so, and is separated in the vapour chamber into vapour and liquid, the latter returning down to the bottom through the wide central fall-tube. Circulation is induced by differences in the density of the liquid in the heated tubes and in the central fall-tube. Apparatus with an external fall-tube are also used. Tube vaporizers with auto-circulation have large throughput capacities.

5.6.2. TUBE VAPORIZERS WITH PUMPED CIRCULATION

Auto-circulation of the liquid in a tube vaporizer stops when the temperature difference between the solution in the vaporizer tubes and the heating liquid

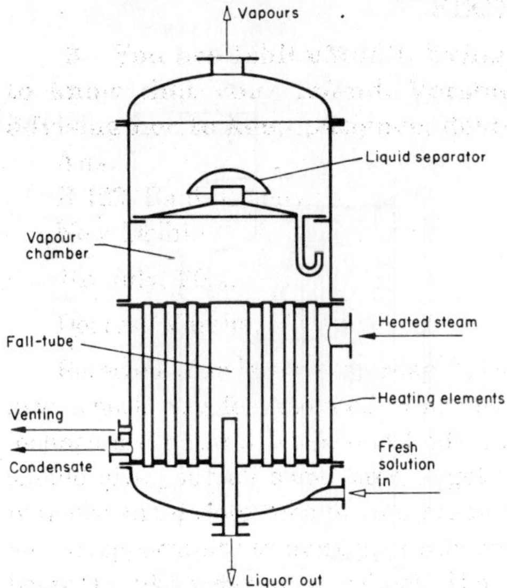

Figure 5.101 Robert auto-circulating vaporizer.

is so small that movement of the liquid no longer takes place. This phenomenon is also affected by the viscosity of the solution, as heat-exchange is poorer in viscous solutions. Forced circulation must be used whenever auto-circulation is too weak or does not take place at all. Figure 5.102 [5.51] shows the principle of a vaporizer with forced circulation.

This type of vaporizer has an external heating unit. Circulation vaporizers with forced circulation of the liquid have higher heat transfer coefficients than auto-circulating vaporizers and hence the material needs to remain in this type of vaporizer for a shorter time.

5.6.3. FALLING FILM VAPORIZERS

The demand for vaporization conditions which help to conserve sensitive materials has led to the development of the falling film vaporizer, also known as the falling flow vaporizer. The term spray film vaporizer which is sometimes also used refers to a special mode of operation where steam is released from a spray film. A spray film exists when the escaping steam only slightly affects the surface and flow of the film. This is the case with low heating surface loading and with low speed of the product vapour or steam flowing along the surface of the film.

Falling film vaporizers are characterized, in comparison with tube vaporizers, by the very short time for which the solution needs to stay in them (on

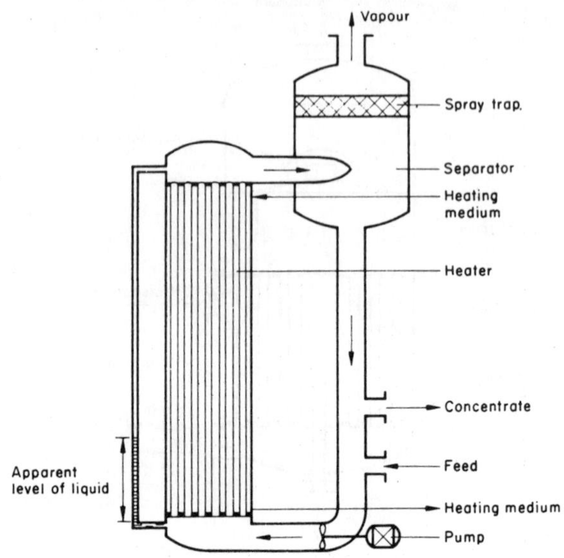

Figure 5.102   Circulation vaporizer with external heater and circulation pump.

average a maximum of 1 min in a single stage plant). Figure 5.103 [5.50] shows a diagramatic representation of such a plant.

An even distribution of the fresh solution over the cross-sectional area is necessary for even vaporization in a falling film vaporizer. Figure 5.104 shows some possible distribution principles.

The liquid should be boiling at input if optimal results are to be obtained. The solution must therefore be preheated, in falling flow vaporizers this is achieved with built-in preheater coils. Figure 5.105 [5.52] shows a scheme of such a vaporizer. Falling film vaporizers are used successfully in the concentration of yeast extract, drug and spice extracts, and in the preparation of coffee and tea extracts and the manufacture of gelatins.

5.6.4.   THIN LAYER VAPORIZERS

Although apparatus which retains the material for only a short time is available in the form of falling-film vaporizers, development did not end there. Whereas the thickness of the film set up in the liquid tube in falling film vaporizers depends on the properties of the solution, the temperatures of the solution and heating medium, and on their flow-rates, the incorporation of rotating devices in thin layer vaporizers has created additional possibilities of influencing the liquid film. Figure 5.106 [5.53] gives a diagrammatic representation of a thin layer vaporizer with a rotary device.

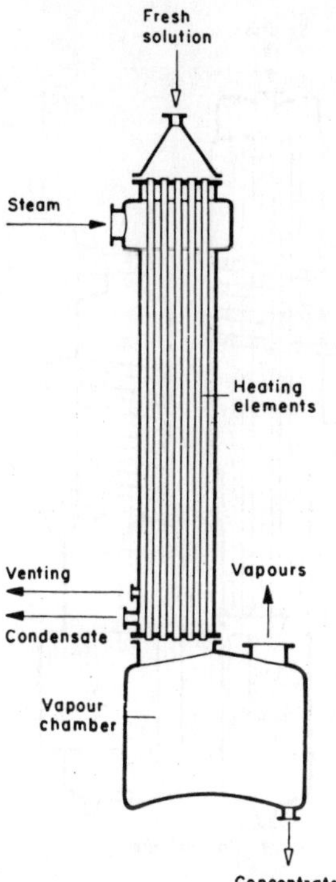

Figure 5.103   Falling film vaporizer (Wiegand Karlsruhe GmbH).

The solution and the heating medium flow from top to bottom, as in the falling film vaporizer. The separation section (vapour chamber) is situated above the heated part. A fast rotor provided with wiper-blades operates inside the vaporizer. The solution is led in above the heating jacket through a distributor ring and is distributed evenly over the wall of the vaporizer by the wiper blades, flowing down the inside of the hot wall in a spiral track, whereupon the more volatile components vaporize. The vapours thus formed flow upwards through the apparatus in the opposite direction and pass into

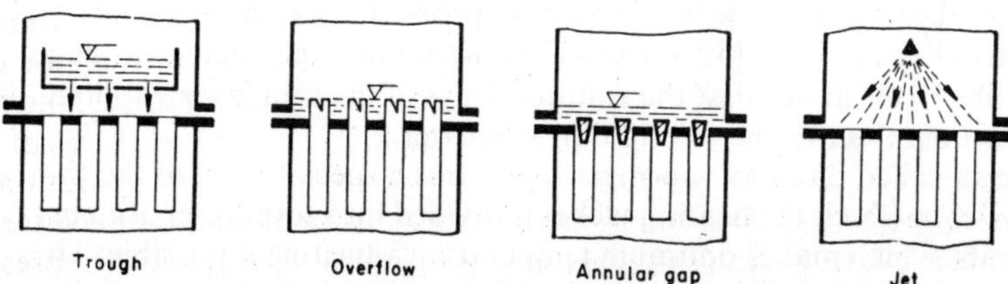

Figure 5.104   Methods for the distribution of the liquid in falling film vaporizers.

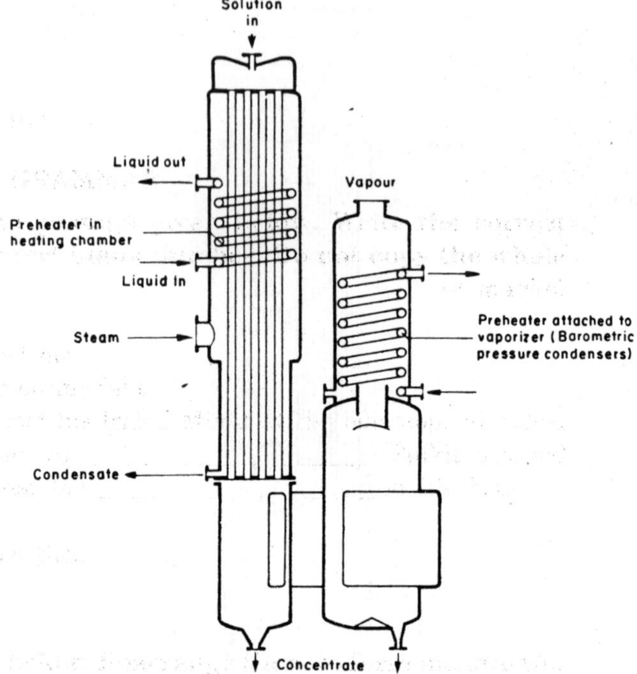

Figure 5.105   Falling flow vaporizer with built-in coiled tubular preheaters.

the separator. The rotating separator ensures the separation of liquid droplets, which flow back into the vaporization zone. The concentrated solution is drawn off at the bottom of the vaporizer. The rotors of the thin layer vaporizer can be provided with wiper blades of various designs. Figure 5.107 shows the various possibilities [5.50].

Pendulum, scraper and sliding wipers are also specified in addition to fixed wiper blades, which are placed at a distance of 0.75–4 mm from the heating jacket. It is possible to concentrate solutions with pendulum and sliding wipers to such an extent that a dry powder is obtained. It is not possible to prevent traces of impurities from the wipers getting into the end product, which makes the use of wiper vaporizers problematical in the pharmaceutical industry. Figure 5.108 [5.53] shows the action of wiper blades in detail.

Figure 5.109 shows that the solution stays in the thin layer vaporizer for a shorter time than in the falling film vaporizer.

The so-called Sambay vaporizer represents a special form of the thin layer vaporizer in which the heating jacket is divided into several separately heated segments, which makes optimum temperature adjustment possible. The rotor has movable wiper blades which are in contact with the heating jacket when the plant is in operation. Materials which tend to form crusts or coatings can

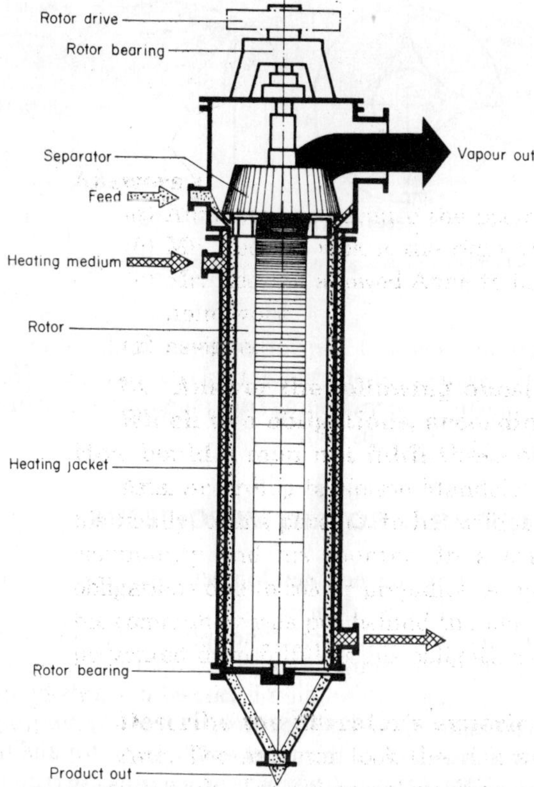

Figure 5.106 Thin layer vaporizer with integral rotor.

be processed in this way. Figure 5.110 [5.50] schematically illustrates a Sambay vaporizer and Fig. 5.111 shows its wiper blades [5.53].

After falling film vaporizers, thin layer vaporizers have gained a firm place in the pharmaceutical industry. They can be operated under vacuum at low temperatures and can thereby also evaporate highly viscous solutions, though of course these reduce the evaporator's output. However, this can be compensated by suitable design of the plant. Aqueous solutions of gelatins,

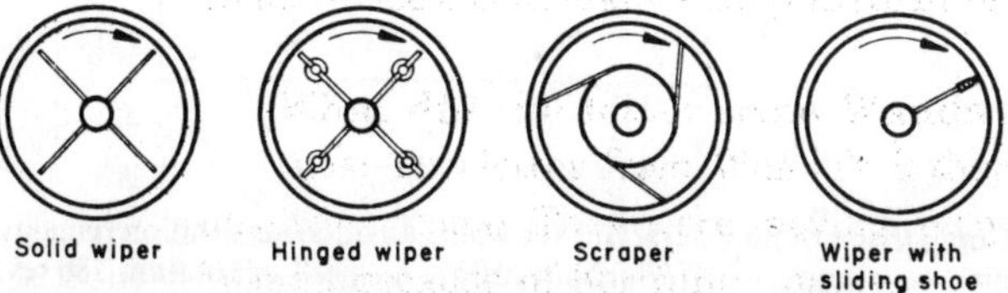

Figure 5.107 Rotary wipers for thin layer vaporizers.

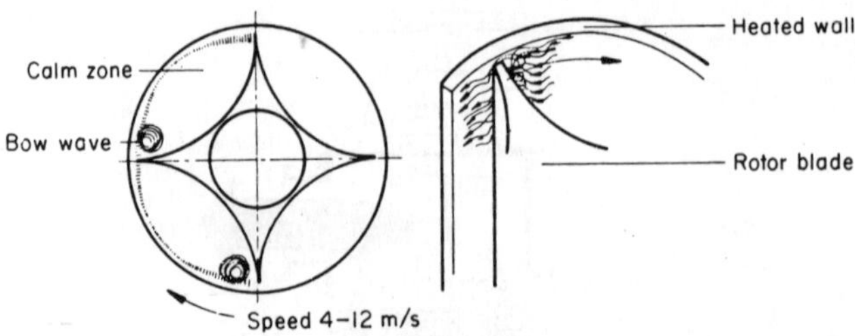

Figure 5.108   Schematic representation of the distribution of liquid on the heated wall.

glucose, fructose and sorbitol are, for example, boiled down in thin layer vaporizers for the recovery of pharmaceutical raw materials. They are used in the foodstuffs industry for dehydrating fruit pulps and concentrating extracts of meat and malt and aqueous roots.

5.6.5.   CENTRIFUGAL ROTARY VAPORIZERS

These represent a completely new development. Figure 5.112 illustrates their principle and Fig. 5.113 their design.

    The solution is fed into the centre of the apparatus from the outside by the steam heated rotor. The rotor revolves at up to 400 or 1600 rev/min, depending on the design. The solution is carried upwards by the centrifugal forces on the heated inside wall of the rotor, thereby evaporating the solvent. The concentrate is drawn off at the upper end of the rotor and the condensed hot steam is led off down through a pipe. The vapours are recycled to preheat

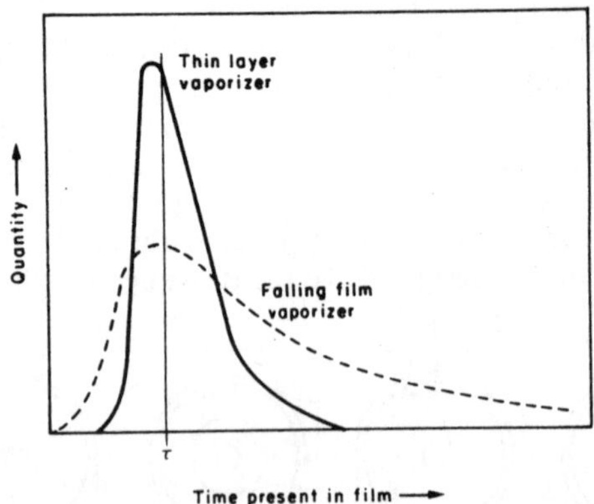

Figure 5.109   Comparison of time durations for which an aqueous solution remains in a falling film and in a thin layer vaporizer. Throughput 2500 L/h; speed of rotation 380 rev/min; mean duration of stay 16.7 s [5.53].

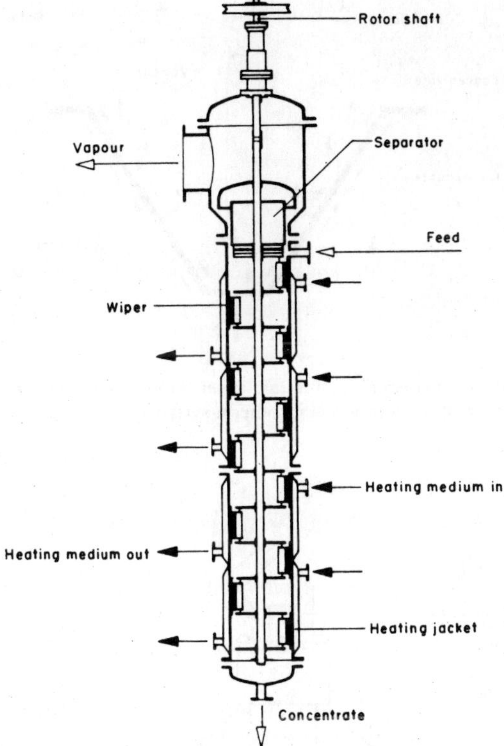

Figure 5.110   Multisegmented rotary thin layer vaporizer (Sambay system).

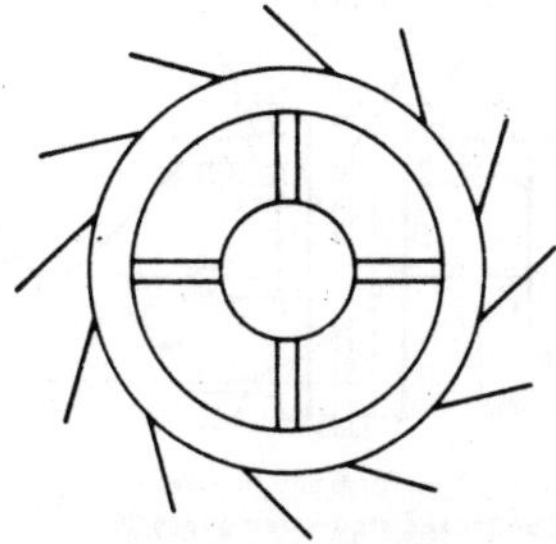

Figure 5.111   B-rotor: a central tube with movable wiper blades (Sambay system) for processing products that tend to form coatings (viscosities up to 5 Pa s).

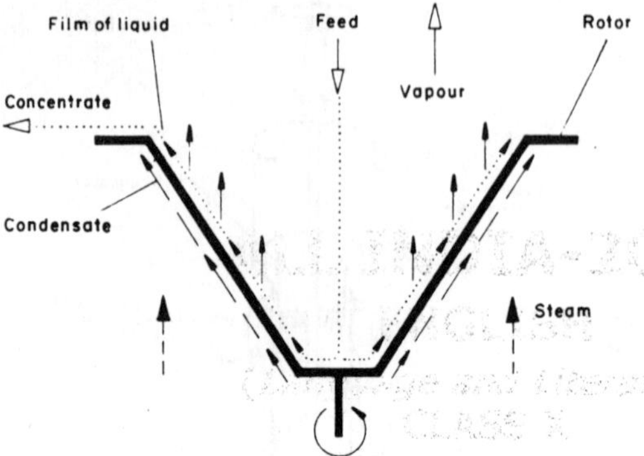

Figure 5.112 Principle of operation of a thin layer vaporizer system in which the product remains for only a very short duration (Liprotherm system).

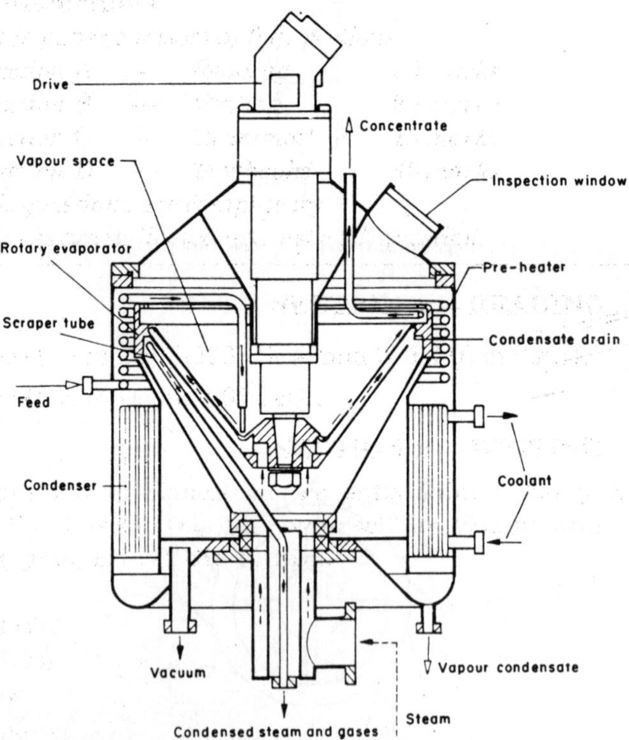

Figure 5.113 Schematic diagram of a Liprotherm system vaporizer in which the product remains for only a very short duration.

the fresh incoming solution and then separated in the vapour condenser. The thickness of the liquid film is only about 0.1 mm and the product takes less than 1 s to pass through the vaporizer.

Centrifugal rotary vaporizers are very suitable for product-conserving vaporization in the pharmaceutical industry and are in use in various parts of the world.

5.6.6. ROTARY THIN LAYER VAPORIZERS

The vaporizer consists of an evacuatable horizontal container which also contains a horizontal lamellar heat-exchanger (Fig. 5.114). Hot steam is passed into this cylinder. The condensed heating steam flows through a stationary pipe which facilitates collection of the condensate. The solution to be concentrated is fed into a receiver at the bottom of the cylinder and the concentrate leaves the vaporizer at the opposite end. The receiver consists of a trough containing a revolving rotor which catches the solution and throws it up against the revolving lamellar cylinder. Figure 5.115 [5.52] shows a cross-section through the plant.

The lamellar construction of the cylinder produces a very large area for heat-exchange. The incoming solution is picked up by the lower rotor and thrown against the lamellar cylinder, where it forms a thin layer from which the solvent vaporizes. After one revolution of the lamellar cylinder the liquid film returns to the pick-up point, where fresh solution is picked up. Excess solution flows back into the trough. Continuous concentration of the solution takes place over the entire length of the cylinder due to the constant

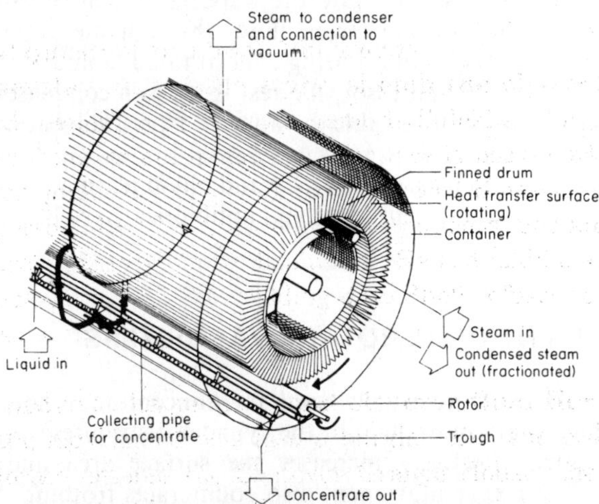

Figure 5.114   Rotary thin layer vaporizer.

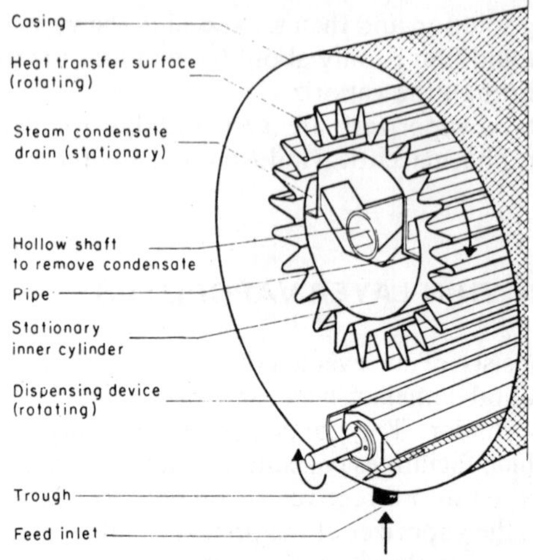

Figure 5.115 Details of the rotary thin layer vaporizer.

alternation of vaporization and re-application. The revolution speed of the lamellar cylinder is 30–90 rev/min and that of the rotor 700–1200 rev/min. The lowest temperature used so far for vaporization of water was 25°C with a temperature difference of ~ 1°C from the heat-exchanger surface. Although the times for which the products stay in the vaporizer are longer than those in the thin layer vaporizer operating by centrifugation, the extremely low temperatures and temperature differences permit conservation of the product. Fig. 5.116 [5.52] shows how the vaporizer is incorporated into a complete system. The heating steam flows through the vaporizer in the opposite direction to the solution being concentrated. Some of the vapour is sucked up by a thermocompressor, the rest is led to a condenser. Compression of the vapour makes it possible to vaporize 1 kg of solvent with ~ 0.4 kg of steam.

Sensitive products such as whole blood and blood plasma, as well as coffee extract, brewer's yeast, nutrient yeast and antibiotics, have been successfully concentrated in this way.

5.6.7. CONVENTIONAL ROTARY VAPORIZERS

The rotary vaporizers are widely used for concentrating solution in the laboratory. The rotary movement of the cylinder distributes solution on the inner wall, which effectively increases the surface area and, therefore, vaporization. The rotary movement also counteracts frothing. Water bath temperature, depth of immersion of the cylinder into the water bath (size of

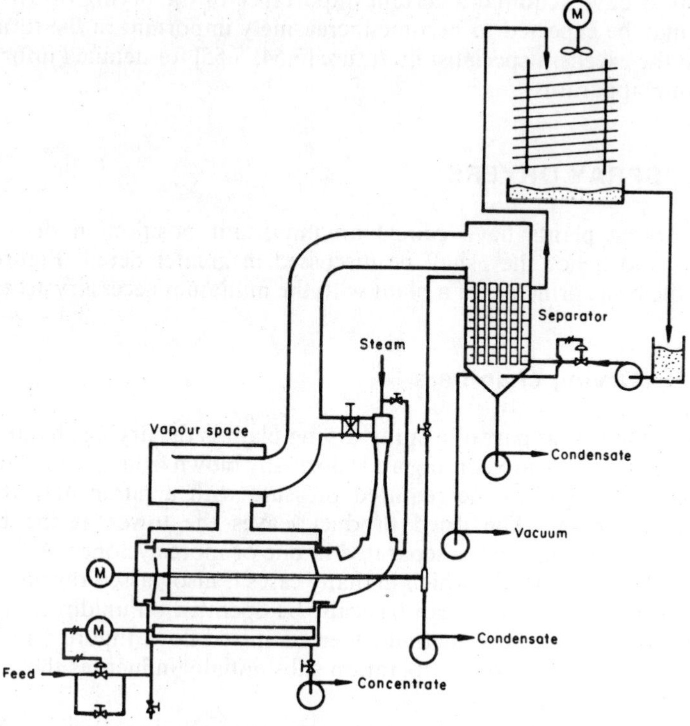

Figure 5.116 Rotary thin layer vaporizer: complete plant.

heated area) and degree of vacuum are parameters which can be regulated to optimize vaporization. Continuous vaporization can be achieved by passing the solution in through an inlet tube at the same rate as distillate flows out. Only small amounts of the solvent should be in the cylinder in order to keep the hydrostatic pressure low.

Rotary vaporizers are characterized by their great adaptability to a wide variety of problems. Apparatus is available not only for vaporization on a laboratory scale but also on a technical and development scale and, hence, even small production batches can be run.

5.7. Apparatus for drying extracts

The many different types of drying apparatus in existence make it impossible to give a comprehensive survey of them within the confines of this section. We shall therefore restrict ourselves to describing only those apparatus which in the past have acquired a certain importance in the drying of extracts or which may be expected to become increasingly important in the future. We refer to the excellent specialist literature [5.54; 5.55] for detailed information on drying apparatus.

5.7.1. SPRAY DRYERS

Spray drying plants have gained an important position in drying drug extracts and hence they shall be discussed in greater detail. Figure 5.117 shows the basic principle of a plant with the minimum necessary accessories.

5.7.1.1. Drying chambers

The most important part of a spray drying plant is the drying chamber, into which the filtered, heated drying air is normally blown from above. The spray solution is brought to the required pressure with a pump and atomized through a sprayer. The dried product leaves the tower at the bottom, together with the air, and is separated in one or more cyclones. A ventilator removes the outgoing air, which in many cases is also passed through a filter and a scrubber. The plants can basically be operated on unidirectional flow or counter-current, and sometimes even on mixed flow. Figure 5.118 shows the differences. The mode of operation substantially influences the design of the dryer.

The ideal unidirectional spray dryer consists of a drying tower in which the air and the liquid sink evenly to the bottom on parallel paths. Figure 5.119 [5.54] shows the flow scheme of such a tower.

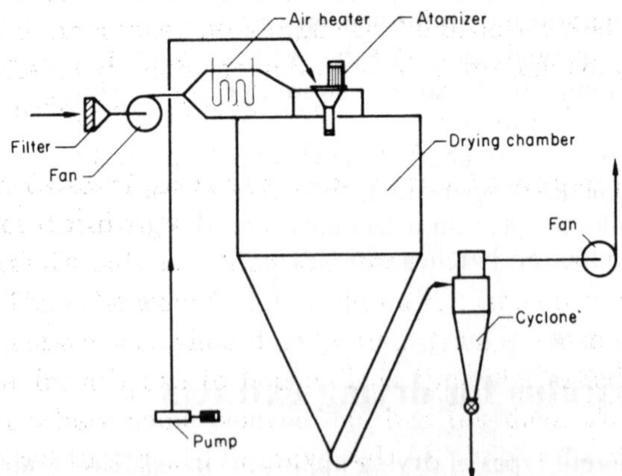

Figure 5.117   Schematic diagram of a spray drying plant.

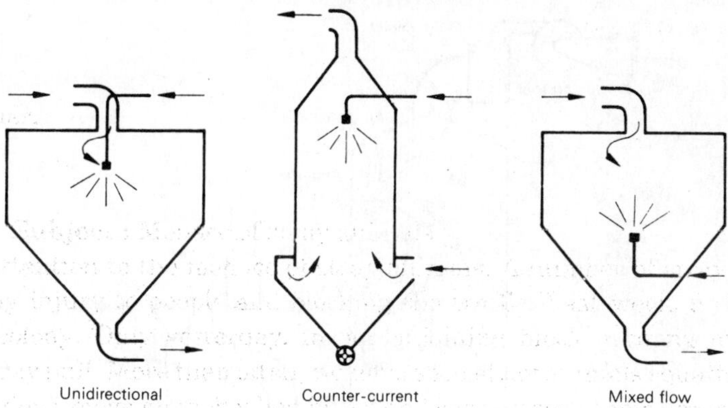

Figure 5.118 A unidirectional, a counter-current and a mixed flow spray drying plant.

This ideal state is not attained in practice, as the individual droplets remain in the tower for different lengths of time as a result of their various sizes, distribution and speed, and of the unevennesses in the flow of the air.

Despite this, there are drying towers which come close to the ideal concept. These towers are of the design described above, i.e. high drying towers containing long drying tracks which permit drying in laminar flow and which are 15–25 m high. Particular care must be taken to ensure even distribution of the drying air over the entire cross-section of the dryer. The laminar airflow is achieved by built-in distributor grids.

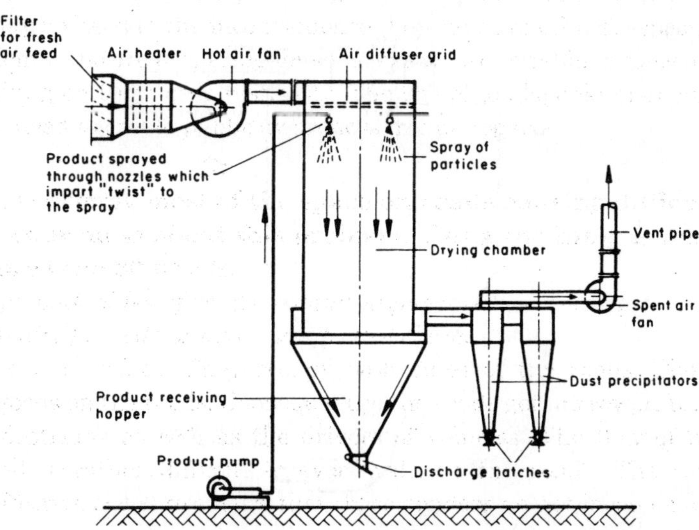

Figure 5.119 Flow scheme of a unidirectional spraying tower.

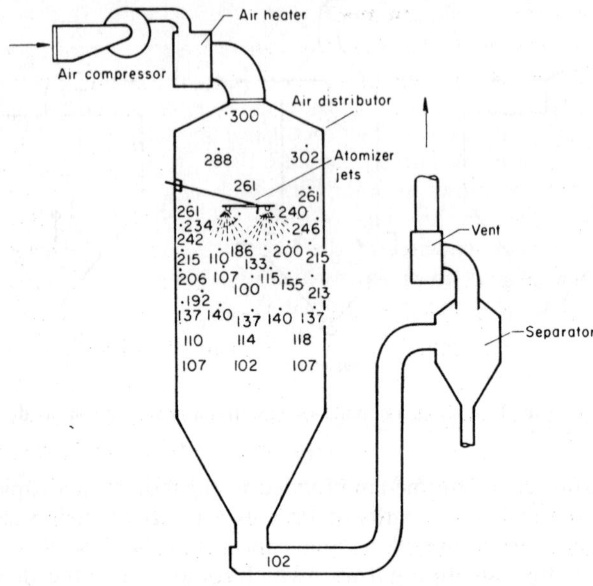

Figure 5.120   Temperature distribution in a spray tower.

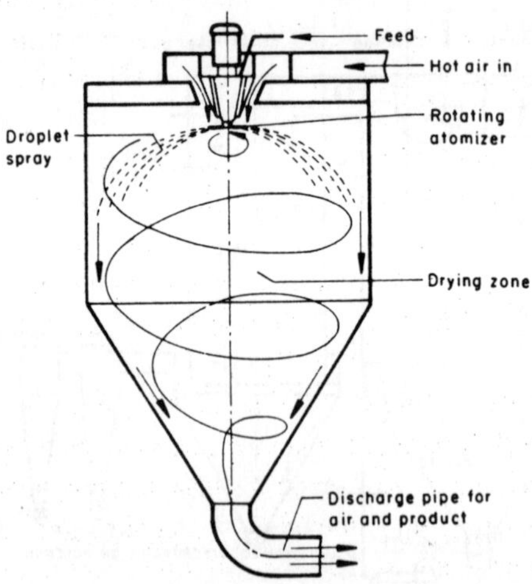

Figure 5.121   Unidirectional flow spray dryer with rotary atomizer, spiral air movement, and a common discharge outlet for product and air.

The speed of the airflow in such plants is 0.1–0.5 m/s. These dryers are generally operated with spray nozzles, one component nozzles being preferred because of the high throughput and the low disturbance of the laminar flow pattern. Spray dryers with a laminar unidirectional airflow are used only for large-scale production. The temperature distribution in this type of dryer has been investigated by Turck [5.56] and the results are shown in Fig. 5.120. Figure 5.121 [5.54] shows another form of airflow in a unidirectional flow dryer. The air enters in the shape of a ring around the sprayer. Some of it is directed vertically downwards by the spray jet and the rest is set into a spiral motion by baffles and flows downwards. The particles are turned from the vertical airstream into the spiral flow and thus take a relatively long path to the floor of the dryer. The design of these drying chambers, and the way in which the air is supplied, permit the towers to be of small heights and diameters. They are therefore also used on the technical and laboratory scale as well as for industrial production. Rotating discs and central nozzles have proved to be suitable spraying systems.

In the counter-current spray dryer the material flows against the air being used to dry it. The air enters at the bottom and the solution to be dried is normally sprayed against it at the top. Figure 5.122 [5.54] shows this diagrammatically. In addition to this type of material and air supply, spraying from below and air supply from above is also commonly used, as is shown in Fig. 5.123.

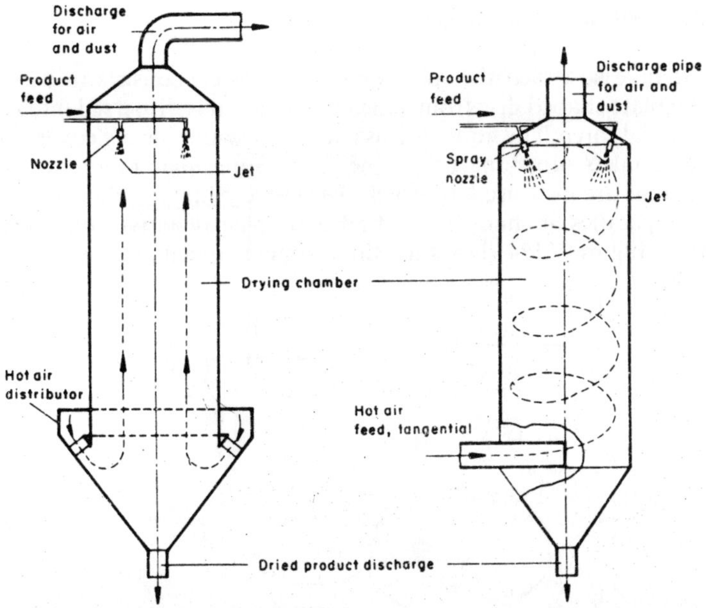

Figure 5.122 Counter-current spray dryer with nozzle sprayer (left) with approximately straight air supply and (right) with spiral air motion.

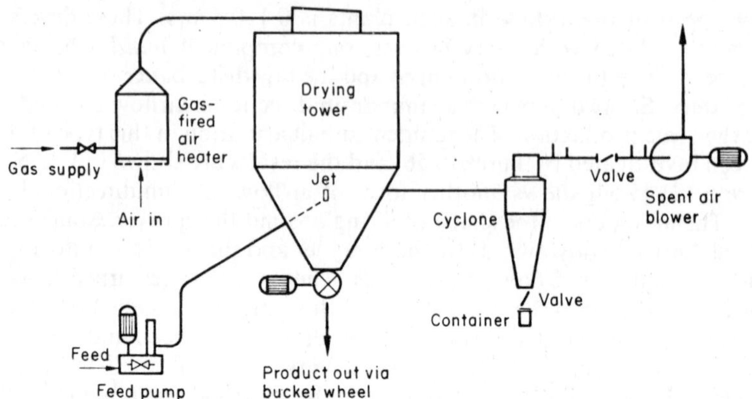

Figure 5.123 Counter-current spray drying with material supplied from below and air supply from above.

5.7.1.2. Atomization systems

Three systems are used for dispersing the liquid into droplets.

5.7.1.2.1. Centrifugal atomization

This system, also called the spinning disc atomizer, consists in its simplest form of a plate shaped disc from which the material is dispersed into droplets by centrifugal force. The liquid forms a film which completely covers the disc, and the droplets are flung off at the edge. Somewhat more cumbersome constructions are now used, in which the simple plate is replaced by a wheel fitted with paddles or channels, or the atomizer-head consists of several rows of nozzles. Figure 5.124 shows the three systems schematically.

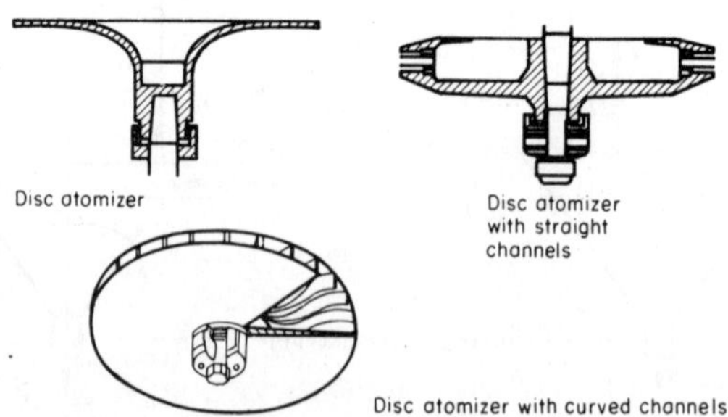

Figure 5.124 Various designs of atomizer disc.

The size of the droplets produced is influenced by the following parameters:

Flow-rate of the solution.
Design of the disc.
Surface tension of the solution.
Rotation speed of the disc.
Viscosity of the solution.

Spinning or rotating discs produce a narrow range of particle sizes. They rotate at speeds of 1500–50 000 rev/min and have diameters of 10–30 cm. They are also suitable for the conveyance of highly concentrated suspensions. They must be made of special materials if they are to be used in the processing of abrasive substances. Spray dried products prepared with atomizer discs are formed as spherical particles.

5.7.1.2.2. One component nozzles

In this spraying device, which is also known as a centrifugal pressure nozzle or high-pressure nozzle, the solution is sprayed or atomized through a nozzle under high pressure. Pressure nozzles usually have a vortex chamber and tangential supply of the material (Figure 5.125) and nozzles which have obliquely grooved heads are fitted in front of a small vortex chamber (Fig. 5.126) [5.55].

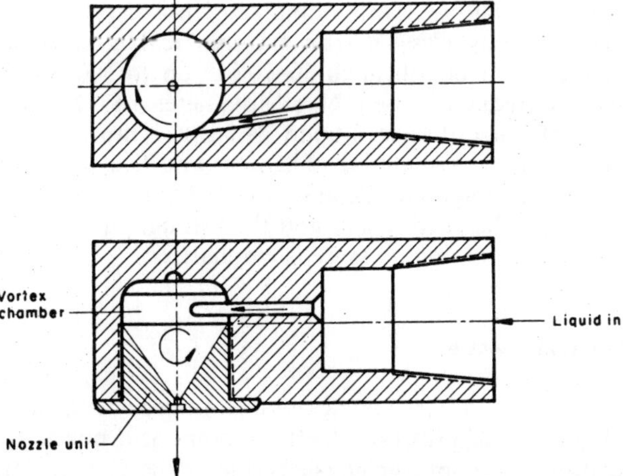

Figure 5.125 Pressure nozzle with vortex chamber.

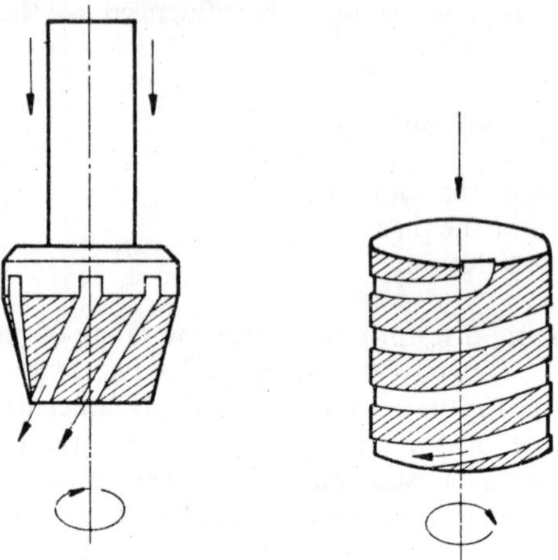

Figure 5.126 Hollow cone (left) and filled cone (right) nozzles.

The size of the droplets produced is influenced by the following parameters:

Flow-rate of the solution.
Internal diameter of the nozzle.
Spraying pressure.
Surface tension of the solution.
Viscosity of the solution.

Particle sizes of 8–800 µm can be obtained with one component nozzles.
The size range of the particles is narrow and spherical in shape. The pressures
used range from 10–30 bar through 30–70 bar up to 500 bar. Pressures of
around 30 bar are frequently used. Nozzle diameters of 0.3–0.5 mm are used
for atomization of water at rates of up to 5000 L/h. It must be remembered in
the processing of abrasive materials that wear and tear on the nozzle has a
direct effect on the throughput rate and output. Fluctuations of the spraying
pressure affect both the particle size and the throughput rate and hence the
need for a constant spraying pressure is emphasized.

5.7.1.2.3. Two component nozzles

The solution is atomized in the two component nozzle with the aid of a gas,
usually air. Figure 5.127 illustrates the two component nozzle diagrammatically. The action of the sprayer nozzle is due to the fact that the thread of
liquid issuing from it cannot withstand the acceleration forces of the air,
which is flowing at a speed of 100–200 m/s, and are consequently torn apart.

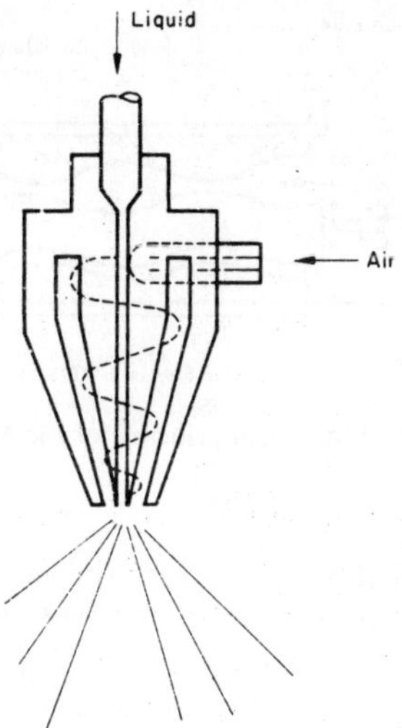

Figure 5.127 A two component nozzle.

The droplets formed are mixed with the air to form a mist. The liquid should flow into the nozzle under slight excess pressure to guarantee good atomization. The droplets produced arc influenced by the following parameters:

Flow-rate of the solution.
Bore of the nozzle.
Pressure of the atomizer air.
Viscosity of the solution.

Two component nozzles give particle sizes of ~ 3–$250\,\mu\mathrm{m}$. This size range is considerably broader than that with one component nozzles or rotating discs. The operating pressure of two component nozzles is 0.5–7 bar; although the lower limit of 0.5 bar given by the makers of the specific machine quoted here seems very low.

When the airstream tears off the droplets the particles produced are more irregular than those produced by one component nozzles and rotating discs. Some of the particles are cracked and fragmented, though spherical ones are also present.

5.7.1.3. Pumps

The pumps necessary for conveying the solution which is to be sprayed fulfil various functions, depending on the atomizer system. Rotating discs and two component nozzles can be supplied with solution by simple conveyor pumps.

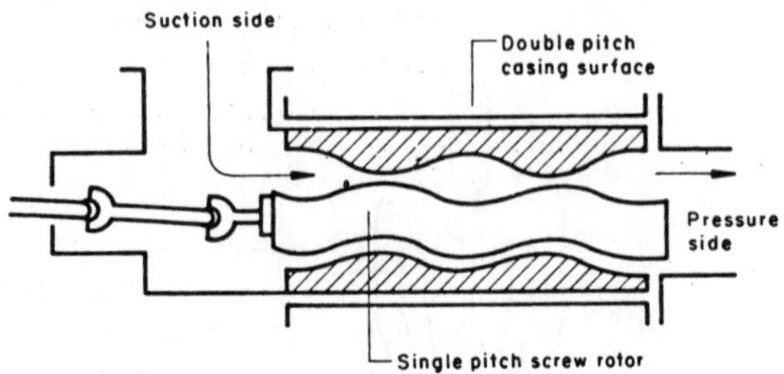

Figure 5.128 Functional principle of the Mohno pump.

Pressures of 30–70 bar must be built up for one component nozzles and hence high-pressure pumps are used. Mohno pumps have proved very useful here, see Fig. 5.128 [5.57].

5.7.1.4. Separation of spray dried products and auxiliary equipment

Spray dried products are always separated from the drying air in cyclones. Two variants are possible here (1) some of the product falls into a receiver at the bottom of the spraying tower and only the fine fraction goes into the cyclone with the air; or (2) all the product is taken into the cyclone by the airstream and separated there.

The first variant is mainly used in large production plants, where two cyclones are generally used. This means that three different particle fractions are produced; the fraction from the second cyclone, which contains the finest product, is usually led back again to the tower in the vicinity of the sprayer nozzle where, with fresh droplets, it forms larger agglomerates. For reasons of economy this fraction should be kept as small as possible.

The second variant is used in smaller spraying towers operating with a spiral air supply, as the speed of the air here is too high for separation at the bottom of the tower. The first variant can, however, also be used in larger towers operating with a tight spiral (vortical) air supply.

Products with a low softening-point may stick to the walls of the tower. The risk of adherence can be reduced in certain types of towers by the use of an 'air broom' (Fig. 5.129 [5.58]).

The air broom is supplied with compressed air, which can be heated if necessary, and blows the air against the wall of the drying chamber through nozzles. The rotating broom runs slowly on the inner wall of the tower, thus cleaning the wall of the dryer.

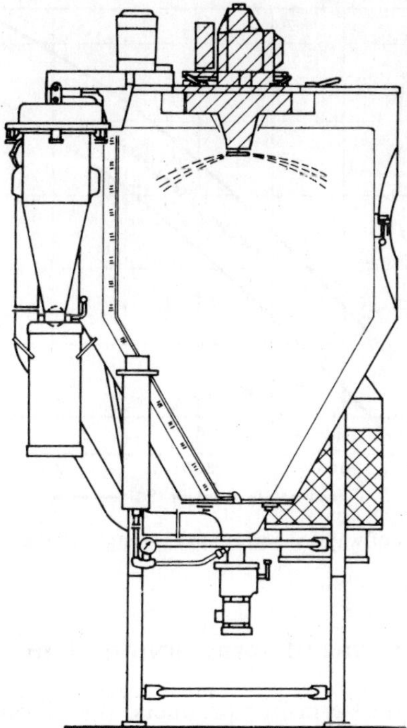

Figure 5.129 Spraying tower with air broom.

5.7.1.5. Dimensions of spray dryers

The size of spray dryers now available ranges from laboratory apparatus
with throughputs of ~ 1 kg solution vaporized per hour, through to technical plants with throughputs of 10–50 kg/h up to production units with
throughputs of ~ 80 tons/h. The conditions which apply to the given
throughput are important and complete statement of the throughput should
contain the following parameters:

> Solution used (for comparative data usually water).
> Temperature of the incoming air.
> Temperature of the effluent air.

Figure 5.130 [5.59] shows the alteration of water vaporization capacity in
relation to the temperature of the incoming air for the effluent temperature
range of 80–120°C in a small production dryer with two component nozzle
atomization.

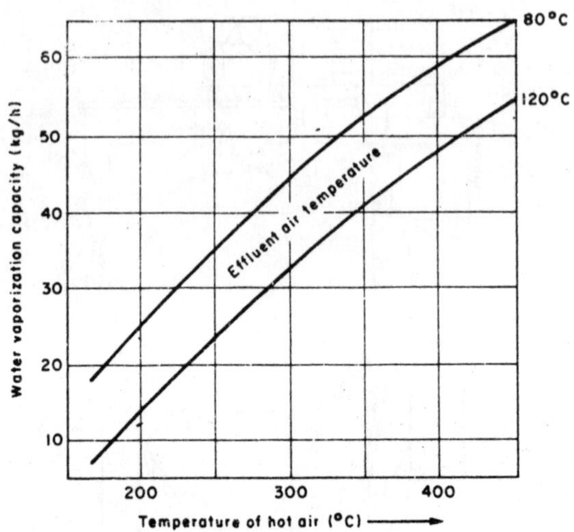

Figure 5.130   Variation of water vaporization capacity with the incoming air temperature.

5.7.1.6. Special forms of spray drying plants

The demand for sterile germ-free products, the necessity in some cases of using organic solvents instead of water and the energy crisis have led to the development of special forms of spray dryers.

5.7.1.6.1.   Aseptic spray drying

This type of plant (Fig. 5.131) operates with a two component nozzle. Both the atomization air and the drying air are filtered to remove particles. The solution to be sprayed also passes through a sterile filter before going into the nozzle. The product is removed from the drying tower by laminar airflow in a sterile room, where it is also packed.

5.7.1.6.2.   Closed circulation spray drying

Although the use of organic solvents has steadily diminished in the last few years because of the risk of explosion and pollution of the environment, there are still cases where the use of organic solvents cannot be avoided. Plants operating by closed circulation were developed for this purpose. Figure 5.132 [5.59] shows a diagram of such a plant.

Nitrogen is normally used as an inert gas, both as a drying medium and as a spraying gas. The spray dried product is drawn off at the lower end of the

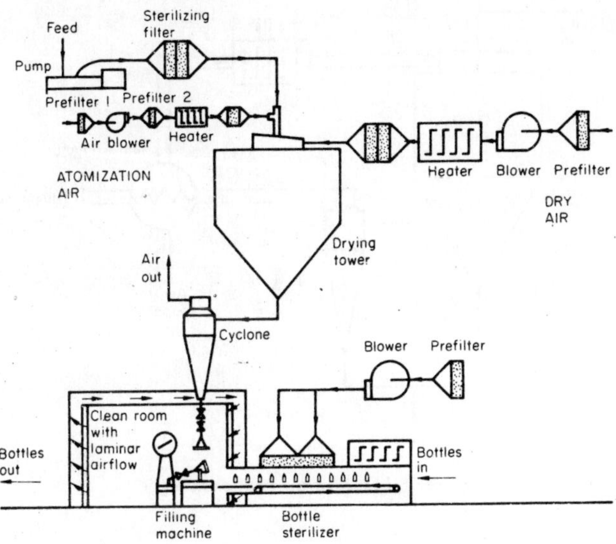

Figure 5.131   Aseptic spray drying plant.

tower and smaller particles are separated by a filter. The nitrogen laden with solvent vapour is led into a condenser, in which the solvent is separated. The gas, usually heated indirectly over steam or hot oil, is led back to the tower via a heater.

5.7.1.7. Recovery of heat

The heat balance of a spray dryer can be improved by recycling the drying air to the tower. The moisture content of the effluent air sets narrow limits on

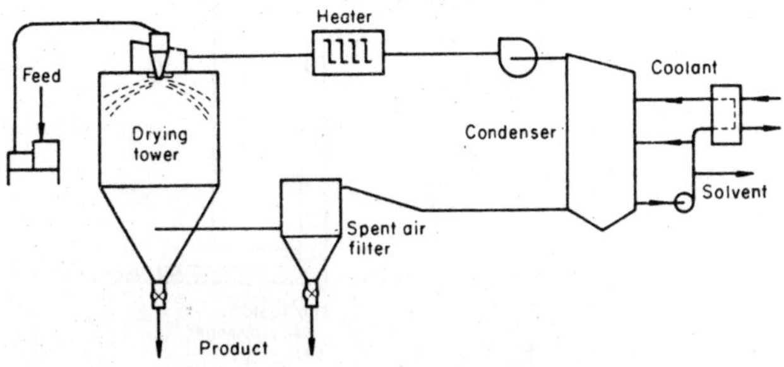

Figure 5.132   Closed circulation spray dryer.

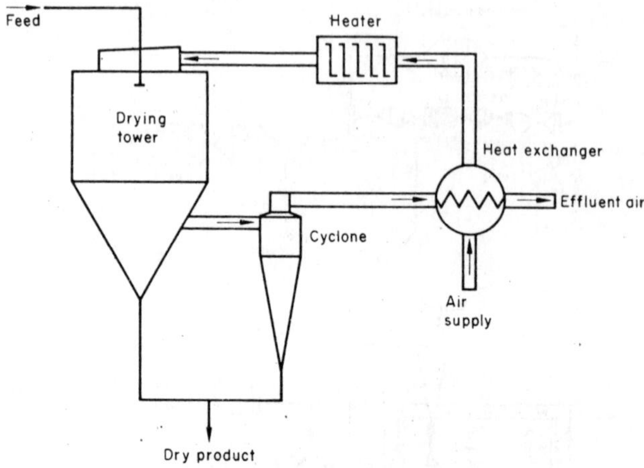

Figure 5.133　Air/air heat exchanger.

this process. On no account must so much effluent air be recycled that the moisture content of the spray dried product is greatly increased.

The simplest way of conserving the heat of the effluent air is by means of heat exchangers in the effluent airstream. Two types are commonly used, namely the air/air heat exchanger (Fig. 5.133) and the air/liquid/air heat exchanger (Fig. 5.134).

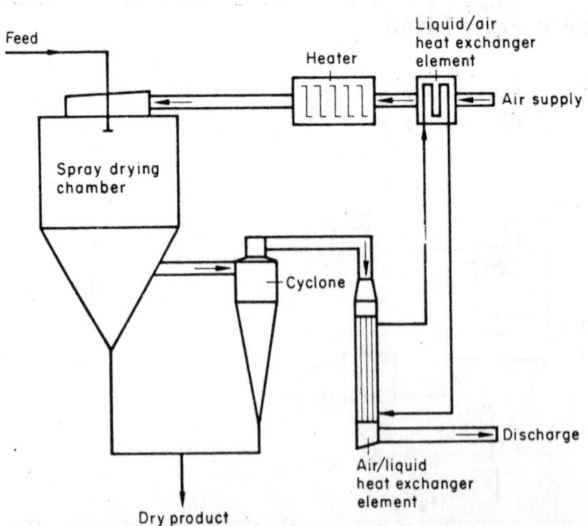

Figure 5.134　Air/liquid/air heat exchanger.

The air/air heat exchanger is the simplest one to incorporate into the design of a new plant. It is possible to incorporate it into existing plants if the conduiting permits, i.e. if the heat exchanger and the air heater are not too far apart for the air supply; an air/liquid/air exchanger is better if these are a long way apart. In the latter system the heat is transferred to a liquid in a first heat exchanger, which then gives it up again to the incoming air in a second heat exchanger near the air heater. Longer distances can also be overcome with this system.

Common to both systems is the fact that the effluent air, from which the heat is to be removed, is contaminated. The dust contained in it coats the exchanger bottles, resulting in poorer heat transfer. Scrubbing systems which permit rapid purification, if possible during operation, must therefore be incorporated from the outset.

5.7.2. VACUUM BELT DRYERS

Belt dryers operating at atmospheric pressure, and therefore requiring correspondingly high temperatures for the removal of, for example, water, have become considerably more widely used through the introduction of vacuum belt dryers (Fig. 5.135).

The air to be dried flows through an inlet into a vessel equipped with a stirrer, from which it is pumped to the dryer by a doser pump, via an attachable filter if necessary. The feeder device, which can be either a swinging arm or nozzle, distributes product evenly on to each belt of the dryer. The material immediately froths up under the action of the vacuum and forms a loose, bubble-filled layer, which is carried along by the conveyor belt. The belt consists of a synthetic plastic material or steel webbing and

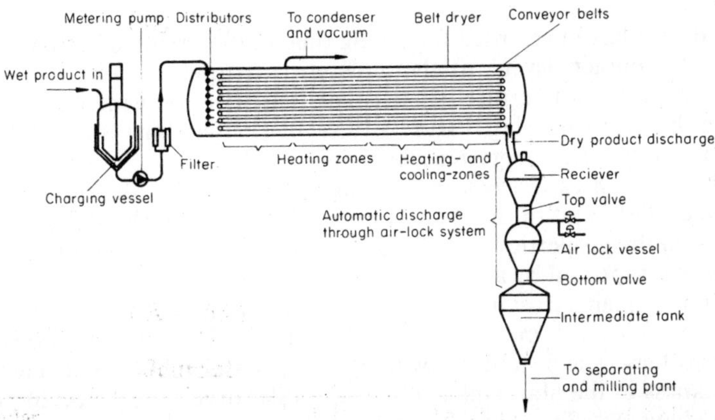

Figure 5.135 Schematic diagram of a vacuum belt dryer.

slides over plates heated with hot water, steam or oil. The dryer is divided into a number of heating zones, namely four to six segments, depending on the design. A cooling zone can also be formed at the end.

The temperature in each zone can be adjusted to the optimum for the drying process. The dried material is preliminarily coarsely comminuted through a grating at the end of the belt and then goes into a sluice consisting of two separate chambers. The upper chamber is always under vacuum, whereas the lower one can be connected alternately to vacuum or to air at atmospheric pressure so that the product is discharged in portions. The solvent vapours are drawn off and condensed.

The following parameters affect the drying:

> Length of the apparatus.
> Speed of the belt.
> Temperature of the heating plates.
> Degree of vacuum.
> Number and adjustment of the individual heating zones.
> Initial moisture content of the material to be dried.
> Desired final moisture content.
> Flow rate of the material.

The advantages of vacuum belt drying are the great flexibility of the process and good conservation of the material. The necessity for subsequent comminution of the material is a disadvantage.

In addition to numerous areas of use in the food industry, pharmaceutical extracts of valerian, horse chestnut, aloes, alder buckthorn bark, etc., have been dried by the vacuum belt method.

5.7.3. ROLLER DRYERS

Roller dryers have been used for a long time in the drying of extracts. They are purely contact dryers, as the moisture is evaporated or vaporized by contact with the hot surface of the roller. Vaporization generally predominates in this type of drying. Figure 5.136 shows possibilities for the design of this machinery taken, from Kneule [5.55].

In the simplest case a heated rotating roller dips into the solution to be vaporized. The solution forms a film on the roller, the thickness of which depends on the properties of the solution. The liquid film dries and the dried product is scraped off by a scraper just before that part of the roller dips into the solution again. In contrast to these single roller dryers are two roller dryers, in which the material is fed from above between two downward rotating rollers. The distance between the rollers is adjustable and determines the thickness of the film. Unlike the single roller dryer, the properties of the solution have no effect on the thickness of the film in these dryers.

In double roller dryers the material is fed between upward rotating rollers

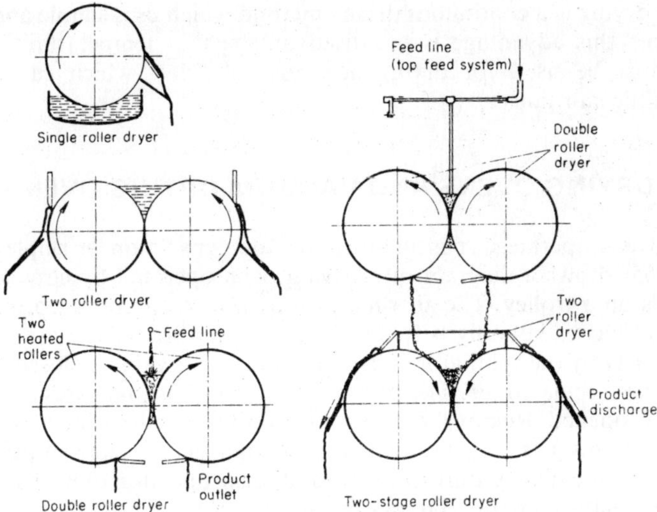

Figure 5.136   Examples of roller dryers.

and the thickness of the film adjusted by small pressure rollers. Two-stage roller dryers finally combine the two roller dryer with the double roller dryer, which results in the material undergoing a longer treatment.

The essential parameters which affect roller-drying are:

Temperature of the rollers.
Revolution speed of the rollers.
Thickness of the film of solution.

The last factor has limited variability, since there is the risk of the partially dried solution forming a crust on the roller if the film is too thick.

Direct contact between the hot roller and the solution makes it necessary to keep the contact time short with heat sensitive substances, by using a thin film and a high roller rotation speed.

The time t for which the material remains on the roller is calculated by the formula

$$t = \frac{\text{Angle between point of application and point of removal}}{360 \times \text{number of revolutions of roller/min}}$$

As well as simple dipping, the solution can also be applied to the roller by another cooled roller which dips into the solution and transfers the film which it has collected to the drying roller. Further possibilities are the use of a so-called toothed roller or a spray-roller. In both cases an auxiliary roller produces a mist of spray which is flung against the drying roller.

Roller drying is a continuous drying method which uses simple apparatus. Set against this advantage is the disadvantage that temperatures of 100–110°C must be used for drying aqueous solutions, which can damage thermolabile substances.

5.7.4. DRYING OVENS AND VACUUM DRYING OVENS

Drying ovens operate discontinuously on the convection principle. Figure 5.137 [5.55] shows a diagram of a drying oven operated by circulating air with grids on a trolley. The air sucked in by the ventilator is led through a heater and passes laterally between the perforated trays. The air is kept in motion solely by the ventilator. When the effluent air valve is closed the oven operates on circulated air. The proportion of fresh air increases as the inlet air valve is opened. Removal of moisture requires a high proportion of fresh air, though this has the disadvantage of high energy consumption. The quantity and the temperature of the air and the proportion of fresh air can be used as regulating factors. Drying ovens generally operate with circulating air at 80–95°C.

Although drying ovens are widely used and easy to operate, they have a number of disadvantages:

> They are only suitable for drying stable substances.
> When the trays are partially filled the product near the edges of the trays is not completely dried, as drying air is deflected ineffectually over the edge zones. In systems such as the one shown in Fig. 5.137 there is also the risk of the product in the upper trays being dried more thoroughly than that in the lower ones, as the warm air initially passes though the upper trays. This effect is particularly noticeable with low drying temperatures and small throughputs of air. After completion of drying of a granulate the relative moisture contents was 15% on the highest tray, whereas it was 95% in the lowest tray.
> Only granular materials, pastes, etc. can be processed in tray-drying ovens.

Vacuum drying ovens operate discontinuously under reduced pressure. Heat transfer takes place here not by convection of heated air but by contact of the material with heated plates or with the floor of the drying oven. They are therefore in principle contact dryers. The drying temperature can be reduced by the application of a vacuum, which is good for heat sensitive materials. The moisture is led away by opening a needle valve so that the pressure rises by about 10 Torr. The resultant airstream removes the moisture, which must then be bound to adsorbents.

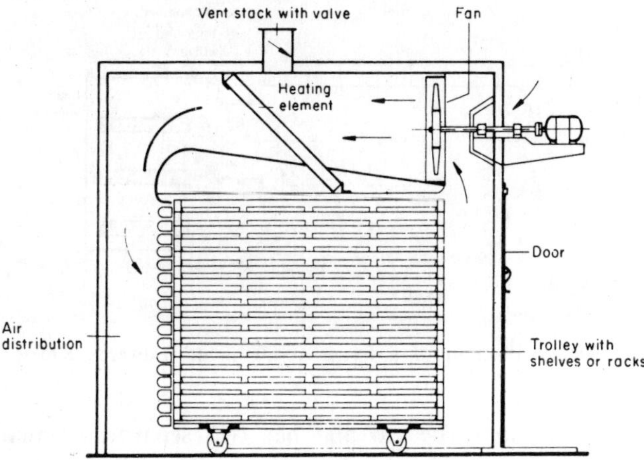

Figure 5.137 Circulating air (ventilated) oven dryer with trays on a trolley (Buttner-Schilde-Haas Company, Krefeld).

5.7.5. FREEZE DRYERS

Freeze drying plants carry out three technical process stages, namely freezing, main drying and further or secondary drying. Such a plant has three parts:

A drying chamber with a temperature controlling device.
A pump for sucking up water vapour and air.
A condenser or absorber.

In the simplest case a desiccator connected to an oil pump can be used as a drying chamber. The desiccator contains concentrated sulphuric acid, calcium chloride or phosphorus pentoxide to absorb the moisture. The deep frozen material (at least 20°C below the eutectic point) is put into the desiccator and vacuum applied. The radiant energy of the surrounding room is sufficient to vaporize small quantities of ice; with larger quantities the vaporization energy can be supplied with an infrared lamp [5.18].

Industrial purpose-built freeze drying plants usually contain freezing equipment integrated into a complete plant. These plants are designed to operate on scales from the laboratory, through the technical, up to the industrial scale, on which tons of material are produced. Figure 5.138 [5.18] shows a laboratory scale operation used for discontinuous operation.

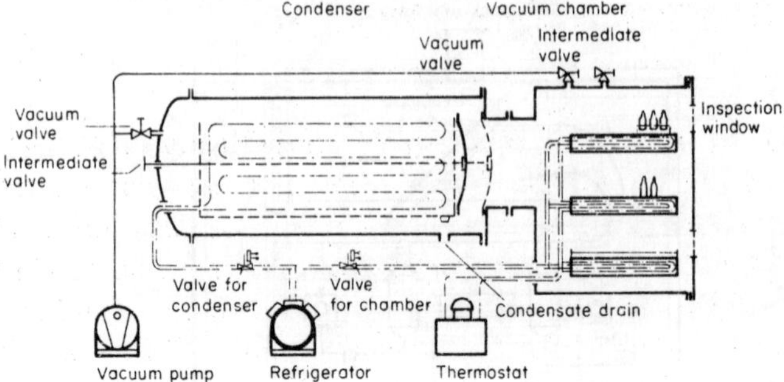

Figure 5.138 Schematic diagram of a freeze drying plant (Edwards Kniese, Marburg-Michelbach).

The apparatus is liquid heated and has two separate circulations for cooling and heating liquid. The moisture which is extracted is separated in a condenser.

Industrial freeze drying plants are made both for batch and for continuous operation. Figure 5.139 [5.55] shows a batch operating plant consisting of eight chambers which can be filled and emptied individually and all of which open into a common cooling chamber.

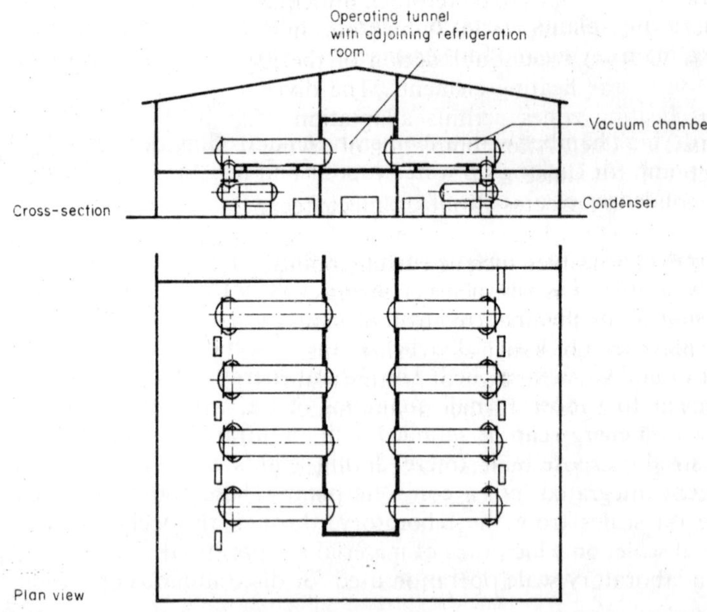

Figure 5.139 Freeze drying plant with eight chambers for batch operation (Leybold-Heraeus design).

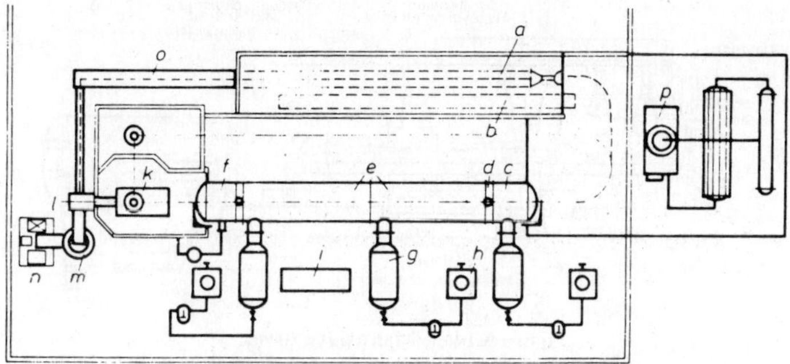

Figure 5.140 Freeze drying tunnel for continuous operation (Leybold-Heraeus design). a, Freezing tunnel and freezing belt; b, loader face; c, conveyor sluice; d, vacuum-tight slide valve; e, drying tunnel; f, outlet sluice; g, condensers; h, vacuum pump sets; i, switchboard; k, trolley discharge; l, emptying of dishes; m, mill; n, doser and packing machine; o, automatic dish washing machine; p, cooler. Dashed line, dish transport; dot dashed line, trolley transport.

Figure 5.140 shows a continuously operating plant [5.55].

Apart from larger throughputs, the advantage of a continuously operating plant is seen in the maintenance of constant operating conditions over a long period. In addition to vacuum chambers, these plants each have an inlet and an outlet sluice. They are therefore in principle tunnel or channel dryers. The product is frozen automatically on plates in the inlet sluice and transported into the vacuum zone. Sublimation of the ice is induced by lowering the plates on to the heating elements. The division of the drying tunnel into various heating zones permits adaptation of the drying conditions to a particular product. After completion of drying the material leaves the tunnel through the exit sluice.

The following process variables affect freeze drying:

Product variables such as eutectic point and solid content.
Apparatus variables such as freezing chamber, plate heating system in the vacuum chamber, removal of moisture.
Process variables such as freezing time, freezing temperature, strength of vacuum, temperature of heating plates, time for which the material stays in the plant, further drying time.

The numerous parameters make it difficult to optimize freeze drying and reliable control of the process requires considerable experience.

5.7.6. MICROWAVE DRYERS

Microwave drying has become important wherever very rapid drying is required. It is thus frequently used in the drying of thin foils in foil

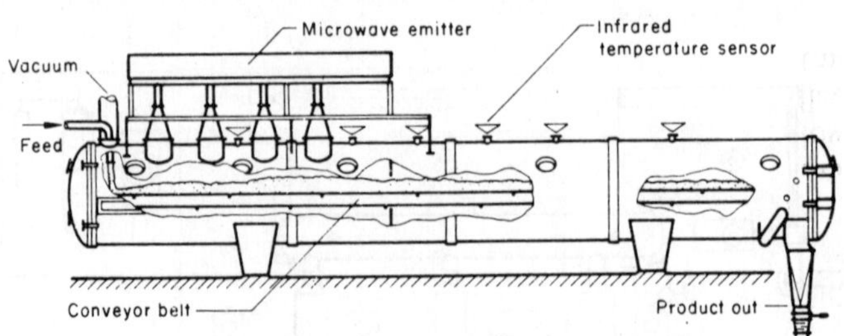

Figure 5.141 Microwave dryer.

manufacture. The frequency range given for this industrial application is 2400–2500 MHz.

Continuously operating microwave dryers are similar in design to vacuum belt dryers. The division into various heating zones and the ability to connect various zones as a cooling zone are also the same. Fig. 5.14 shows a microwave dryer [5.61].

Microwaves induce vibration of polar molecules. The simultaneous application of a vacuum causes the product to froth up in a manner similar to that in vacuum belt drying. The structure of the dried product is also similar. The advantage of microwave drying lies in the shorter drying time. Its application extends, for example, to the preparation of desiccated orange and grapefruit concentrates.

Chapter 6

Quality assurance of phytopharmaceuticals

6.1. Terms and definitions

Standardization and regulatory measures are used for the quality assurance of phytopharmaceuticals. Both terms are used in diverse ways in the published literature, creating the impression that they are mutually interchangeable. The fact that this is not the case becomes apparent from the Pharmacopoeia monograph 'Standardized drug powders. Pulveres normati'. This dosage form, described under Section 1.3.3, is reserved for potent drugs, according to *DAB 8* and *Ph. Eur.* When they are determined chemically or biologically they are adjusted to the required value by addition of a calculated quantity of inert excipient substances or of the powdered drug of minimum content. In every case the standard values lie within narrow limits precisely fixed by the pharmacopoeias. On the one hand, the Pharmacopoeia Commissions appear to have tacitly assumed that this must always be a case of the 'dilution' of drug materials with too high a content of active substances, as nowhere is it mentioned that a drug of low content should be adjusted by mixing with material of high percentage or even with pure active substance. On the other hand, the addition of low content drug cannot strictly speaking take place, as a minimum content for the drug is fixed in the pharmacopoeia. This is illustrated by the example of 'Belladonna leaves, *Ph. Eur. I*' and 'Standardized (Adjusted) Belladonna powder, *Ph. Eur. I*'. A minimum content of 0.30% total alkaloids, calculated as hyoscyamine, is required for Belladonna leaves, whereas adjusted Belladonna powder must have a total alkaloid content of 0.28–0.32%.

USP XX takes this difficulty into account, as it states under 'Vegetable and Animal Drugs':

'The requirements for vegetable and animal drugs apply to the articles as they enter commerce; however, lots of these drugs intended solely for the manufacture or isolation of volatile oils, alkaloids, glycosides or other active principles may depart from such requirements.'

This regulation is obviously made for the preparation of active substance concentrates or the recovery of pure active substances, and should also apply to all potent medicinal preparations. This means that standardization is carried out on drugs containing active substances which can be determined chemically or biologically and even then only on commercial products.

Regulation is preceded by standardization, which is primarily used for verifying the identity of the investigated sample (drug material or preparation). In every case where the active substances are not known and hence cannot be determined chemically or biologically, standardization is at the same time the method used for quality assurance. The strict separation of the terms 'standardization' and 'regulation' has still not been fully brought about. The statement of the composition of a purgative given in the Rote Liste

> '1 dragee contains: 85.75–100 mg of dried extract of deresinified *Alexandria senna* fruits adjusted to 20 mg hydroxyanthracene derivatives, calculated as sennoside B',

is complete and correct. It contains both a statement of the quantity of extract contained in the preparation and its origin and also a statement on the regulation and the basis of its calculation. It must be emphasized that the formulation 'adjusted to 20 mg hydroxyanthracene derivatives...' represents regulation, even though no quantity range as defined in the 'Pulveres normati' of the pharmacopoeias is given. It can, however, be considered as a mean value of a quantity range within the rules applying to solid medicinal dosage forms for total content and content uniformity. In contrast:

> '1 dragee contains: Extr. Fol. Sennae sicc. 155.6–200 mg (standardized to 14 mg sennosides A + B)',

is incorrect, although it has the same stated content, as far as clear terminology is concerned, as here the term 'standardization' is used instead of the term 'regulation'.

Statements such as '1 tablet contains: sennosides A + B from follic. Sennae 5 mg' are incomplete, as they do not indicate any relationship between the quantity of extract used, and hence the quantity of accompanying substances, and the active substance content.

The many different statements on ready-to-use medicines are certainly attributable to the existing lack of a uniform terminology. Every effort should however be made to standardize the terminology for new registrations and licensing of medicines and for renewal of these.

6.2. Standardization

Drugs are generally standardized according to the ten points listed under Section 2.2.1 and with consideration of the remarks made under Section 6.1.

If these are drugs for which monographs are given in the current pharmacopoeias, their individual examinations should be carried out in accordance with the directions given. If no such monographs exist, it can be helpful to consult older editions of the pharmacopoeias or other literature, especially publications which are up to date with the present state of knowledge.

Generally speaking, no difficulties occur with medicines containing potent constituents and those with biologically or chemically determinable components. Standardization is more difficult with drugs which have no pharmacological action (measurable pharmacological action is not to be equated with efficacy or activity) and the active principles of which are not known. The tests listed under Section 2.2.1, with the exception of point 8, must be referred to here for quality assurance. Thin layer chromatographic investigations (point 2), with which a so-called fingerprint of the drug can be obtained, are particularly suitable for identification of these drugs.

The need to search for one or more principal indicative substances is debatable, especially when the activity of the drug is to be ascertained by their quantitative determination. It must not be expected that the indicative substance content of the drug will be proportional to the content of active substances. The custom of selecting indicative substances of a drug from the class of substances to which the active principle is assumed to belong has often led to disappointments. Worldwide endeavours to isolate and discover the chemical nature of the active substances in such drugs have necessitated the replacement of the previously chosen indicative substances by other substances. Striking examples of this are camomile flowers, valerian roots and hawthorn leaves.

Whereas the present opinion is that the cardioactivity of *Crataegus* preparations is principally attributable to procyanidines, *DAB 8* permits the identity of the drug to be checked using thin-layer chromatography via chlorogenic acid and flavonoids (comparative substances: chlorogenic acid and rutin); the flavonoids are determined with the prescribed content determination and calculated as hyperoside.

Figure 6.1 shows a chromatogram of the flavonoid fractions of various species of *Crataegus*. The chromatogram was developed with a running solvent which differs from that given in *DAB 8*. The illustration was taken from the thesis by Adelheid E. Hecker-Niediek, Marburg, 1983.

6.2.1. CORRECT AND MEANINGFUL USE OF INDICATIVE SUBSTANCES

6.2.1.1. Indicative substances for identification

Thin layer chromatography is a simple and reliable aid for supporting the macro- and microscopic identification of a plant drug/drug plant and also for identification of an extract and of any medicinal dosage forms which may be prepared from it. If no tested, authentic drug plant or extract material is

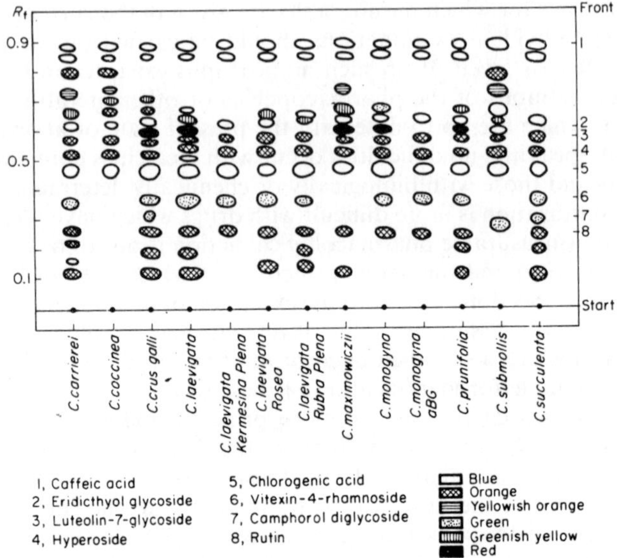

Figure 6.1 Thin layer chromatogram of flavonoid fractions of various *Crataegus* species and subspecies on HPTLC silica gel. Solvent: Flavonoid—FM I (ethyl acetate+ethyl methyl ketone+formic acid+water, 50:30:10:10). Natural product reagent A (Neu, Roth Co.) and Merckofix; UV detection at 366 nm.

available, indicative substances are also chromatographed. They can be selected from among the constituent substances of the drug plant, irrespective of whether these are active or accompanying substances. They should however be characteristic of the drug plant under investigation and easily demonstrable. Analytical data other than those obtained by thin layer chromatography can also be used for identification and quality assurance. The total nitrogen content or a qualitative amino acid analysis can, for example, be used for Urticae herba or other drug plants.

However, it is not sufficient to demand that identically coloured zones or spots from the investigated solution appear in the chromatogram at the level of the indicative substances. Rather, the colour and R_f value of the zones appearing in addition to the indicative substance spots must be described.

It is interesting that *DAB 8* permits the use of so-called external indicative substances which are not contained in the drug plant itself. '*Arnica* flowers, *DAB 8*' may be quoted as an example. A comparative solution of rutin in methanol is also applied to the thin layer plate in addition to the solution under investigation which is prepared by extraction of the drug powder with methanol. This flavonoid (rutin), which does not occur in *Arnica* flowers, is used to recognize any possible adulteration by marigold (*Calendula*) flowers, as it is stated:

'No zones with a yellowish to green colour should appear in the chromatogram of the investigated solution at the same level or below the zone in the chromatogram of the comparative solution.'

6.2.1.2. Indicative substances for content determination

Only those substances which actually occur in the material being evaluated can be considered in the selection of indicative substances for quantitative evaluation of a drug plant material or of a product prepared from this. An 'external' indicative substance cannot be used.

The choice of the indicative substance is also governed by the problem in question. The quantitative determination of any substance regularly occurring in the drug plant for ensuring the constant action of the drug or of its preparations is meaningless if this substance cannot be demonstrated to be the active principle. It can, however, be meaningful if it is carried out, for example, on an extract which occurs up to a certain percentage in a tablet preparation, as the extract content in the ready-to-use preparation can be ascertained via determination of the indicative substance. In such cases the choice of the indicative substances is made on an analytical basis, i.e. they must be easy to isolate and to determine exactly in small quantities without great effort and expense.

6.2.1.3. Indicative substances for stability testing

The shelf life or the stability limits of plant drugs and their products containing active substances which cannot or cannot yet be detected and determined chemically or biologically also cannot be determined. The use of indicative substances as a substitute for the unknown active substance can be regarded as justified insofar as, by definition, no one substance or group of substances but the whole range of extractable constituent substances is considered to be responsible for the action of phytopharmaceuticals. Difficulties occur in the selection of the indicative substance. The multiplicity of constituents, their qualitative and quantitative differences through the growing period and the numerous influences, e.g. time of harvesting, further processing of the drug plant material, etc., make the demand that the most labile compound must always be chosen as the indicative substance for testing of stability hardly logical. Furthermore, each type of preparation requires its own choice of indicative substance, as its lability differs greatly under various environmental conditions. A compound susceptible to hydrolysis would not be a suitable indicative substance in dry preparations. The same applies to substances sensitive to oxidation or to light. In this situation it seems more logical to check phytopharmaceutical preparations and their precursors both organoleptically for taste, smell and appearance and chemically and physically for any alterations at regular intervals during storage.

Determination of the density, refractive index and alcohol content of liquid preparations and especially the comparison of thin layer chromatograms of fresh material with those prepared upon completion of manufacture of the test preparation provide definite indications of the stability limits.

It can be stated in conclusion that quantitative determination of one or more indicative substances permits no definitive statement on the stability of the activity when the active principle is not known. Care should therefore be taken not to interpret every little alteration as a serious instability.

6.3. Regulation

The adjustment of phytopharmaceutical preparations to a certain active substance content, known as regulation, can easily be carried out in liquid preparations such as fluid extracts, tinctures, etc., by further cautious product conserving concentration or by dilution with the menstruum. Dried extracts should be adjusted as described under Section 1.3.4.6.

In certain cases the active constituents are enriched by removal of so-called ballast substances. The active substances can also be concentrated by extraction methods other than the usual ones, such as, for example, extraction with supercritical gases (see Section 4.2.4). However, it must not be forgotten that we are constantly concerned here with the manufacture of phytopharmaceutical preparations, the demands for the quality of which are not based solely on the exactly determinable active substance content (see Section 1.1). Hefendehl and Lander [6.12] emphasized that the extract or the tincture is regarded as the active component of a medicine designated as a pharmaceutical preparation, not an individual substance contained in it. One must therefore proceed in such a way in any regulatory adjustment, irrespective of whether it is concentration or dilution, that the original ratio of active substances to accompanying substances of the same drug plant material is preserved within acceptable limits.

Although the regulatory adjustment of dried extracts ultimately depends on the selection of drug materials of higher content (in the following called strong or stronger) and lower content (in the following called weak or weaker) drugs, possibilities are discussed here for the adjustment to the required content at various stages of preparation.

(1) Drug materials of various content are mixed as already described in Section 6.1 in a ratio which, depending on their contents, results in an end-product with the required content when they are extracted together. This method is difficult and is rarely used in practice.

(2) Each part of a drug plant is extracted by a fixed method. This produces extracts of differing contents. The mixture ratio for two or more components can be determined by the rules of mixture calculation (S. Meyer, *Grosser Rechenduden* (*Handbook of Calculation*), Bibliographisches Institut, Mannheim, 1964).

Example:
  Extract A has an alkaloid content of 1.55%.
  Extract B has an alkaloid content of 1.30%.

The adjusted extract shall have a total alkaloid content of 1.30–1.45%, calculated as hyoscyamine. Extract A is therefore too strong. The mixture ratio calculation is shown in Table 6.1.

It should be noted that mixing of the two extracts must result in a homogeneous product, which can only be satisfactorily achieved when the starting-extracts have approximately the same particle-size ranges.

(3) Inhomogeneities can be prevented by mixing extracts of different content when the extracts are still in liquid form. The outlay of analytical apparatus and materials is very much greater here, as not only the active substance contents of both of the liquid extracts, but also their total quantities which are to be processed and their dry residues with respect to residual moisture, have to be determined. The calculation is then made as described in Section 1.3.4.6, where x signifies not the quantity of inert adjuvant substance which is to be added but the quantity of weaker extract

Table 6.1

Extract	Content (%)	Mean adjusted content (%)	Content difference	Crossover	Proportion
A	1.55		+0.15		0.10
B	1.30	1.40	−0.10		0.15

The mixture ratio A:B is therefore 10:15.

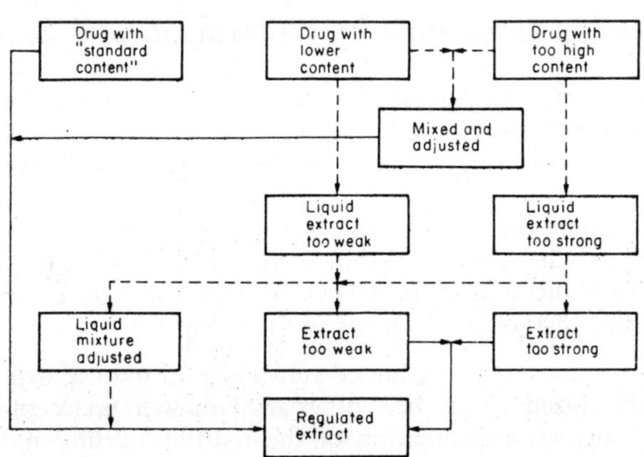

Figure 6.2   Possible ways to prepare adjusted dried extracts.

which is to be used. This method would be too cumbersome in practical industrial preparation.

6.4. Physical quality assurance

Quality assurance of phytopharmaceutical products has so far been discussed solely from the chemical and physiological points of view. The physical quality however plays an equally important role for the manufacturer and processor of plant extracts. Without the addition of suitable adjuvant substances many plant extracts occur in a form which makes further processing considerably more difficult, often even impossible. Hence extracts of *Crataegus* fruits, *Curcuma* extracts and many others cannot be processed to more manageable dry products either by roller, belt or spray drying. One particular example is the male fern extract already mentioned in Section 1.3.4.4, which is produced as a solvent-free thin extract. In all such cases the manufacturer cannot handle this without considerable additions of inert adjuvant substances. Before drying, therefore, a proportion of Aerosil, lactose, maltodextrin, glucose syrup or starch constituting up to 50% of the end-product is added to such plant extracts. As the ratio of active substances to accompanying plant substances remains unaltered here, the manufacturer has only to declare the measures he has taken, for example as follows:

> '100 g of the preparation contains 45–55 g of Extractum ..., corresponding to 5 g of active substance group XY (e.g. total alkaloids. glycosides and others), calculated as Y. Maize starch was used as excipient.'

6.5. Quality assurance by cultivation and breeding

Although medicinal plant cultivation and breeding are not in the province of pharmaceutical technology, we mention here a development which may well have a great influence on the use of phytopharmaceuticals as useful medicines.

The two alternatives of gathering wild medicinal plants and cultivating them have already been mentioned in Section 2.1.2. For various reasons the cultivation of medicinal plants will become increasingly dominant in the future. C. Franz [6.8] cites a number of these reasons:

(1) The fact that many medicinal substances of natural origin cannot be synthesized, or can be synthesized only with unacceptably great effort, necessitates creation of the natural starting-material, i.e. cultivation of the medicinal plant.

(2)   The unreliability of supply of drug plants gathered from the wild shows the need for their cultivation. The quantities available in the often widely scattered gathering areas are limited. Expert collectors are becoming increasingly difficult to find. All this results in the increasing occurrence of mistaken identity and adulteration of drug plant materials.

(3)   The increased demands for safety of medicines in general have also led to increasing demands for purity and quality of phytopharmaceuticals and of the drug plant materials from which they are obtained. The fundamentally heterogeneous composition of wild gathered medicinal plants does not meet the demands for consistent and reproducible quality.

(4)   Legal regulations such as the 1973 Washington Protection of Species Agreement and the more recent 1980 West German Nature Protection Order will considerably hinder the trade and processing of wild plants.

Franz draws the conclusion that not only should cultivation of medicinal plants be intensified and renewed but also much more consideration should be given to the factors affecting the quality of the product. Medicinal plants must be cultivated with phytochemical aspects in view, as the success of such cultivation depends less on the quantity of plants produced and much more on their quality. The active substance content of a cultivated medicinal plant can be affected by various factors:

(a)   Genetic variation and hereditary transmission of the secondary substances,

(b)   Morpho- and ontogenetic variability, i.e. differences in the active substance contents in various parts of the plant and during its growth,

(c)   Environmental influences (location, fertilization).

Agricultural methods are now available which permit the grower to produce high quality drug plants relatively quickly in large quantities. The reader is referred to Refs 6.9 and 6.10 for further considerations.

6.6.   Stabilization and stability

6.6.1.   TERMS AND DEFINITIONS

The stabilization of a phytopharmaceutical preparation consists of measures which guarantee its keeping quality or stability. Stability means the unaltered state of the product under customary or precisely defined storage and transport conditions. Alterations which reduce the quality and shelf-life can be of a physical, chemical or microbial nature. They can therefore be detected in comparison with the original state organoleptically (= sensorily):

Physically (e.g. via the refractive index, measurement of turbidity, colorimetrically, etc.),
Chemically (e.g. by determination of one or more of the constituents),
Biologically and microbiologically (bacterial count determinations, etc.).

Adjusted preparations, the lower and upper limits of the active substance content of which are stated in pharmacopoeias or other published directions, must first of all comply with these limit values. Even if this is guaranteed, it does not necessarily mean that no other alterations may have also occurred.

Stability is not to be understood as an absolute term. Alterations take place gradually under the customary storage conditions or even upon strict adherence to storage instructions until finally a state is attained which could be described as 'unstable' or 'unkeepable'. Here, the active substance content has fallen to such an extent, or the physical state has altered so much, or microbial contamination has assumed such proportions, that the professional use and the activity of the product can no longer be guaranteed to be harmless to the patient.

The keeping quality of products which have simply been standardized (see Section 6.2) is ascertained by comparing the properties of the fresh product used for the standardization procedure with those found after certain periods of storage.

As nearly every pharmaceutical product undergoes some kind of alteration process, the speed of which depends on the nature of the active substances, the properties of the excipients and additives, the nature and size of the contact surfaces and many other factors, it is customary to take the stability limit at 90% of the declared active substance content. This assumes that apart from loss of active substance no other detrimental alteration has taken place and that the decomposition of the active substance has not led to the formation of toxic products.

It is also customary to add a so-called manufacturing or safety 'margin' to medicinal products in which the active substances undergo relatively rapid decomposition and in which the therapeutic breadth (the margin between therapeutic dose and toxic dose) permits this, so that the product appearing on the market contains about 110% of the declared active substance content. This is seldom considered for phytopharmaceutical preparations.

6.6.2. METHODS OF STABILIZATION

6.6.2.1. Drying

Physical, chemical and microbial alterations take place in the liquid state, especially in aqueous medium, more easily than in dry products. It follows from this that the drying of phytopharmaceutical preparations, and especially keeping such preparations dry, is the simplest and best method of protection. The pharmacopoeias have recognized this for a long time, as they no longer accept spissum extracts as official dosage forms. The residual moisture content in dried extracts is generally limited to a maximum of 5%. The demand in *DAB 7* for a maximum of 3% has proved to be impracticable, as extracts dried to such an extent readily attract water from the surrounding atmosphere. An equilibrium of 6–7% residual moisture is established in extracts which are not hermetically sealed during storage under the average relative atmospheric humidity prevailing [FRG]. The storage of dried extracts over blue silica gel (*DAB 8*) is useful only if the partial vapour pressure above the protected extract is greater than that above the drying agent at a given atmospheric humidity, in other words, if the drying agent is more hygroscopic than the extract. The rate at which the extract absorbs moisture depends very much on its specific surface area. This means that spray dried and freeze dried (lyophilized) extracts are more sensitive than roller or oven dried ones. This fact must also be taken into account when grinding dried extracts, i.e. this should not be done in rooms where there is a high atmospheric humidity. There should also be no noticeable downward temperature difference from the air in the room to the material being milled, as otherwise water vapour will condense on the cold extract.

Large quantities of extracts cannot be stored over a drying agent, as uneconomically large quantities of the latter will be required to ensure an adequate adsorption capacity. Only moisture tight packing can be used here (see Sections 1.3.4.6 and 1.4.16).

Physical alterations scarcely occur at all under the above-mentioned conditions. Chemical processes such as enzymatic reactions, which are detectable down to a residual moisture content of about 10%, oxidations, hydrolysis, etc., proceed extremely slowly when storage in a cool place protected from light, as well as adequate drying, are guaranteed. Entry of oxygen must be restricted by the choice of suitable packing materials. In extreme cases the material must be vacuum-packed or sealed under an inert gas. The impermeability of the packing material to oxygen should be checked if this packing material is a synthetic (plastic) substance.

Microbial alterations, i.e. generally multiplication of the bacteria present in the product to a no longer acceptable number (see Section 2.2.1.9), are also dependent on the residual moisture in the product. The dependence of microbial growth on the so-called a_w value (a_w = water vapour pressure above the substrate/vapour pressure of pure water) or the water activity is shown in Table 6.2.

6.6.2.2. Stabilization of liquid preparations

As mentioned above, alternative processes proceed more rapidly in liquid phytopharmaceutical preparations than in dry ones. The following alterations are relatively easy to recognize:

Table 6.2.   Influence of the a_w value of the medicament on growth of bacteria present in it

a_w	Organisms totally inhibited by lowest a_w	Preparations with such an a_w value
1.00–0.95	Gram-negative bacilli, spores of Bacillaceae, certain yeast species	Preparations with ~40% (w/w) sucrose or 7% (w/w) NaCl
0.95–0.91	Most cocci, Lactobacillaceae and vegetative cells of Bacillaceae, certain fungal moulds	Preparations with ~55% (w/w) sucrose or ~12% (w/w) NaCl
0.91–0.87	Most yeasts	Preparations with ~65% (w/w) sucrose or 15% (w/w) NaCl
0.87–0.80	Most fungal moulds, *Staphylococcus aureus*	Flour, milled products with ~17% water content
0.80–0.75	Most halophilic bacteria	Aqueous solutions with ~26% (w/w) NaCl
0.75–0.65	Xerophilic fungal moulds	Flour, milled products with ~10% water
0.65–0.60	Osmophilic yeasts	Supersaturated sugar solutions with ~10% water content

Physical alterations, such as the formation of sediments, colour changes, etc.

Alterations due to microbial growth, recognizable by the formation of a pellicle of mould, cloudiness or a sediment which can be easily disturbed.

Chemical alterations, when these can be detected organoleptically by smell, taste or appearance.

Chemical alterations such as hydrolytic decomposition, racemization, oxidation, etc., can only be detected with difficulty, i.e. with analytical apparatus and agents.

Measures such as are described under Section 7.2 must be taken to prevent, or at least delay, such processes.

The prevention of any microbial growth in a liquid phytopharmaceutical preparation by diethylpyrocarbonic ester (diethylpyrocarboxylate) is unfortunately no longer permitted, as this substance may possibly form carcinogenic urethanes with free amino acids or amines in the substrate. This discovery should really be thoroughly checked again in case this elegant method could be reinstated.

Chapter 7

Further processing to the finished medicine

According to Section 4 of the German Medicines Act, finished ready-to-use medicines are premanufactured medicines which are put on the market in a defined packing suitable for sale to the consumer. In this respect phytopharmaceutical preparations such as packaged tea or infusion mixtures, ready-to-drink active substance extracts and certain fluid extracts or tinctures are regarded as finished, ready-to-use medicines. However, in most cases they are intermediate or semi-finished products which must be further processed to liquid or solid dosage forms. The difficulties arising here are often greater than those encountered in the processing of individual substances of natural or synthetic origin.

7.1. Preparation of solid dosage forms from dried extracts

The processing of plant extracts into soft gelatin capsules does not usually present any particular difficulties. Most extracts, despite being hydrophilic, can easily be converted into pumpable suspensions with lipophilic dispersants when they are finely powdered and then encapsulated by the Scherer process.

In contrast, direct filling into hard gelatin capsules is made difficult by the poor flow properties of most dried extracts. Freeze and spray dried extracts are moreover far too voluminous, i.e. their bulk density is too low for them to be able to fill a capsule of the normal size with a single dose. The dried extracts must therefore be granulated as in the manufacture of tablets (see below).

In addition to poor flow properties and the low bulk density, extracts existing as multicomponent systems usually exhibit pronounced hygroscopicity and often have a very low eutectic point. These very properties make tabletting difficult and direct compaction usually impossible.

Table 7.1 Formulae for extract tablets [7.1]

Formula and directions (example)			
A		Plant extract	150.0
		Aerosil purum	20.0
		Amylum solani	20.0
	(a)	Mix well and sieve	
	(b)	Thoroughly moisten with 2% glycerine-alcohol	
	(c)	Granulation with cetyl-isopropanol 10% ca. 20.0	2.0 dry substance
	(d)	Dispersion	
	(e)	Drying	
B		External phase lubricant	8.0
			200.0

1000 cores each of 0.2

AESCULUS COMP. 0.2			
A	I	Aneurin vitamin B1	2.0
		Extract Aesculi hippocast.	123.0
		Extract, Hamamelis	25.0
		Aerosil pur.	20.0
		Amylum solani	18.0
	(a)	Moisten with glycerine-isopropanol and dry	
	(b)	Moist granulation with cetyl-isopropanol 10% ca. 20.0	2.0 dry substance
	(c)	Dispersion and drying	
B		External phase lubricant	10.0
			200.0

1000 cores each of 0.2

CARDIOTONICUM 0.26			
A	I	Rutin	20.0
		Extract. Crategii 1 + 1	90.0
		Extract. Visci	35.0
		Extract. Allii	35.0
		Amylum solani	38.0
		Aerosil pur.	30.0
	(a)	Mix well in a mixer and sieve	
	(b)	Soak with glycerine-isopropanol 2% in sufficient quantity	
	(c)	Drying	
	(d)	Granulation with cetyl-isopropanol 10% ca. 20.0	2.0 dry substance
	(e)	Dispersion and drying	
(B)		External phase lubricant	10.0
			260.0

1000 cores each of 0.26

Table 7.1 Continued.

SEDATIVUM
A	I	Extract. Valerianae	150.0
		Aerosil pur.	20.0
		Amylum solani	20.0
	(a)	Thoroughly moisten with glycerine-isopropanol and dry	
	(b)	Granulate with cetyl-isopropanol 10% 20.0	2.0
	(c)	Dispersion and drying	
B		External phase lubricant	8.0
			200.0

HEPATICUM CHOLAGOGUM 0.29
A	I	Extract. Aloes	60.0
		Fel. tauri	20.0
		Rhiz. curcum.	50.0
		Extract. gentian	15.0
		Extract. chelidonii	10.0
		Extract. chinae	10.0
		Extract cynarae	50.0
		Aerosil pur.	35.0
		Amylum solani	27.0
	(a)	Mix well in mixer and sieve	
	(b)	Soak with glycerine-isopropanol, sufficient quantity	
	(c)	Moist granulation with cetyl-isopropanol ca. 30.0	3.0 dry substance
	(d)	Dispersion and drying	
B		External phase lubricant	10.0
			290.0

1000 cores each of 0.29

LAXATIVUM COMP. 0.21
A	I	Extract. Aloes	60.0
		Extract. rhei	60.0
		Extract. cascar. sagrad.	25.0
		Diacetyldioxyphenylisatin	5.0
		Fel. tauri sicc.	15.0
		Aerosil pur.	24.0
	(a)	Thoroughly moisten with glycerine-isopropanol and dry	
	(b)	Granulate with cetyl-isopropanol 10% ca. 10.0	1.0 dry substance
	(c)	Dispersion and drying	
B		External phase lubricant	10.0
			210.0

1000 cores each of 0.21

At a very early stage the Arbeitsgemeinschaft für Pharmazaeutische Verfahrenstechnik (APV) (The German Pharmaceutical Industrial Process Technology Association) had endeavoured to establish practical directions for carrying out these processes with dried extracts. The formulae used in a training course in 1961 [7.1] are reproduced in Table 7.1. These are characterized by the addition of highly disperse silicic acid (Aerosil®) which reduces the hygroscopicity of the extracts and loosens the powder. The use of hydrophobic (water repellant) Aerosil® R 972 was also described. Satisfactory tablets were obtained with the following formula [7.2]:

Dried extract (+ 5% Aerosil® R 972)	45%
Potato starch	10%
Aerosil® R 972	2%
Talcum siliconisatum	3%

Hydrophobic Aerosil® R 972 is mixed with the extract. This protects the hygroscopic extract from penetration of water and hence from agglutination. This premixture is mixed with the other constituents and made into briquettes, which are then ground up and admixed with the following further constituents:

Granulatum simplex	35%
Talcum siliconisatum	3%
Aerosil compositum	2%

The resultant tablets have a breaking strength of $2 \, kg/cm^2$, a friability of 0.75% and a disintegration time of less than 6 min.

Crippa [7.3] has exhaustively investigated formulation factors with the example of a dried cascara extract. He was able to show that using water instead of chloroform as granulating liquid and polyvinylpyrrolidone as binder, but with an otherwise unaltered formula, results in a greatly increased disintegration time.

The addition of colloidal silica and methylene caseine, which reduce the disintegration time, is also suggested for the production of acceptable tablets.

A suggested formula is reproduced in Table 7.2.

Crippa [7.3] suggested binding water by the addition of adsorbants such as highly disperse silicic acid and thus of imparting a certain porosity to the tablet. This effect is further intensified by the use of microfine cellulose as a filler and disintegrating agent, as Graf and Bornkessel [7.4] were able to demonstrate with the example of tablets containing a freeze dried valerian extract. Table 7.3 shows their results.

Table 7.2 Tablet formula for cascara extract

Composition	Quantity per tablet
Dried, granulated cascara extract	125 mg
Lactose	60 mg
Magnesium stearate	1.5 mg
Colloidal silica	5 mg
Microcrystalline cellulose	8.5 mg
Tablet weight	200.0 mg
Hardness	30–35 N
Friability	0.25%
Disintegration time	10–15 min

Table 7.3   Tablet formula for valerian extract

Composition	Formula	
	No. 1	No. 2
	(Quantities in %)	
Freeze-dried valerian extract	20.0	40.0
Colloidal silica[a]	1.0	1.0
Granulated microfine cellulose[b]	59.0	40.0
Powdered microfine cellulose[c]	20.0	9.4
Rice starch	—	9.6
Tablet weight	250 mg	122 mg
Disintegration time	18 min	4 min 48 s
Breaking strength (in a Heberlein tester)	170 N	58 N
Friability (Roche Friabilator)	0.05%	0.05%

[a]Aerosil 200.   [b]Elcema P100.   [c]Elcema G250.

The altered ratio of valerian extract to Elcema, together with the lower compaction pressure (smaller breaking strength), produces tablets which, for products containing extract, have very good disintegration times and friabilities which are also maintained upon storage.

We were able to confirm the positive effect of microfine and microcrystalline cellulose in our own experiments [7.5] with tablets containing extract. It was also confirmed here that the hydrophobization of the extract originally suggested by Keymer [7.2] is preferable to loosening or disintegration with colloidal silicic acid. An extract ground with added magnesium stearate exhibits a hygroscopicity which is definitely lower than that of an extract treated with colloidal silica. Grinding on a high-speed pin mill with about 10–15% magnesium stearate evidently reduces the wettability of the dried extract. The resultant product is a loose, free-running powder. These measures, combined with the use of microcrystalline or microfine cellulose, can be recommended for the tabletting of numerous extracts. The use of magnesium stearate naturally raises the question of bioavailability. It is well known that hydrophobized substances can have a detrimental effect both on disintegration time and on rate of dissolution. A paper by Burger and Dialer [7.6] is of interest in this connection. These authors investigated a number of commercial preparations containing valerian root with respect to their disintegration time and dissolution rate.

This series of preparations was chosen as a rapid commencement of the effect of the drug is desired with soporifics and sedatives and hence one is interested in finding such substances which have short disintegration times and a rapid release of active substance. All the preparations were sugar-coated dragees. The weight of the sugar coating was about 400–800% of that of the core, which is evidently necessary for masking the smell. Only one of all the eight investigated commercial preparations fulfilled the pharmacopoeia requirements for disintegration time. The authors concluded from the determined dissolution rates that solid medicinal preparations containing dried extracts should have their own monographs in future pharmacopoeias. Only the disintegration time has any relevance for the bioavailability of such preparations.

7.2.   Preparation of liquid dosage forms from extracts

Liquid phytopharmaceutical products can be prepared from tinctures, fluid extracts, spissum (thick) extracts and dried extracts. The difficulties arising here can often be greater than those in the preparation of solid dosage forms.

DAB 8 states that solutions of dried extracts must not be stored as such for any length of time, as precipitations due to differences in the solubility of certain constituents in the original menstruum used in the preparation of the extract and in the liquid used for reconstituting (redissolving) it occurs upon redissolution of concentrated extracts, particularly of dried extracts. Precipitations, however, still occur fairly rapidly even when the same mixture as was used for extraction of the drug plant is used for reconstitution of the dried extract, as the extractive substances dissolved in the miscella will have undergone partial physicochemical alterations due to concentration and drying which impair their solubility.

The following can be regarded as measures for counteracting the formation of precipitates:

(1)   Solvents with compositions corresponding to the menstruum used for preparing the plant extract are the best for redissolving a dry extract or a spissum extract or for diluting a fluid extract or a tincture.

(2)   pH shifts and other inconsistencies must be avoided when reconstituting extracts and especially when mixing them with other medicinal substances or their preparations. This applies particularly to products containing alkaloids [see 7.7].

(3)   In some cases addition of cosolvents helps to stabilize the solutions. Polyvalent alcohols such as 1,2-propyleneglycol, glycerol, sorbitol, low molecular weight polyethyleneglycols and glucose syrup are mostly used.

(4)   The stabilizing addition of non-ionogenic surfactants has occasionally also been described. Crippa [7.3] has reported the solubilizing action of polyoxyethylene-20-sorbitan monostearate (Tween 60) in the dissolution of a spissum extract of *Ruscus aculeatus*.

When any precipitates have been removed by filtration it should be checked that the activity of the preparation has not been diminished. This can generally be assumed to be the case in adjusted preparations, though even here a certain degree of alteration of the phytopreparation may have to be accepted.

There is no procedural rule applicable to all cases. The measures to be adopted must be established and optimized for each individual case.

7.3. Preservation of liquid phytopharmaceutical preparations

Guidelines as have been published by Nürnberg [7.8] and others for non-phytopharmaceutical preparations are generally applicable. The view that an alcohol concentration of more than 15% has an adequate preservative effect must be treated with caution in the case of liquid preparations containing extract. Preparations are known in which microbial growth was still found even with a 40% alcohol content. It must not be forgotten that phytopharmaceutical preparations often represent important nutrient media for microorganisms, particularly for yeasts and fungal moulds.

Adequate conservation by (a) the alcohol present in the mixture or (b) the added preservative should be ascertained in each individual case in the conservation loading test [see 7.9]. Some phytopharmaceutical preparations are well known for their low bacterial content and even 'autosterility'. This has been found to be particularly striking in drug plant extracts containing anthraquinone, the active substances of which possess definite bactericidal activity. Other extracts, however, appear to be particularly susceptible to microbial growth, one example being extracts of Fructus Cynosbati. Addition of preservatives is indispensable here.

p-Hydroxybenzoates, sorbic acid and benzoic acid active against yeasts and fungal moulds are still the most important preservatives available. Here it should be noted, particularly for the conservation of alcoholic extract solutions, that the stability of *p*-hydrozybenzoate esters increases in the order methyl, ethyl, propyl and butyl ester. The solubility of the butyl ester in alcoholic preparations is adequate and hence it should be preferred to the otherwise more customary mixture of methyl and propyl ester when mixed with the propyl ester, because of the particularly good antimicrobial action of this combination.

References

To simplify the citation and listing of literature references, the individual references are numbered in sequence within each chapter using the chapter number. The presentation in this listing is again in order of chapter then reference.

Chapter 1

1.1 P. Dilg and G. Jüttner, *Pharmazeutische Terminologie*, 3rd Edn, Govi-Verlag, Frankfurt/M (1983).

1.2 H. G. Menssen (Ed.), *Moderne Aspekte der Phytotherapie*, Pharm & medical inform. Verlags-GmbH, Frankfurt/M (1981).

1.3 P. H. List, W. Schmid and E. Weil, *Arzneim. Forsch.* **19**, 181 (1969).

1.4 S. Heinzl, *Dtsch. Apoth. Ztg* **122**, 2370 (1982).

1.5 *"Pharma jahresbericht 1981/82" des Bundesverbands der Pharmazeutischen Industrie e. V.* Frankfurt/Main.

1.6 H. Schilcher, *Pharm. Ztg* **127**, 2174 (1982).

1.7 J. Langerfeldt, *Dtsch. Apoth. Ztg* **122**, 1417 (1982).

1.8 Ch. Franz and A. Wünsch, *Planta Medica* **24**, 1 (1973).

1.9 C. Weichan, *Pharmazie* **5**, 351 (1977).

1.10 J. Weichmann, *Landwirtsch. Forsch.* 30/I, Sonderheft (1977).

1.11 K. Gierschner, *Chem. Rundschau* **34**, No. 45; No. 48 (1981).

1.12 R. A. Baker, *J. Food Sci.* **41**, 1198 (1976).

1.13 P. H. List, *Arzneiformenlehre*, 3rd edn. Wissensch. Verlagsgesellschaft, Stuttgart (1982).

1.14 *Hagers Handbuch der Pharmazeutischen Praxis*, 4th Edn, Springer-Verlag, Heidelberg (1967, 1980).

1.15 H. Schilcher, *Pharm. Ztg* **127**, 2178 (1982).

1.16 G. Kassis and P. H. List, *Pharm. Ind.* **43**, 1036 (1981).

Chapter 2

2.1 H. Flück, *Acta Pharm. Helv.* **43**, 1, (1968).

2.2 L. Hörhammer, in *Hagers Handbuch der Pharmazeutischen Praxis*, Vol. VI B, p. 72. Springer-Verlag, Heidelberg (1979).

2.3 H. Schilcher, *Pharm. Ztg* **126**, 2119 (1981).

2.4 W. Schier, *Dtsch. Apoth. Ztg* **121**, 323 (1981).

2.5   H. Schilcher, *Dtsch. Apoth. Ztg* **111**, 818 (1971).
2.6   G. Maas, *Deut. Pflanzenschutz.* **34**, (8) (1982).
2.7   E. Kühle, in *Hagers Handbuch der Pharmazeutischen Praxis*, 4th Edn, Vol. II, p. 420. Springer-Verlag, Heidelberg (1969).
2.8   L. H. Grimme, *Pharmuz* **2**, 51 (1973).
2.9   K. H. Wallhäusser, *Pharm. Ind.* **34**, 549 (1972).
2.10 H. Böhme and K. Hartke, *Europäisches Arzneibuch*, Vols. I, II, p. 185. Wissensch. Verlagsgesellschaft, Stuttgart (1976).
2.11 Th. Lammers and R. Gewalt, *Z. Hygiene* **144**, 350 (1968).
2.12 M. H. Bucca, *J. Bacteriol,* **71**, 491 (1956).
2.13 L. Z. J. Toth, *Arch. Microbiol,* **32**, 409 (1959).
2.14 J. L. Friedl, L. F. Ortenzio and L. S. Stuart, *J. Ass. Offic. Agr. Chemists* **39**, 480 (1956).
2.15 J. B. Opfell, J. B. Hohmann and A. B. Latham, *J. Am. Pharm. Ass.,* Sci. Ed. **48**, 617 (1959).
2.16 C. R. Phillips, *Bacteriol. Revs* **16**, 135 (1952).
2.17 H. Horn, M. Privora and W. F. Weuffen, *Handbuch der Desinfektion und Sterilisation*, Vol. II, p. 190. VEB Verlag, Berlin (1973).
2.18 H. Sucker, *APV-Informationsdienst* **17**, 59 (1971).
2.19 G. Retzlaff, G. Rust and J. Waibel, *Statistische Versuchsplanung*, 2nd Edn. Verlag Chemie, Weinheim (1978).
2.20 F. Bandermann, in *Ullmanns Encyklopädie der Technischen Chemie*, 4th Edn, Vol. 1. Verlag Chemie, Weinheim (1968).
2.21 H.-A. Trübenbach, Optimierung der Ethylenoxid-Begasung von Drogen. Dissertation, Marburg (1980).
2.22 H. Schilcher, *Planta Med.* **44**, 65 (1982).
2.23 P. H. List and S. Hanafi, *Dtsch. Apoth. Ztg* **103**, 1314 (1963).

Chapter 3

3.1   *Lösemittel Hoechst,* 6th Edn. Hoechst AG, Frankfurt/M. (1976).
3.2   H. Böhme and K. Hartke, *Europäisches Arzneibuch*, Vols. I, II, p. 99. Wissensch. Verlagsgesellschaft, Stuttgart (1976).
3.3   P. H. List, in *Hagers Handbuch der Pharmazeutischen Praxis*, 4th Edn, Vol. II, p. 128. Springer-Verlag, Heidelberg (1972).

Chapter 4

4.1   H. Rumpf and K. Schönert, in *Ullmann's Enzyklopädie der Technischen Chemie.* 4th Edn, Vol. II. Verlag Chemie, Weinheim (1972).
4.2   G. Aldolphi, *Lehrbuch der chemischen Verfahrenstechnik.* VEB Verlag, Leipzig (1969).
4.3   F. C. Bond. Trans. Am. Ind, Mining Met. Engrs, Techn. Pub. No 3308-B, *Mining Eng.* **4**, 484 (1952).
4.4   W. Beushausen, *Chem. Ing. Techn.* **31**, 553 (1959).
4.5   O. Lauer, *APV-Info* **19**, (2/3), 91 (1973).
4.6   J. Wessel, in *Ullmann's Enzyklopädie der Technischen Chemie.* 4th Edn., Vol. II, p. 43. Verlag Chemie, Weinheim (1972).
4.7   K. Leschonski, in *Ullmann's Enzyklopädie der Technischen Chemie.* 4th Edn., Vol. II, p. 35. Verlag Chemie, Weinheim (1972).
4.8   DIN 66 142.
4.9   H. Paul, *Glückauf* **74**, 277 (1938).

4.10   H. Paul, *VDI-Zeitschrift* **82**, 1197 (1938).

4.11   *Prüfsiebung – Probenahme und Siebanalyse*, 6th Edn, technical publication by Siebtechnik, Mülheim/Ruhr (1969).

4.12   O. Lauer, *CZ-Chemie-Technik* **1**, 363 (1972).

4.13   F. Patat and K. Kirchner, *Praktikum der Technischen Chemie*. 2nd Edn. Walter de Gruyter, Berlin (1968).

4.14   J. Flesselles and M. Gayet, *Beiträge Tabakforsch* **9**, 161 (1977).

4.15   H. Smit, *Beiträge Tabakforsch.* **3** (1), 39 (1965).

4.16   A. J. Kruozynski and H. Larsen, *Beiträge Tabakforsch.* **3** (5), 395 (1966).

4.17   L. Hübschen, *Beiträge Tabakforsch.* **3** (5), 371 (1966).

4.18   Alpine–Information 4/72, Pfeffer (schwarz und weiß).

4.19   Alpine–Information 3/74, Feinzerkleinerung von Lebensmitteln mit dem Cryogen-Kaltmahlverfahren.

4.20   O. Lauer. Personal communication.

4.21   P. H. List, *Arneiformenlehre*, 3rd Edn. Wiss. Verlagsgesellschaft, Stuttgart (1982).

4.22   I. A. Muravev, V. D. Ponomarev and U. G. Pshukov, *Pharm. Chem. J. (UdSSR), Engl. Edn.* **5**, 102 (1971).

4.23   C. Srinivasulu, S. R. S. Sastri, K. S. Narasimhan and S. N. Mahapatra, *Res. Ind.* **23**, 224 (1978).

4.24   M. Melichar, *Pharmazie* **17**, 290 (1962).

4.25   K. Bauer and K. Heber, *Pharmaz. Zentralhalle* **71**, 513 (1930).

4.26   F. Müller, *Dtsch. Apoth. Ztg* **103**, 87 (1963).

4.27   H. Palme and G. Winberg, *Arch Pharm.* **254**, 537 (1916); **256**, 223 (1918).

4.28   J. Büchi and K. Feinstein, *Acta Pharm. Helv.* **11**, 334 (1936).

4.29   F. Müller, *Arzneim. Forsch.* **13**, 551 (1963).

4.30   O. E. Schultz and J. Klotz, *Gesetzmäßigkeiten der Perkolation*. Editio Cantor, Aulendorf/Württ (1955).

4.31   M. Melichar, *Arzneim. Forsch.* **18**, 1341 (1968).

4.32   T. Voeste, and K. Wesp, in *Ullmann's Enzyklopädie der Technischen Chemie*, 4th Edn, Vol. II, p. 722. Verlag Chemie, Weinheim (1972).

4.33   K. Schoenemann and T. Voeste, *Fette and Seifen* **54**, 385 (1952).

4.34   D. F. Boucher, J. C. Brier and J. O. Osborn, *Trans. Am. Inst. Chem. Engn, p. 722.* **38**, 967 (1942).

4.35   G. Karnowsky, *J. Am. Oil Chem. Soc.* **26**, 570 (1949).

4.36   O. E. Schultz and J. Klotz, *Arzneim. Forsch.* **3**, 471 (1953).

4.37   F. Müller, *Pharm. Zentralhalle* **107**, 270 (1968).

4.38   F. Müller and J. B. Mielck, *Arch. Pharmaz.* **301**, 631 (1968).

4.39   M. Melichar, *Arzneim. Forsch.* **18**, 1482 (1968).

4.40   K. G. Goncharenko, *Pharm. Chem. J. (UdSSR) Engl. Ed.* **13**, 764 (1979).

4.41   G. K. Goncharenko, N. Niskanov and I. A. Gengrinovich, *Pharm. Chem. J. (UdSSR) Engl. Ed.* **11**, 981 (1977).

4.42   O. Isaac, Verfahren zur Herstellung von Kamillenextrakten. DOS 2 316 363, 17 Oct. 1974.

4.43   O. Isaac, Verfahren zur Herstellung von Kamillenextrakten. DOS 2 227 292, 20 Dec. 1973.

4.44   M. S. Khagi, *Pharm. Chem. J. (UdSSR) Engl. Ed.* **5**, 437 (1971).

4.45   R. Madaus, Verfahren zur Gewinnung von Silymarin aus Pflanzen. DOS 29 14 3 30, 9 April 1979.

4.46   F. Blaich, Verfahren zum Herstellen eines Wirkstoffkonzentrates aus Sennesdrogen. DAS 1 050 506, 1 March 1958.

4.47   L. Hörner, W. Rösch and K. Deuschel, Verfahren zur Extraktion von Morphin aus Mohnkapseln. DOS 27 26 925, 15 June 1977.

4.48   M. I. Lakoza et al., *Pharm. Chem. J. (UdSSR) Engl. Ed.* **9**, 237 (1975).

4.49   M. Melichar, V. Rusek and G. Solisch, *Ceskoslav. Farmazie* **2**, 338 (1953).

4.50   M. Melicher, V. Rusek and G. Solisch, *Ceskoslav. Farmazie* **3**, 336 (1954).

4.51   M. Melicher, V. Rusek and G. Solisch, *Ceskoslav. Farmazie* **4**, 512 (1955).

4.52   U. Bogs, *Dtsch. Apoth. Ztg.* **98,** 917 (1958).
4.53   U. Bogs and K. Damm, *Pharmazie/Pharmazeutische Praxis* **17** (1), 1 (1962).
4.54   U. Bogs and K. Damm, *Pharmazie/Pharmazeutische Praxis* **21,** 149 (1966).
4.55   W. Walter. Untersuchungen über die Turboextraktion von Chinarinde, Diss. Nr. 3165, ETH-Zürich (1961).
4.56   V. G. Ainshtein, L. B. Shagalov and Y. P. Vainberg, *Pharm. Chem. J. (UdSSR) Engl. Ed.* **6,** 523 (1972).
4.57   P. H. List, *Arzneiformenlehre*, 3rd Edn, p. 262. Wiss, Verlagsgesellschaft, Stuttgart.
4.58   R. Haul, H. H. Rust and J. Lützow, *Naturwissenschaften* **37,** 523 (1950).
4.59   W. Specht, *Z. Lebensm. Unters und Forsch.* **94,** 11 (1952).
4.60   O. E. Schultz and J. Klotz, *Arzneim.Forsch* **4,** 325 (1954).
4.61   W. F. Head, H. M. Beal and W. M. Lauter, *J. Am. Pharm. Ass. Sci. Ed.* **45,** 239 (1956).
4.62   F. O. Schmitt, C. M. Johnson and A. R. Olson, *J. Am. Chem. Soc.* **51,** 370 (1928).
4.63   S. C. Liu and H. Wu., *J. Am. Chem. Soc.* **56,** 1005 (1934).
4.64   A. Henglein, *Naturwissenschaften* **43,** 277 (1956).
4.65   A. Henglein, *Naturwissenschaften* **44,** 179 (1957).
4.66   W. Süss, W. and H. Hanke, *Pharm. Zentralhalle* **106,** 86 (1967).
4.67   W. Süss and H. Hanke, *Pharmazie* **24,** 270 (1969).
4.68   P. E. Wray and L. D. Small, *J. Am. Pharm. Ass.* **47,** 823 (1958).
4.69   P. C. Bose, T. K. Sen and G. K. Ray, *Indian J. Pharm.* **23,** 222 (1961).
4.70   A. E. De Maggio and J. A. Lott, *J. Pharm. Sci.* **53,** 945 (1964).
4.71   M. E. Ovadia and D. M. Skauen, *J. Pharm. Sci.* **54,** 1013 (1965).
4.72   C. Srinivasulu, S. C. Srivastava and S. N. Mahapatra, *Indian J. Pharm.* **32,** 90 (1970).
4.73   W. Süss, *Pharmazie* **27,** 615 (1972).
4.74   N. S. Shchepilov, P. N. Makarenko and A. S. Chernyak, *Pharm. Chem. J. (UdSSR) Engl. Ed.* **6,** 240 (1972).
4.75   V. J. Thakkar, A. N. Saoj and V. K. Deschmukh, *Indian Pharm.* **36,** 136 (1974).
4.76   D. Thompson and D. G. Sutherland, *Ind. Engng Chem.* **47,** 1167 (1955).
4.77   N. G. Bozhko, *Pharm. Chem. J. (UdSSR) Engl. Ed.* **4,** 640 (1970).
4.78   Y. Z. Rakhman-Zade et al. *Pharm. Chem. J. (UdSSR) Engl. Ed.* **6,** 789 (1972).
4.79   I. Issaev and D. Mitev, *Farmacija (Sofia)* **1968** (3), 1 (1968).
4.80   I. Issaev and D. Mitev, *Pharmazie* **27,** 236 (1972).
4.81   V. D. Boiko and I. V. Mizinenko, *Pharm. Chem. J. (UdSSR) Engl. Ed.* **4,** 504 (1970).
4.82   M. K. Bojadschieva and C. T. Dimov, *C.R. Acad. Bulgare Sci.* 26, No. 8, 1049 (1973).
4.83   M. K. Bojadschieva and C. T. Dimov, *C.R. Acad. Bulgare Sci.* 27, No. 6, 823 (1974).
4.84   M. K. Bojadschieva and C. T. Dimov, *C.R. Acad. Bulgare Sci.* 27, No. 8, 1101 (1974).
4.85   C. T. Dimov, *Pharmazie* **33,** 105 (1978).
4.86   N. A. Gromova et al., *Pharm. Chem. J. (UdSSR) Engl. Ed.* **8,** 709 (1974).
4.87   O. E. Schultz and J. Klotz, , *Arzneim.Forsch.* **3,** 471 (1953).
4.88   F. Müller et al., *Acta Pharm. Technol.* **26,** 240 (1980).
4.89   B. Cuntze, Dissertation, Kiel (1969).
4.90   T. V. Astakhova and S. A. Minina, *Pharm. Chem. J. (UdSSR) Engl. Ed.* **11,** 256 (1977).
4.91   J. C. Ferrada et al., *Contrib. Cient. Tecnol. (Univ. Tec. Estado, Santiago)* **25,** 35 (1977).
4.92   A. S. Lipkovskii, *Pharm. Chem. J. (UdSSR) Engl. Ed.* **9,** 448 (1975).
4.93   U. Osmanov et al., Optimization of extraction of alkaloids from Vinca erecta roots using a mathematical experiment planning method, Deposited Doc. 1973, Viniti 7061–73 (russ).
4.94   A. Murav'en and T. P. Zyubr, *Pharm. Chem. J. (UdSSR) Engl. Ed.* **6,** 798 (1972).
4.95   M. B. Zinko et al. *Pharm. Chem. J. (UdSSR) Engl. Ed.* **11,** 529 (1977).
4.96   J. K. Figlmüller, *Oesterr. Chem. Ztg* **56,** 221 (1956).
4.97   J. K. Figlmüller, *Oesterr. Chem. Ztg* **56,** 261 (1955).
4.98   W. Kehse, *Chem. Ztg* **94,** 56 (1970).
4.99   D. F. Othmer and W. A. Jaatinen, *Ind. Engng Chem.* **51,** 543 (1959).
4.100   D. F. Boucher, Paper presented at the Massachusetts Meeting May, 11–13, 1942.
4.101   R. Dousse and E. Ugstad, *Lebensm. Wiss. und -Technol.* **8,** 255 (1975).
4.102   K. Masters, Continuous Extractor for Liquorice, *Process Biochemistry*, 18 (March 1972).

4.103 M. I. Lakoza, *Pharm. Chem. J. (UdSSR) Engl. Ed.* **9**, 519 (1975).

4.104 A. M. Plyashkevich and M. D. Zamyshlyaeva, *Med. Prom- St.'USSR*, No **12**, 22 (1964).

4.105 W. Sirtl, Grundlagen der Hochdruckextraktion, Nova-Swiss, Symposium, 29–31 Aug, p. 12 (1979).

4.106 Buse-Kohlensäure-CO^2, Räumliches Zustandsdiagramm für Kohlensäure CO_2, Publication by Buse, Bad Hönningen, FRG.

4.107 G. Brunner and S. Peter, *Chem. Ing. Technik* **53**, 529 (1981).

4.108 P. Hubert and O. G. Vitzthum, *Angew. Chem.* **90**, 756 (1978).

4.109 H. Brogle, Komprimierte Gase als Lösungsmittel, Publication by Nova-Swiss, Effretikon, Switzerland.

4.110 Krupp-Forschungsinstitut. Personal communication.

4.111 H. Coenen, H. et al., *Techn. Mitt. Krupp-Forsch. Ber.* **40**, 1 (1982).

4.112 F. Cramer, Neue Zustandsdiagramme von Kohlendioxid-Kohlensäure bis zu 12 000 at. Publication by Buse, Bad Hönningen, FRG.

4.113 W. Sirtl, Verfahren und Anlagen der Hochdruckextraktion, Nova-Swiss Symposium 28–31 Aug, p. 15 (1979).

4.114 H. D. Brannolte, Hochdruckextraktion mit CO_2 – Eine neue Stofftrennmethode, Lebensmitteltechnik-Fachzeitschrift für Lebensmitteltechnologie 14, 5 May 1982.

4.115 Various authors, *Angew. Chem.* **90**, 747 (1978).

4.116 G. M. Schneider, E. Stahl and G. Wilke, *Extraction with Supercritical Gases*. Verlag Chemie, Weinheim (1980).

4.117 K. Zosel, *Angew. Chem.* **90**, 748 (1978).

4.118 E. Stahl, *Parfümerie Kosmetik* **63**, 117 (1982).

4.119 E. Stahl, *Angew. Chemie* **90**, 778 (1978).

4.120 E. Stahl, Verfahren zur Extraktion von Kamille mit überkritischen Gasen, DOS 2 709 033, 2 March (1977).

4.121 E. Stahl and E. Schütz, *Arch. Pharm.* **311**, 992 (1978).

4.122 E. Stahl and E. Schütz, *Planta Med.* **40**, 12 (1980).

4.123 E. Stahl and E. Schütz, *Planta Med.* **40**, 262 (1980).

4.124 E. Stahl and E. Willing, *Pharm. Ind.* **42**, 1136 (1980).

4.125 R. Stahl and E. Willing, *Planta Med.* **34**, 192 (1978).

4.126 T. P. *Maslob.-Zhirovaya Prom.* **24**, No. 6, 34 (1958).

4.127 H.-A. Kurzhals and P. Hubert, Verfahren zur extrativen Bearbeitung von pflanzlichen und tierischen Materialien. Eur. Pat. 79 10 3931.6 (12. 10. 1979).

4.128 E. Stahl, *Fette-Seifen-Anstrichmittel* **84**, 444 (1982).

4.129 H. Trawinski, in *Ullmann's Enzyklopädie der Technischen Chemie*. 4th Edn., Vol. II, p. 204. Verlag Chemie, Weinheim (1972).

4.130 H. Hemfort, Separatoren. Publication by Westfalia Separator AG, Oelde, FRG (1979).

4.131 C. Alt, Filtration, in *Ullmann's Enzyklopädie der Technischen Chemie*. 4th Edn, Vol. II, p. 154. Verlag Chemie, Weinheim (1972).

4.132 Seitz-Filtermedien. Publication by Seitz-Asbest-Werke, Bad Kreuznach; PAW 611002 8 670 GK.

4.133 O.-B. Genius, Hypericin from St. John's wort. Deutsches Patent 15 69 849, 26 July (1967).

4.134 W. Winkler and P. Patt, *Naturwissenschaften* **47**, 83 (1960).

4.135 G. Kamphius et al., *Österr. Apoth. Ztg* **24**, 535 (1970).

4.136 P. Hietala and J. Sundman, Verfahren zur Isolierung von 1-Dopa aus Pflanzenextrakten. DOS 2241 414. 23 August (1972).

4.137 R. Madaus, Aesculus-Saponin und Salze. DOS 25 30 741, 11 July (1975).

4.138 P. A. Yavich et al., *Pharm. Chem. J. (UdSSR) Engl. Ed.* **10**, 215 (1976).

4.139 E. Bombardelli, Verfahren zur Gewinnung eines Ginseng-Extraktes und diesen enthaltende pharmazeutische Zubereitungen. DOS 2724 947. 2 June (1977).

4.140 N. Diding et al., *Acta Pharmac. Suec.* **5**, 177, 183, 199, 205 (1968).

4.141 M. Nikbacht, Mikrobiologische Verunreinigungen pflanzlicher Arzneizubereitungen und mögliche Maßnahman zur Keimzahreduzierung. Diss. Marburg (1974).

4.142 R. Billet, *Verdampfung und ihre technische Anwendung*. Verlag Chemie, Weinheim (1981).

4.143   E. Weder, in *Einführung in die thermische Verfahrenstechnik*, ed. by P. Grassmann, p. 32. Walter de Gruyter, Berlin (1967).
4.144   VGI-Wasserdampftafel. Springer-Verlag, Berlin (1968).
4.145   Wiegand-Verdampfer, Technical booklet No. P 73/la/VII/-78/4000 d E + B.
4.146   A. V. Visher, *Pharm. Chem. J. (UdSSR) Engl. Ed.* **9**, 121 (1975).
4.147   P. H. Stahl, *Feuchtigkeit und Trocknen in der pharmazeutischen Technologie*. Dr. Dietrich Steinkopff-Verlag, Darmstadt (1980).
4.148   G. Schepky, *Dtsch. Apoth. Ztg* **112**, 87 (1972).
4.149   F. Kneule, *Das Trocknen*. Verlag H. R. Sauerländer, Aarau und Frankfurt/Main (1975).
4.150   O. Krischer and K. Kröll, *Trocknungstechnik*, Vol. I. *Die wissenschaftlichen Grundlagen der Trocknungstechnik*, 2nd Edn, Springer-Verlag, Göttingen (1963).
4.151   P. Grassmann, *Einführung in die thermische Verfahrenstechnik*, Walter de Gruyter, Berlin (1967).
4.152   E. Henne, *Luftbefeuchtung*, 2nd Edn. Verlag C. F. Müller, Karlsruhe (1976).
4.153   R. Rinkel, Sprühtrocknung pharmazeutischer Produkte, APV-Informationsdienst (4), 170 (1968).
4.154   K. Bullock and J. W. Lightbrown, *J. Pharm. Pharmacol.* **15**, 228 (1942).
4.155   P. Prokofièv and D. Solonouz, *Farmatsiya* **4**, 22 (1941).
4.156   A. M. Realdon, *Boll. chim. Farm.* **97**, 87 (1958).
4.157   V. I. Ganchukov and V. Davituliani, *Pharm. Chem. J. (UdSSR) Engl. Ed.* **5**, 423 (1971).
4.158   S. G. Barashkov et al., *Pharm. Chem. J. (UdSSR) Engl. Ed.* **8**, 364 (1974).
4.159   H. Junginger, *Pharm. Ind.* **36**, 333 (1974).
4.160   E. Nürnberg, *Pharm. Ind.* **38**, 74 (1976).
4.161   K. Kröll, *Trockner und Trocknungsverfahren, 2. Trocknungstechnik*. Springer-Verlag, Mannheim (1978).
4.162   H. A. Troesch, *Chem. Ing. Techn.* **26**, 311 (1954).
4.163   B. Wahl, *Chem. Ing. Techn.* **46**, 351 (1974).
4.164   Technical documentation of VBT-Prozeßtechnik AG, Zürich, Switzerland.
4.165   J. Flink, in *Freeze Drying and Advanced Food Technology*, ed. by S. A Goldblith et al., p. 143. Academic Press, New York (1975).
4.166   H. Rahm, Fortbildungskurs der Gesellschaft Schweizerischer Industrie-Apotheker (GSIA) 1979/1980: Lesson 6, p. 37.
4.167   E. Graf and B. Bornkessel, *Dtsch. Apoth. Ztg* **118**, 503 (1978).
4.168   G. V. Chebak, *Pharm. Chem. J. (UdSSR) Engl. Ed.* **7**, 386 (1973).
4.169   N. E. Chernov et al., *Pharm. Chem. J. (UdSSR) Engl. Ed.* **11**, 538 (1977).
4.170   R. K. Himmel, Grundlagen und Technologie der Mikroverkapselung, APV-Info (4), 206 (1968).
4.171   E. Nürnberg and M. Krieger, *Acta Pharm. Technol.* **25**, 49 (1979).
4.172   E. Nürnberg and E. Graf, *Pharmazie in unserer Zeit* **3**, 1 (1974).
4.173   W. A. Ritschel, *Pharm. Ind.* **33**, 612 (1971).
4.174   W. A. Ritschel, *Pharm. Ind.* **33**, 532 (1971).
4.175   R. Hegenauer, Dragoco-Report 10/1978, p. 203 (1978).
4.176   W. Treibs, in *Die ätherischen Öle*, ed. by E. Giedemeister and F. Hoffmann, Vol. I, Akademie-Verlag, Berlin (1956).

Chapter 5

5.1   W. Schaub, Electronic Grading in the Food Industry. Publication by Gunson's Sorts Ltd. 20–21 St. Dunstan's Hiil, London EC3, UK.
5.2   H. Samans, *CZ-Chemie-Technik* **2**, 345 (1973).
5.3   Alpine AG, Augsburg, Leaflet 18/2 d.
5.4   H. Samans, *Lebensmitteltechnik* (8), 424 (1976).
5.5   Alpine AG, Augsburg, Leaflet 14/5.

5.6 Mahlen mit flüssigem Stickstoff bei geregelten Temperaturen, Die chemische Produktion, Oct. 1981, p. 32.

5.7 Tabakaufbereitung – Alpine Information 4/72.

5.8 Alpine AG, Augsburg, Leaflets 16/1; 6 675.

5.9 O. Lauer, *CZ-Chemie-Technik* **1**, 363 (1972).

5.10 K. Leschonski, W. Alex and B. Koglin, *Chem. Ing. Techn.* **36**, 901 (1974).

5.11 Alpine AG, Augsburg, Leaflet 50-52/1 d.

5.12 Alpine AG, Augsburg, Leaflet 314/1 d.

5.13 J. Engelsmann AG, Leaflet 5.79 RD.

5.14 J. Wessel, in *Ullmann's Enzyklopädie der Technischen Chemie*. 4th Edn, Vol. II, p. 43. Verlag Chemie, Weinheim (1972).

5.15 F. Mogensen, Theorie des Sizer-Verfahrens. Bergbau-Wiss. Part 6, 136 (1960).

5.16 F. Mogensen, *The Quarry Manager's J.* H. 10, 409 (1965).

5.17 J. Wessel, *Aufbereitungstechnik* **10**, 492 (1969).

5.18 P. H. List, *Arzneiformenlehre,* 3rd Edn. Wiss. Verlagsgesellschaft mbH, Stuttgart (1982).

5.19 J. K. Hartmann, Verfahren zur Kaltgewinnung von Extrakten. DAS 1 124 637, 27 Aug. (1959).

5.20 T. Voeste and K. Wesp, in *Ullmann's Enzyklopädie der Technischen Chemie*. 4th Edn., Vol. II, p. 726. Verlag Chemie, Weinheim (1972).

5.21 H. List, Extraktion. Leaflet from List-Industrielle Verfahrentechnik, Pratteln, Switzerland.

5.22 K. Kipke, Ekato-Rührtechnik. Publication by Ekato-Rühr- und Mischtechnik, Schopfheim, FRG (1980).

5.23 Gebr. Lödige, Operating instructions for the Lödige Plowshare mixer, Gebr. Lödige, Paderborn, FRG.

5.24 Nautamix, Publication by Nautamix Europa, B.V., Haarlem, The Netherlands.

5.25 IKA-Werke, Janke & Kunkel, Bad Krozingen, FRG.

5.26 NIRO Atomizer, Techn. Bulletin B-Na-150, Niro-Atomizer A/S, 305 Gladsaxevy, Soeborg, Denmark.

5.27 W. Kehse, *Chemiker Ztg* **95**, 54 (1970). Manufacturers: Extraktionstechnik, Gesellschaft für Anlagenbau mbH, D-2000 Hamburg 76, FRG.

5.28 U-Extraktor, Ungar. Pat. 159 977.

5.29 Hoelscher-Technik-Gorator, Herne, FRG.

5.30 F. J. Supraton, Zucker GmbH, Neuss, FRG.

5.31 K.-H. Brunner, Separatoren und Dekanter für die kontinuierliche Extraktion, Westfalia, Technisch wissenschaftliche Dokumentation No 4 (1982).

5.32 K. Theiler and T. Paschedag, *Seifen-Öle-Fette-Wachse* **105**, 269 (1979).

5.33 S. A. Minina et al., *Pharm. Chem. J. (UdSSR) Engl. Ed.* **10**, 407 (1976).

5.34 Eries, F-69740 Genas, France.

5.35 Nova-Werke AG, Effretikon, Switzerland.

5.36 HDA-Anlagenbau, Konstanz, FRG.

5.37 H. Coehnen et al., *Krupp-Forsch. Ber.* **40**, 1 (1982).

5.38 E. Stahl, *Parfümerie und Kosmetik* **63**, 117 (1982).

5.39 K. Zosel, Verfahren zur Desodorierung von Fetten und Ölen. DAS 23 32 038, 23 June (1973).

5.40 K. Zosel, Verfahren zur Entkoffeinierung von Kaffee. DOS 2905 078, 10 Feb. (1979).

5.41 C. Alt, in *Ullmann's Enzyklopädie der Technischen Chemie*, 4th Edn., Vol. II, p. 199. Verlag Chemie, Weinheim (1972).

5.42 N. A. Gromova et al. *Pharm. Chem. J. (UdSSR) Engl. Ed.* **8**, 709 (1974).

5.43 W. Hemming, *Verfahrenstechnik*, 2nd Edn. Vogel-Verlag, Würzburg (1980).

5.44 W. R. A. Vauck and H. A. Müller, *Grundoperationen chemischer Verfahrenstechnik*, 2nd Edn. Steinkopf, Leipzig (1966).

5.45 H. Hemfort, Separatoren, Zentrifugen für Klärung, Trennung, Extraktion. Westfalia Separator AG (1979).

5.46 H. Hemfort, *Chemiker Ztg* **90**, 247 (1966).

5.47 Deutsche Garap, Gesellschaft für Apparatebau. Düsseldorf, FRG.

5.48   Carl Padberg, Zentrifugenbau GmbH, Lahr, FRG.

5.49   J. L. Stein and K. Imhof, in *Ullmann's Enzyklopädie der Technischen Chemie*. 4th Edn, Vol. II. Verlag Chemie, Weinheim (1972).

5.50   R. Billet, *Verldampfung und ihre Technischen Anwendungen*. Verlag Chemie, Weinheim (1981).

5.51   F. Schaefer, in *Ullmann's Enzyklopädie der Technischen Chemie*. 4th Edn, Vol. II. Verlag Chemie, Weinheim (1972).

5.52   E. Wandel, RTC-Verdampfer zur Hochkonzentration wärmeempfindlicher Produkte. Publication by Niro Atomizer, Soeborg, Denmark.

5.53   Luwa, Dünnschicht-Technik. Publication by Luwa SMS GmbH, Butzbach, FRG.

5.54   K. Kröll, *Trockner und Trochnungsverfahren – Trocknungstechnik*, Vol. 2, 2nd Edn. Springer-Verlag, Heidelberg (1978).

5.55   F. Kneule, *Das Trocknen*, Verlag Saurländer, Frankfurt/Main (1975).

5.56   E. Turke, *Z. Aerosol Forsch.* **6**, 505 (1955).

Chapter 6

6.1   H. G. Menssen, *Therapiewoche* **21**, 2648 (1971).

6.2   W. Spaich and L. Gracza, *Pharm. Ztg* **125**, 1706 (1980).

6.3   B. Siegfried, *Schweiz. Apoth. Ztg* **106**, 561, 589, 661, 714 (1968).

6.4   H. G. Menssen, *Pharm. Ztg* **124**, 233 (1979).

6.5   H. G. Menssen, *Dtsch. Apoth. Ztg* **121**, 1255 (1981).

6.6   H. G. Menssen, *Dtsch. Apoth. Ztg* **122**, 2317 (1982).

6.7   H. Schilcher, DIAITA Part 6, p. 7 (1968); Part 3, p. 18 (1969).

6.8   C. Franz, *Dtsch. Apoth. Ztg* **122**, 1413 (1982).

6.9   K. Ebert, *Arnei- und Gewürzpflanzen. Ein Leitfaden für Anbau und Sammlung*. Wiss. Verlagsgesellschaft mbH., Stuttgart (1982).

6.10   A. M. Schwarzenbach, *Biol. i. u. Zeit* **11**, 107 (1981).

6.11   E. Sprecher, *Dtsch. Apoth. Ztg* **122**, 2717 (1982).

6.12   F. W. Hefendehl and C. Lander, Quality assurance of plant medicines – requirements for registration and licensing. Lecture in APV course. Qualitätssicherung pflanzlicher Arzneimittel, Nov. 1982.

6.13   F. W. Hefendehl, *Planta Med.* **42**, 111 (1981).

Chapter 7

7.1   Various authors, Rezepturen für Extrakt-Tabletten. APV-Info (2), 44 (1961).

7.2   R. Keymer, *Pharm. Ind.* **31**, 721 (1969).

7.3   F. Crippa, *Fitoterapia* **49**, 257 (1978).

7.4   E. Graf and B. Bornkessel, *Dtsch. Apoth. Ztg* **118**, 530 (1978).

7.5   P. C. Schmidt, unpublished results.

7.6   A. Burger and R. Dialer, *Sci. Pharm.* **49**, 461 (1981).

7.7   H. Flück, *Pharm. Weekbl.* **100**, 1361 (1965).

7.8   E. Nürnberg, *Acta Pharm. Technol.* **23**, 111 (1977).

7.9   S. Urban, *Acta Pharm. Technol.* **22**, 247 (1976).

Index